ACTIVE GALACTIC NUCLEI

WILEY-PRAXIS SERIES IN ASTRONOMY AND ASTROPHYSICS
Series Editor: John Mason, B.Sc., Ph.D.

Few subjects have been at the centre of such important developments or seen such a wealth of new and exciting, if sometimes controversial, data as modern astronomy, astrophysics and cosmology. This series reflects the very rapid and significant progress being made in current research, as a consequence of new instrumentation and observing techniques, applied right across the electromagnetic spectrum, computer modelling and modern theoretical methods.

The crucial links between observation and theory are emphasised, putting into perspective the latest results from the new generations of astronomical detectors, telescopes and space-borne instruments. Complex topics are logically developed and fully explained and, where mathematics is used, the physical concepts behind the equations are clearly summarised.

These books are written principally for professional astronomers, astrophysicists, cosmologists, physicists and space scientists, together with post-graduate and undergraduate students in these fields. Certain books in the series will appeal to amateur astronomers, high-flying 'A'-level students, and non-scientists with a keen interest in astronomy and astrophysics.

ROBOTIC OBSERVATORIES

Michael F. Bode, Professor of Astrophysics and Assistant Provost for Research, Liverpool John Moores University, UK

THE AURORA: Sun–Earth Interactions

Neil Bone, School of Biological Sciences, University of Sussex, Brighton, UK

PLANETARY VOLCANISM: A Study of Volcanic Activity in the Solar System, Second edition

Peter Cattermole, formerly Lecturer in Geology, Department of Geology, Sheffield University, UK, now Principal Investigator with NASA's Planetary Geology and Geophysics Programme

DIVIDING THE CIRCLE: The Development of Critical Angular Measurement in Astronomy 1500–1850 Second edition

Allan Chapman, Wadham College, University of Oxford, UK

THE DUSTY UNIVERSE

Aneurin Evans, Department of Physics, University of Keele, UK

COMET HALLEY - Investigations, Results, Interpretations
Volume 1: Organization, Plasma, Gas
Volume 2: Dust, Nucleus, Evolution

Editor: John Mason, B.Sc., Ph.D.

ELECTRONIC AND COMPUTER-AIDED ASTRONOMY: From Eyes to Electronic Sensors

Ian. S. McLean, Department of Astronomy, University of California at Los Angeles, California, USA

URANUS: The Planet, Rings and Satellites

Ellis D. Miner, Cassini Project Science Manager, NASA Jet Propulsion Laboratory, Pasadena, California, USA

THE PLANET NEPTUNE: An Historical Survey Before Voyager, Second edition

Patrick Moore, CBE, D.Sc.(Hon.)

ACTIVE GALACTIC NUCLEI

Ian Robson, Director, James Clerk Maxwell Telescope, Director Joint Astronomy Centre, Hawaii, USA

THE HIDDEN UNIVERSE

Roger J. Tayler, Astronomy Centre, University of Sussex, Brighton, UK

Forthcoming titles in the series are listed at the back of the book.

ACTIVE GALACTIC NUCLEI

Ian Robson BSc., Ph.D., F.R.A.S., C.Phys., F.Inst.P.
Director James Maxwell Telescope,
Director Joint Astronomy Centre, Hilo, Hawaii

JOHN WILEY & SONS
Chichester • New York • Brisbane • Toronto • Singapore

Published in association with
PRAXIS PUBLISHING
Chichester

The White House,
Eastergate, Chichester,
West Sussex, PO20 6UR, England

Published in 1996 by
John Wiley & Sons Ltd
in association with Praxis Publishing Ltd

Wiley Editorial Offices

John Wiley & Sons Ltd, Baffins Lane,
Chichester, West Sussex PO19 1UD, England

John Wiley & Sons, Inc., 605 Third Avenue,
New York, NY 10158-0012, USA

Jacaranda Wiley Ltd, G.P.O. Box 859, Brisbane
Queensland 4001, Australia

John Wiley & Sons (Canada) Ltd, 22 Worcester Road,
Rexdale, Ontario M9W 1L1, Canada

John Wiley & Sons (SEA) Pte Ltd, 37 Jalan Pemimpin 05-04,
Block B, Union Industrial Building, Singapore 2057

A catalogue record for this book is available from the British Library

ISBN 0-471-95853-0 Cloth
ISBN 0-471-96050-0 Paperback

Printed and bound in Great Britain by Hartnolls Ltd, Bodmin

This book is dedicated to my wife, Chris

Photograph of the author taken on November 12th 1985, setting up the Wilfred Hall Telescope at one of the Observatories of Lancashire Polytechnic (now the University of Central Lancashire) prior to public observations of Halley's comet. Ian Robson was the Director of the Observatories and the Head of Department of Physics and Astronomy prior to being seconded to Hawaii.

Photograph by Don McPhee and courtesy of Guardian Newspapers.

Table of contents

Foreword

There is no more daunting task for any author than to try to make sense of the plethora of data which has accumulated over the last 35 years on Active Galactic Nuclei. It is startling to recall that it was only in 1963 that the large redshifts of quasars were established and that the first X-ray survey of the whole sky only took place in 1970. In the same year, infrared astronomy was in its infancy and we had to wait until the 1990s before the high energy gamma-ray sky was opened up for the study of the most exotic Active Galactic Nuclei. The astronomical landscape has changed for good, and all disciplines have benefited from the opening up of the whole of the electromagnetic spectrum for astronomy.

Some wavebands have certain astronomical disciplines as their birthrights. The optical waveband is still the primary waveband for the study of the stars and the stellar content of galaxies. The infrared and millimetre wavebands are the prime tools for studies of the processes of star formation. The study of the hot Universe is the province of X- and gamma-ray astronomy. Interstellar molecules are the speciality of the centimetre, millimetre and submillimetre wavebands. One might have thought that the problems of understanding Active Galactic Nuclei might have been the birthright of one of these wavebands, but that is most certainly not the case. As each waveband has been opened up, active galactic nuclei have featured prominently among the most important discoveries of all of them. This is not just the hyperbolae of the proponents of particular wavebands. This is how active galactic nuclei actually behave—their emissions are quite extraordinarily broadband and the full range of the remarkable phenomena for which they are responsible can only be appreciated by an all-waveband study.

This is difficult enough, but what makes the story so much more complicated is the fact that many of the observed phenomena have no counterpart in laboratory or solar system physics. There are no nearby black holes to study, as we can, for example, the astrophysics of our Sun. There are no jets of relativistic material ejected locally. The physics of the presumed thick discs about the most luminous quasars have no counterpart in our locality. If it is any reassurance, at least all the problems are associated with the single family of objects which we classify under the general title Active Galactic Nuclei.

But it is an unruly family. Many bold attempts have been made to bring some order into the many different classes of object which fall within the category of Active Galactic Nuclei and there has been some modest success in doing this. As soon as something begins to work, however, exceptions are always found and, whilst this does not necessarily mean going back to the drawing-board, models which try to encompass everything end up looking rather messy and contrived. There must be something fundamental we do not yet understand. I am reminded of the state of the classification of stellar spectra at the end of the 19th century, when there were 23 different classification schemes which were only rationalised with the work of Annie Cannon and her colleagues who studied large enough samples systematically for a proper statistical evaluation to be made of the significance of the different schemes. Perhaps we are at the corresponding stage in the evolution of the taxonomy of Active Galactic Nuclei.

However, it turns out, we can be certain of one thing. There is new physics to be understood in the astrophysics of Active Galactic Nuclei. I find the arguments in favour of the presence of supermassive black holes in these nuclei wholly convincing. This then gives us the opportunity of studying the physics of matter in the presence of very strong gravitational fields which is excluded in terrestrial laboratories. The environments of the black holes are unlike anything in our normal experience. If the interpretation of the recent X-ray observations of the broad iron lines in MGC -6-30-15 is correct, we can now observe matter in its death throes as it falls inevitably within the event horizon of a massive black hole. Who would have thought 35 years ago that we would even dare to make such a claim? The physics of jets and their interaction with the environment is a central problem for many aspects of high energy astrophysics and yet we really have little secure understanding how they are produced or their astrophysical role in the evolution of active galaxies. The studies go far beyond their local significance within galaxies. The fact that extraordinarily powerful quasars are observed at a time when the Universe was less than 20% of its present age shows that the formation of supermassive black holes had already taken place by then, before the largest scale structures we know of had condensed out of the expanding substratum.

There are few areas of astrophysics which have such wide physical and astrophysical significance and yet which are so difficult to condense into a readable volume of manageable length than the subject of Active Galactic Nuclei. I enjoyed enormously reading the typescript of this book by Ian Robson, not only because of his great achievement in assembling all the material into a coherent volume, but also because of the evident enthusiasm and enjoyment for the subject which comes through on every page. This is what astronomy is really like. He has been in the enviable/unenviable position of being the Director of the Joint Astronomy Centre in Hawaii where much of the forefront astrophysics described in this book has been carried out. He has had daily contact with the many astronomers who now pass through the island and has caught the excitement of the many different approaches to these thorny problems.

This is a wonderful book to put into the hands of any undergraduate. It contains a comprehensive account of many of the most important areas of research into active galactic nuclei without being so technical that the end goal of the exercise is lost. I hope it will give many young people some of the flavour of research in modern astronomy and encourage them to delve further for themselves. Ian's splendid up-to-date account of a vast array of important observations and their interpretation will certainly make my own efforts in trying to master the literature on active galaxies that much easier. As they say in American bookshops, 'Read and enjoy!'

Malcolm Longair, Cambridge, October 1995.

Author's preface

The study of active galactic nuclei is a very hot topic in astrophysical research and occupies a significant fraction of the work of the entire astronomical community. Why is this research so appealing? There are a number of obvious reasons, and some not so obvious. The need to explain the very high rates of energy generation, requiring supermassive black hole power sources, jets of relativistic particles blasting out of the nucleus of a galaxy and exotic particle–radiation interactions are clear challenges for the application of the laws of physics. Another reason is that to obtain a full understanding of what makes an AGN and why AGNs come in different varieties and flavours requires observations over the entire electromagnetic spectrum, from radio waves to gamma-rays. This is another challenge, to build the instruments to probe the secrets of AGNs. For the theorist, the conditions in the central engines are extreme and are found in only a small number of astrophysical contexts, therefore it is a fertile proving ground for new theories and models of particle and radiation interactions. Finally, there is something else. It is just so very exciting and therefore fun.

I commenced this book a number of years ago, while I was teaching at the University of Central Lancashire in the UK. Being a Head of Department and Director of Observatories left little time for work on the book and progress was slow, with weekends and nights being the only time available during normal working times. The improvement of the laptop computer proved a boon and numerous flights to Hawaii on observing visits, plus accompanying my wife and me on skiing holidays moved things along. I must have gone through at least five brands of laptop and became one of the gurus for lightweight notepads. The move to Hawaii to be Director of the James Clerk Maxwell Telescope and the Joint Astronomy Centre, gave the final push to complete the book. This was compounded by the fact that because my wife remains in the UK with her work, the evenings and weekend provide additional time to bring it to a close. She will be delighted when the book is finished and we can revert to normal holidays, where the accompanying laptop becomes only a means of communication back to the office to keep up with the email, rather than a tool to absorb any free moments. However, the subject of AGN research is so fast moving that actually bringing the book to a 'final' conclusion has been more of a challenge than I anticipated as there always seemed to be that little extra something about to come out in the next journal that just had to be included. In the end I set a deadline that nothing beyond the end of September 1995 would be included.

In facing the daunting task of writing a book on AGNs, the publishers and I decided at the outset that this book should do two things, it should appeal to the student of astrophysics, especially AGNs, but it should also appeal to the layreader who has a general interest in an exciting topic of astronomical research. Therefore this book is not full of equations, although it would have been easy to make it so. I have found from over twenty years of teaching astrophysics that for the lecturer, writing down the equations is the easy part. Giving the background words that allows the student to soak up the information and put it in context is the hard part, especially without a course text book. AGNs have not had a course text, and this has made it very difficult for teachers.

This book aims to provide the background information and describe the *physical principles* at work in the study of AGNs. It also seeks to provide a wealth of detailed information regarding not only the most modern of the observations, some just completed in the last week as this book was going to press, but also a historical context. I feel strongly about this. The history of science is fascinating and as a physicist one realizes that the foibles of history have formed our view of the Universe and our physical laws. The strange quirks of fate that brought widely differing people like Tycho Brahe and Johannes Kepler together, shape our destiny. Therefore I have unashamedly given a historical flavour to the first three chapters of the book, to set the scene for the modern reader. Standing on the shoulders of giants is one of the most apt expressions I have heard, each generation stands on the shoulders of brilliant scientists of the previous generation. The field of AGN research has been blessed with a number of giants, both observationally and theoretically. Although it is always invidious to name these individuals, because one always manages to omit the obvious person who should be listed, for me the Martin Rees, Roger Blandford and Rick Begelman's of this world have proved to be inspirational and fill me with awe for their grasp of the physical processes and how they can conceive of the details of esoteric situations with which I can only struggle to comprehend the rudiments. I have always enjoyed discussions and arguments with Geoffrey Burbidge, for whom I continue to have a tremendous respect. The master of observational spectroscopy, Don Osterbrock has always made me realize what a great observer really is, and the fascination of the historical picture. Then we come to people like Michael Rowan-Robinson who had far-sighted ideas in the early 1970s and never really got the credit he deserved. Finally we come to the next generation, the 'Ski' Antonucci's of this world. They move the scene on and boldly go where no astronomer has gone before. I look forward to their continued use of the telescopes that I am charged to manage in order that we can extend our knowledge and understanding of AGNs.

For the general layreader, I have given numerous section overviews where necessary and have tried to avoid the most complex of the processes and jargon, although it is inevitable that some have slipped through. The message to all is not to panic. The layreader can safely skip the mathematical parts, it is not absolutely essential to know precisely how synchrotron radiation works (and how many of us really understand the intricate details of the electrodynamics) but to know what it is, when it happens and what it does. This is the flavour I have tried to incorporate, along with the details (to some depth) for the student who does need to know how it works. This has been a challenge as an educationalist. I have not been afraid to repeat topics. In my twenty years of teaching experience I have learned that one should not assume that because something has been said once, it will have been absorbed by the student. Sensible repetition of key points rarely does anyone harm.

For all readers I have given extensive further reading sections and these have been broken down into two sections, the general section which is mostly review articles and the specialized section containing research papers from the main astronomical journals. I have two piles of reprints and preprints at my home in Hawaii, each of which would be over six feet high if stacked together (which is not wise in this earthquake zone). In selecting my choices for inclusion I have looked at papers which I found both rewarding and important. I used four criteria for inclusion, the sheer scientific importance of the paper, the demonstration of a key observation, an excellent source of references on the subject, or because it was very topical. I have tried to cover most of the key papers which have shaped the modern picture of AGN research, this includes discovery papers and theory papers. Part of the great satisfaction of writing this book has been the excuse to find the time to read the famous papers of long ago, some before I was born. Anyone who reads the journals of the 1940s will see how vastly different astronomy is today compared to what it was then. It is a salutary experience but also vividly shows the success of our field. Nevertheless, trying to condense these into a list appropriate for a book rather than a library catalogue will inevitably mean that there are important papers I have omitted, I apologize to the authors, no slight is intended.

This book is not meant to show that the topic is easy. It is not. Neither does it set out to present a neat picture, life is not like that and certainly scientific investigation is no exception. This book shows the difficulties, the problem areas, the potential selection effects, the areas in which we have no data, potential speculation and observational tests for the future. This is a book for those who wish to understand the working of the field rather than a picture book of answers. I make no apology for this. The study of active galaxies is at the very forefront of astronomical investigation with a significant fraction of the world's astronomers working in this field and a large fraction of telescope time devoted to it.

Although our knowledge and understanding has improved dramatically since the discovery of active galaxies, nevertheless we cannot claim to have solved everything. We do not have an agreed global picture of what makes active galaxies, but we are now close to being able to build models which seem to fit the observations. We can start to ask what ingredients are necessary to make a quasar, a Seyfert, a BL Lac and so on. But a cook-book it is not. This book contains a wealth of information of diverse nature, but I hope that this information is presented in a coherent manner so that the reader can be taken through the stages in simple steps.

For me this book has been a big challenge and trying to find the quality time to do the research has been the hardest part. On the other hand, it has made me concentrate on understanding the 'big picture', at least as much of it as we know today. It has driven me to understand areas of science or observational techniques that I was not as familiar with as I might have been, especially the esoterics of X-ray astronomy and the complexities of the reflection models which dominate current thinking on the central regions around the supermassive black hole. As a teacher, this is always an ongoing process in preparing for new courses, but as a telescope director it is easy to slide back into management and one's own speciality and start to lose out on the physics of the 'big picture'. For this part alone, writing the book has been well worthwhile.

Due to the requirement to keep costs down, a factor that I was most insistent on in order to make this book affordable to students, the number of colour plates has, by necessity, been restricted. There are some obvious pictures which are absent but this could not be avoided. On the other hand, I would also like to point out the huge amount of astronomical information and pictures now available to everyone on the *World Wide Web.* All self-respecting astronomy organizations have a WWW page and the public pages from the HST and NASA are wonderful for the picures they contain. Use the web, it is the medium of the future. http://www.jach.hawaii.edu is our site on the web. Information on the JCMT, UKIRT and links to other sites can be found there.

Finally a tip to potential authors—use the excuse of writing a book to obtain a sabbatical and enjoy the quality time so precious these days. Never, ever, volunteer to produce camera-ready copy, no matter what the inducements, unless you are a LateX guru. It is a nightmare. Having written the words and selected the figures, the page layout and all the sub-editing associated tasks take much of the pleasure out of the writing. On the other hand, one has control over the final production and attempting this 8,000 miles from the publisher would inevitably have slowed final production down by many months. Thank goodness for long flights from Honolulu to the UK, the use of Admirals Clubs turned out to be a great boon for completing this process.

In conclusion, I sincerely hope the readers enjoy this book, this is its prime purpose.

Ian Robson
Hilo, September 24th 1995

Acknowledgements

This is the good part, to say thank you to all those who have helped me in this task. The numerous discussions I have had with colleagues over the years, at conferences and meetings, at telescopes, at my university department and other universities in the UK, have all shaped what has appeared in this book. Even more, those astronomers over the many years who have always encouraged and helped others in understanding the secrets of the Universe have been a source of inspiration from which I have benefited. Malcolm Longair has been a guiding light in this respect and has kindly provided a foreword to this book. There is no doubt that astronomy as a discipline is really full of 'good people'.

I would particularly like to take this opportunity to thank all those astronomers who went out of their way to provide me with special figures. Some made extensive changes to their electronic versions to furnish the figures appearing here and this is very much appreciated. At all times, I have been amazed at how extremely busy astronomers have continued to do that little extra to provide their best pictures. The following deserve special thanks: Norbert Bartel, Tom Herbst, David Hughes, Robert Goodrich, James Graham, Lincoln Greenhill, Prab Gondhalekar, John Hutchings, John Kormendy, Dave Leisewitz, Alan Marscher, Corinna von Montigny, Patricia Orr, Brad Peterson, Dave Sanders, Jason Stevens, Clive Tadhunter and Geoff Taylor. I would also like to acknowledge the journals and books for their permission to reproduce material and in this context I note that the *Astrophysical Journal* is published by the University of Chicago Press.

In writing this book, which contains such a wide coverage of techniques and physics it is paramount to ensure that the information is correct. I would like to thank my colleagues, Ian Butchart, Bob Carswell, Tim Cawthorne, Thierry Courvoisier, Robert Laing and Ian McHardy who gave a detailed and critical reading to individual chapters. The book has benefited immensely from their conscientious efforts in this regard. Above all, Alex Filippenko proved absolutely brilliant in reading the manuscript to give it a thorough shake-down in terms of facts, physics and grammar. I have been tremendously impressed with his ability in all areas and he certainly puts the 'gramm checker' on my PC to shame, one of them is superior and my money is on Alex. Any errors which remain have resulted from late changes and the responsibility is mine alone.

I would like to thank the publisher, Clive Horwood of Praxis, for being so positive at a time when I was really fed-up after seeing the scientific publishing industry go though trauma with take-overs and closures. I am also grateful to Clive for agreeing to my demands to produce a paperback version. Clive seeks to promote quality in a book, both in content and also presentation. I wish him well in his future endeavours. Lisa Briaris did an excellent job of keeping track of the permissions to publish, while Mike Shardlow did an excellent job of editing the final manuscript. I learned all about the 'em' and 'en' rules, but I'm not at all sure I'm converted to his use of 'ize' instead of 'ise'; maybe I'm just old-fashioned. John Mason sorted out many of the colour images and galaxy pictures and remained always enthusiastic.

Finally I would like to take this opportunity to dedicate this book and to thank my wonderful and long-suffering wife of twenty-five years. She has been incredibly supportive of all my enterprises in life, rock climbing, sports cars, research, travel, skiing, and this book. Indeed, she also gets thanks for her expertise (or at least fast learning ability) to sort out the complexities of the 'headers' and 'indexing' for this camera-ready copy along with numerous other quirks of Microsoft Word. Other observing astronomers will recognize the phrase 'astronomy widow', given the fact that we observers seem to be away from home so much and I'm afraid this has been a fair description. On the other hand, she has been treated to seeing some of the world's best observatories in some of the most exotic locations which I figured was at least some compensation. Even with her high-pressure career, she has always found the time to help out whenever I get stuck or multiple deadlines loom large. She is the ultimate in support and a brilliant back-up team to have at home, even though that happens to be 8,000 miles away while I am posted to Hawaii. With the completion of this book, she has been promised a real holiday (without laptop), the only problem for both of us is finding the time to take it somewhere.

List of illustrations, Plates and tables

Chapter 7

Chapter 8

Chapter 9

Colour plates

Tables

1

The Universe—an overview

1.1 INTRODUCTION

The study of active galaxies lies at the forefront of current astronomical research. Astronomers investigating these most powerful objects in the Universe use the latest state-of-the-art detection techniques along with the largest and most sophisticated telescopes in the world. To understand the secrets of active galaxies astronomers need to make observations over the entire electromagnetic spectrum, from radio waves through the millimetre/submillimetre, far-infrared, infrared, optical, ultraviolet, extreme ultraviolet, X-rays and finally gamma-rays. To cover this vast spectral range, telescopes in space as well as those on the ground are used. Herein lies one of the great beauties of the study of astronomy: it is all encompassing and to make progress in the field one needs to acquire a wide range of skills. As a topic of research, it is challenging, sometimes frustrating (particularly when precious telescope time is lost to bad weather or instrument problems) but above all, it is tremendously exciting.

The Universe is made up of galaxies, many containing more than a hundred thousand million stars. We live in such a Galaxy, called the Milky Way Galaxy. Now imagine a galaxy possessing an amazing power source at its nucleus, pouring out the same energy as that emitted by a thousand galaxies like the Milky Way. Then add another touch of incredulity by requiring that this fantastic power source occupies a volume of space only the size of the Solar System! This sounds too fantastic to be true. Nevertheless, with the discovery of quasars, these were precisely the bizarre facts facing astronomers in the early 1960s. Quasars are the most powerful of the active galaxies. Some quasars are highly luminous radio emitters, others are very bright in X-rays, while all show broad spectral emission lines of highly ionized species. The output of a number of active galaxies is known to vary with time, sometimes on timescales as short as minutes and hours. A small minority of objects have fantastic narrow jets of energy pouring out from their very centres. Active galaxies are truly amazing objects of study.

We shall describe the overall characteristics of active galaxies later in this chapter, but from the very outset we should make it clear that the enormous luminosity emanating from the small volume of the emission source originally presented astronomers with a major puzzle. Indeed, in the most luminous of objects, the quasars, this central luminosity source is so powerful that it easily outshines the light from all the surrounding stars of the host galaxy. Then, we no longer see a galaxy of stars, just an intensely bright central spot. In attempts to explain what kind of energy source could emit the tremendous power from such a small volume, it was not surprising that many competing theories endeavoured to explain this new and intriguing phenomenon. By a process of scientific deduction coupled with more and better observational data, the picture has become much clearer over the last twenty years. It is with some confidence that we can now claim to have a basic understanding of how active galaxies are powered. A vast majority of astronomers are now convinced that they are powered by supermassive black holes located at, or close, to their centres. In this context, supermassive refers to masses greater than about one million times that of the Sun ($M_{BH} > 10^6 M_{\odot}$, and we introduce the subscript '$\odot$' to represent solar quantities such as mass, radius and luminosity.).

These supermassive black holes possess deep gravitational potential wells and work like enormous engines, converting potential energy of infalling material into kinetic energy. In so doing, the material that is being inexorably dragged towards the black hole produces an accretion disk and is heated to temperatures greater than the surface temperatures of the hottest stars. This radiant energy blazes forth as UV or X radiation. Hence the fundamental power source of active galaxies is the gravitational potential well of a supermassive black hole. It is this that is responsible for the enormous luminosities of these objects, making them stand out like bright beacons in the depths of space.

However, before we become euphoric with the idea that everything is solved and there are no remaining mysteries to the workings of active galaxies, we should add a note of caution. Although black holes are now a well-accepted theoretical idea, astronomers have so far failed to identify with *absolute certainty* a single black hole anywhere in the Universe. This is true for solar mass sized black holes right up to supermassive black holes. The concept of a black hole was postulated by the Reverend John Mitchell as far back as 1786 and by Laplace in 1795. This was resurrected in terms of Einstein's general theory of relativity by Schwarzschild in 1916, but conclusively proving their existence by direct observational evidence has remained elusive.

Nevertheless, good progress is being made in this quest; observations from the 1970s through to the present time have produced a number of excellent candidates for solar mass sized black holes residing in binary systems in our Galaxy. The most famous is the X-ray source Cyg X-1, while the best conclusive candidate so far is a Galactic X-ray nova called V404 Cyg. This is believed to be a black hole of mass around 12 $M_{\odot}$ orbiting a giant star which has been stripped of its outer envelope by the gravitational attraction of the black hole. One of the first observations by the refurbished Hubble Space Telescope (HST) in early 1994 provided very strong support for the existence of a supermassive black hole. In this case the black hole resides at the centre of the nearby active galaxy M87 (fig. 1.2) and has a mass of just over $10^9 M_{\odot}$. In chapter 6 we shall take a detailed look at supermassive black holes, accretion disks and the observational evidence supporting their claim to fame as the power source for active galaxies.

Given the lack of conclusive proof, one might wonder why astronomers have so readily accepted this exotic phenomenon. The answer is because the theories and subsequent predictions of the behaviour of how a black hole can act on the surrounding matter seem to fit most of the otherwise unexplained observations. Furthermore, and very importantly, alternative theories have failed to stand the test of time. Although this by itself does not prove the existence of black holes, and it is always possible that new physics yet to be discovered might provide better answers, this process of inter-linking observation and theory shows how science progresses. Powerful theoretical arguments and strong observational evidence have convinced the vast majority of astronomers that supermassive black holes are the ultimate powerhouse, the central engines of active galaxies. We shall support this view throughout this book.

However, the story of black holes is even more complex. Most objects in the Universe are rotating in some manner, and the theories of black hole formation predict that black holes will also be rotating systems along with their surrounding material. Due to the conservation of angular momentum and particle interactions, the infalling matter will not fall directly into the black hole, but, as is very common in astrophysics, will form a flattened system. A prime example of a flattened system is the Solar System, where all the planets orbit the Sun in a plane, another is spiral galaxies (see fig. 1.2). The matter falling into a black hole forms an accretion disk outside the hole, and it is the material of the accretion disk that is extremely hot and radiates into space, not the black hole itself. The theoretical studies on black holes and accretion disks make it readily understandable why no observation has ever directly revealed one; they are far too small for current telescopes to resolve. If the Sun were to shrink to be a black hole, the inner parts of the accretion disk would have a diameter of only 100 km or so! The presence of a black hole is best revealed by its interaction on the surroundings due to its high gravitational mass. As we saw above, for solar mass size black holes, membership of a binary system provides the best observational candidates.

Before leaving this section we should point out that not everyone is totally convinced by the black hole story for all active galaxies. A very small number of astronomers consider that massive clusters of hot and luminous stars might be the powerhouses for the nuclei of some of the active galaxies, especially the less powerful ones. This brings us nicely to an important point of definition. Although we shall use the term active galaxy in this book, we must stress that it is only the very central region of a galaxy that is active, not the entire galaxy. A more precise description for these objects is *active galactic nuclei* (AGN) and this is a term that is widely used by professional astronomers. When we refer to an active galaxy we are just raising a flag to show that the galaxy in question has a substantial luminosity that is dominated at some wavelength by processes other than starlight. We will use the term active galactic nuclei, or AGN, synonymously with active galaxy.

With increasingly powerful telescopes and detectors it is now believed that up to 10% of all galaxies harbour an active nucleus. There are other galaxies that are also very luminous; these are called *super starburst* galaxies and their luminosity is dominated by thermal re-radiation from heated dust. We know that for their less powerful cousins, the far-infrared or *starburst* galaxies, the source of the re-radiated dust emission is due to massive young stars. However, we can also ask what triggers this massive burst of star formation. In the ultraluminous infrared galaxies, there is intense speculation that these are indeed powered by supermassive black holes, their normal signature being hidden by the mass of the surrounding dust. Perhaps these are really buried quasars, a topic we shall return to in chapter 9.

High luminosity means a high rate of energy loss. In terms of the lifetimes of highly luminous AGNs, unless there is an adequate supply of gas to fuel the activity, a galaxy is unable to sustain the observed luminosity for a significant fraction of the age of the Universe. We will see that although this is possible, it leads to a tremendously massive black hole being built up in the galaxy and for which no evidence has been found. Therefore, most astronomers believe that activity is a transient phenomenon which occurs in many galaxies—*active galaxies are not active all the time*. This is a very important statement.

In our quest to understand the workings of active galaxies it is necessary to gain a feel for the overall picture. Indeed, active galaxy research is one of the most complex in modern astronomy. This is not because of its intrinsic difficulty or theoretical requirements, rather it is due to the wealth of detail that needs to be absorbed in order to pull together all the diverse observations and physical processes that provide us with our ultimate understanding. This, of course, is part of the challenge. To make the process of understanding as simple as possible we shall tackle this challenge in steps. The first step may come as a surprise, as we see why the classification system for active galaxies has been a significant barrier to understanding. We shall also find it instructive and informative in constructing the big picture to take a historical look at the discoveries of AGNs. We will see how objects with strange sounding names such as quasars, Seyferts and blazars were discovered. We shall learn how new pieces of observational evidence fitted into the ever-improving picture, culminating in the modern viewpoint of a grand unification, which is described in chapter 9.

In all observational sciences, evidence is obtained from experimental data. For astronomers the experiments usually consist of collecting radiation from the distant reaches of space by using giant telescopes and the best detection systems. These provide the precious clues just like those in a detective story that will hopefully lead to the unravelling of the secrets of the mysteries of space. This book elucidates the secrets of quasars, Seyferts, blazars and other objects, all of which make up the family of active galaxies. We shall even indulge ourselves in speculation about how active galaxies were formed, how they evolved and subsequently changed with time. This leads into the realms of cosmology and the origin of the Universe itself, but here we shall tread warily. A sign that I placed on my office door in the late 1970s reads 'Many spiral galaxies have been quasars'. In later chapters we shall discuss the implications of this profound statement and in so doing discover whether the spiral galaxy in which we live, the Milky Way, was once a quasar.

In keeping with this great detective story, along the way we shall make note of those aspects of technological innovation by which important discoveries were made. Likewise, in the true spirit of Sherlock Holmes or the modern-day equivalent of Miss Marple, we shall investigate and learn how the clues we receive in the form of observational data from telescopes can be interpreted through the laws of physics into deductions about the properties of these most exotic and powerful objects in the Universe. Physics is both fantastic and fun, and in chapter 2 we will discover how astronomers probe the depths of the Universe and through their faith in the laws of physics determine what we can never hope to see or touch in our earthbound laboratories.

Before getting to grips with active galaxies we will first set the scene in section 1.2

by reviewing our place in the Universe. Those readers already familiar with the Solar System and galaxies, and comfortable with astronomical methods of distance measurement, can move on to section 1.3. Throughout this book we shall endeavour to use SI units, unless they are so out of context with mainstream astronomical usage as to be misleading. For readers not used to metric units, other units will occasionally be given. Appendix 1 lists units and conversion factors. As we have already seen, we shall make use of normal scientific notation based on powers of ten. One thousand is written 10^3 and is called kilo (as in kilometre), one million is 10^6 and is prefixed as mega (as in megaparsecs), one thousandth is 10^{-3} (as in millimetre) and one millionth is 10^{-6} (as in micrometre). We define a billion (giga) to be 10^9, a thousand million.

1.2 OUR PLACE IN THE UNIVERSE

1.2.1 The Solar System to the stars

Because distances in the Universe are so vast, the standard units of length, the metre (m) and kilometre (km), are far too small for convenient use. Light, and all electromagnetic (e-m) radiation, travels at a velocity of 2.998×10^8 m s^{-1} (metres per second) in vacuum. For simplicity, we shall take the velocity of light to be 3×10^8 m s^{-1} or 3×10^5 km s^{-1}. The popular unit of astronomical distance measurement, the light-year, is simply the distance travelled by light in one year. There are $60 \times 60 \times 24 \times 365.25 = 3.16 \times 10^7$ seconds in a year, therefore a light-year is equal to $(3.16 \times 10^7) \times (3 \times 10^5) = 9.467 \times 10^{12}$ km. However, although the light-year is easy to understand, it is not the unit of distance used by astronomers. Instead, astronomers generally use a unit called the parsec (pc). Before defining this we should pause for a moment to reflect on the implications of the velocity of light, which although incredibly high, is nevertheless finite and has an absolutely crucial significance for astronomy. When we look out into the depths of space, we see objects at varying distances. The farther away the object, the longer it has taken the light to reach us. Therefore we do not see the Universe as it is now, but as it was sometime in the past. Put another way, as we look farther out into the depths of space we see the Universe when it was younger.

Astronomers can look back into the distant past, to a time long before life existed on Earth. In so doing their telescopes are a form of time machine, probing the history of the Universe. Indeed, it is precisely by studying the most distant objects in the Universe that astronomers have some confidence that they can construct theories of the evolution of the Universe from the very beginning of time to the current epoch. Furthermore, these theories predict how the Universe will evolve in future times.

Let us briefly give one or two examples of the consequences of the finite velocity of light. In everyday life events seem to happen instantaneously as far as a local observer is concerned, because distances are very small. However, the arrival of the space age brought much larger distances into play and the finite velocity of light began to be noticeable. When the Apollo astronauts ventured onto the lunar surface, the radio communication delay between them and the receiving station on Earth was about 1.3 seconds (and vice versa). The light travel time to Mars is 4.3 minutes, to Jupiter almost 35 minutes and over an hour to Saturn. Sending commands to space probes at these distances requires careful planning and the time delay of sending and the spacecraft receiving the command must obviously be taken into account.

Another example of the finite velocity of light can be seen in an interesting thought experiment. Let us postulate the rather highly unlikely event of the Sun catastrophically going out! The Sun provides all the light and heat for the Earth and is at a mean distance of 1.496×10^8 km. If for some unexplained reason, the Sun were to be extinguished uniformly all over its surface at, say noon Earth-time, then we on the Earth would not know anything was amiss until 8.3 minutes after noon. This is the time the light would take to reach us from the nearest point on the solar surface. We would then see a dark spot suddenly appear at the centre of the Sun's disk. This would spread rapidly across the surface until the entire Sun was extinguished just 2.32 seconds later. Apart from creatures existing near undersea volcanic vents, life on Earth would then come to an abrupt end. Luckily this is just a thought experiment, but it shows that even on a very local scale (astronomically speaking) everything is relative and we only see the Sun as it actually existed about 8 minutes ago. Astronomically speaking, the Sun is 8.3 light-minutes away from the Earth.

The finite velocity of light is also crucially important in that the variability of light from an object can tell us about the size of the region that is emitting this light. Consider the above example again. Because the centre of the solar disk as seen on the sky is closer to the Earth than the edge of the disk, the time that the black spot takes to travel outward from the centre of the disk to the edge is just the extra time it takes the light from the limb of the Sun to reach us. This is the radius of the Sun, which by our calculation is some 2.32 light-seconds in length. So if we watched the Sun going out from the Earth, then we would measure that it went from full bright to completely out in 2.32 seconds and we would define this as the variability timescale. We can translate this into a size for the Sun by assuming that the Sun is spherically symmetric (i.e. round, which is a pretty good approximation). Therefore, if 2.32 light-seconds is the radius, then the diameter of the Sun becomes 4.64 light-seconds.

We can extend this concept to measurements of variability, including those seen from active galaxies. For simultaneous brightening of an object, such as the surface of the Sun in the above example, the different arrival times of the light tells us about the size of the object (assuming a particular geometry). For a brightening of, say one day, we can infer that the object is about 1 light-day in radius if we assume it is spherical. This provides a method of determining the approximate size of the emitting region from the variability timescale. This is particularly important for the study of active galaxies because although it does not require the distance to the object in question to be known, it nevertheless puts a limit on the size of the object. Another important use of variability timescales for active galaxy research is subtly different but still uses the same principle that information cannot be communicated between two regions faster than the speed of light. This measures how one region (such as gas clouds surrounding an active nucleus) responds to the brightening (or fading) of the central engine. The time delay between the response of the gas to the stimulus directly gives the distance between the two regions and so allows us to build up a picture of the structure of the AGN. In section 5.4.4 we will see that this leads to a very important technique for active galaxy study known as reverberation mapping.

Let us now return to the issue of why astronomers use the parsec rather than the light-year. 'Parsec' sounds much more complicated than a light-year until one understands how it originates. Parsec is an abbreviation for the parallax second. I suspect this has not made the story any clearer, but it has provided a clue. The 'second' does not refer to the unit of time, but the unit of angular measure, the arcsecond (″). Remember, there are 60 arcseconds in 1 arcminute, 60 arcminutes in 1 degree and 360 degrees in a circle. In passing, the diameter of the full Moon seen from the Earth is about 30 arcminutes. But what about parallax? Parallax is the method of determining distance practised by surveyors using theodolites. They make angular measurements of an object, such as a measuring stick, from the two ends of a baseline. Knowing the length of the baseline and the two enclosed angles, application of simple trigonometry allows the distance of the object in question to be calculated. Such a method was used to make all the early maps and is still used today in the surveying and construction industries.

The parsec is the distance at which the Earth–Sun distance subtends an angle of 1″. The Earth–Sun distance of 1.496×10^8 km (~ 93 million miles) is called the astronomical unit, which is abbreviated to AU. The determination of distances by the method of trigonometric parallax is outlined in fig. 1.1. In terms of simple trigonometry we have $\sin\theta = (1\ \mathrm{AU})/d$. But for very small angles, $\sin\theta$ is nearly equal to θ, and therefore $\theta = (1\mathrm{AU})/d$, or alternatively,

$$d = \frac{1\ \mathrm{AU}}{\theta}. \tag{1.1}$$

The above deduction uses the angular measure of radians and we would like to convert this to more usual units of arcseconds. There are 2π radians in 360° (see section 2.2.3), so $1'' = 1/206{,}265$ radians. Therefore we can now express the displacement angle, θ, in arcseconds (″), and from our definition of the parsec we have

$$d = \frac{1}{\theta''}\ \mathrm{pc}. \tag{1.2}$$

To convert the parsec to kilometres, for $\theta = 1''$ we convert to radians by dividing θ by 206,265, substituting for the AU in km, and we then obtain 1 parsec (pc) $= 206265 \times (1.496 \times 10^8) = 3.086 \times 10^{13}$ km. If the angular displacement, θ, of a nearby star with respect to the very distant background stars is 0.05″ as measured from the radius of the Earth's orbit, then the star lies at a distance of 20 parsecs. Note that we had to divide the angle as seen from the extremes of the Earth's orbital path around the Sun by the factor of two in order to comply with the definition of the parsec using the length of the astronomical unit, the radius of the orbit.

Trigonometric parallax is a direct method of distance determination, involving only the accuracy of measuring the parallax angle and the knowledge of the value of the astronomical unit. It is free of many of the assumptions that are used in other methods to determine stellar distances. The measurement of the distances of nearby stars using trigonometric parallax is of crucial importance and provides a platform on which the rest of the distance scale of the Universe is constructed. We need to know the distance to determine one of the most fundamental intrinsic properties of any radiating body, its luminosity. Unfortunately, because of the distances involved and the difficulty of measuring such small angles, only about 100 stars had very good distance measures derived by trigonometric parallax until the arrival of the Hipparchus satellite, which improved this by several orders of magnitude. To explore the shape and size of our Galaxy and beyond, other methods of determining distance must be used. However, these are all indirect techniques, building on other measurements in the form of a distance ladder. As we venture farther into the Universe, our knowledge of distances and all quantities that depend on this knowledge (such as luminosity, mass and size) become much less certain. We will not pursue this topic further; suitable references for additional study are presented in chapter 1.11.

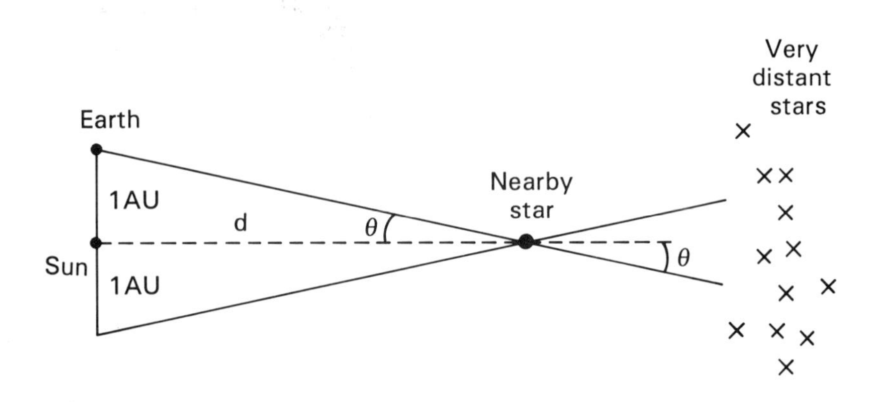

Fig. 1.1. Outline of the method of distance determination by trigonometric parallax. The nearby star changes its position with respect to the distant background stars when seen from the extreme positions of the Earth's orbit six months apart.

To give a simplified overview of distances, separations and sizes, stars in our Galaxy are about 1 pc apart, our Galaxy is about 30 kpc in diameter, the largest galaxy is about 100 kpc in size and galaxies are about 1 Mpc apart in space. Finally, we also note that 1 pc = 3.262 light years. So if you feel more comfortable with light years, just multiply the numbers in parsecs by about 3.3 and you will be in the right ball park.

1.2.2 The Milky Way Galaxy

Now that we are more familiar with the idea of distances we can begin to explore our place in the cosmos. The Universe is a vast sea of galaxies, many shining by the light of more than a billion stars. We live in a spiral galaxy that we call the Milky Way. It is shaped like a flattened disk with a nuclear bulge and a good picture is to imagine two eggs done 'sunny side up' placed back-to-back (plate 1). Our Galaxy is about 30 kpc in diameter, 5 kpc thick in the central bulge and 0.5 kpc thick towards the rim, where the Solar System lies. The Galaxy contains over two billion stars as well as gas and dust;

these latter materials are mostly confined to the plane, or disk of the Galaxy.

We can see the disk of our Galaxy when we look up at the faint band of light in the night sky. This is the Milky Way. It is the accumulated light from the millions of stars lying in the plane of our Galaxy. It was in 1609 that Galileo Galilei, using the first astronomical telescope, discovered that this hazy band of light was due to millions on millions of stars. Go out on a clear dark night, away from street lights, and look for the Milky Way. The light you see will be from stars that are much less than 1 kpc away. View the Milky Way with either a small telescope or binoculars and you will see countless stars, seemingly packed one on top of another. However, this is far from the case. As we noted above, stars are roughly 1 pc apart in the Galaxy and the appearance of overcrowding is just an optical illusion. The Sun lies in the plane of the Galaxy and the stars of the Milky Way stretch a full 360 degrees around the sky, giving the impression that the Solar System might be at the centre. This is incorrect but it turns out to be very difficult to determine exactly where the Sun lies in the Galaxy as we shall now see.

Indeed, until the early part of this century, just as the ancient civilizations had always considered the Earth to be the centre of the entire cosmos, it was believed that the Sun lay at the centre of the Galaxy. The reasoning behind this view is rather obvious. A simple study of the night-time sky reveals a generally symmetric picture with bright stars approximately uniformly distributed and the faint glow of the Milky Way stretching like a narrow band around the sky. From our point of view, we seem to be at the centre of things. Many statistical studies of the distribution of stars in space were undertaken using a technique called star counting. This was begun by Herschel in 1760 and culminated in some highly sophisticated work during the first few years of this century. However, in spite of their sophistication, all these studies produced a heliocentric (Sun-centred) model for the Galaxy, which we now know to be completely incorrect.

Harlow Shapley provided the breakthrough. Shapley was an American ex-journalist turned astronomer who, in 1910, started a detailed investigation of globular clusters. Globular clusters are massive, spherically symmetric assemblies of over a hundred thousand stars. As his studies progressed he was astounded to discover that unlike stars, which had been shown to be approximately symmetrically distributed in the night sky, the globular clusters were far from uniform. He found that over 30% of the globular clusters seemed to be grouped together in only 2% of the sky. This highly pronounced anisotropy could not come about by chance, but must be revealing something about their intrinsic distribution in space. Shapley made the bold assumption that the globular clusters must be telling us something about the shape of our Galaxy. He reasoned that globular clusters, being massive systems in their own right, must be major constituents of the Galaxy. As such they were probably a good pointer to the overall shape and size of the Galaxy and hence of the underlying distribution of stars. Somehow the star-counting method was failing to delineate the correct stellar distribution. He was absolutely correct.

Excited by this discovery, he refined his techniques of determining the distances of globular clusters and thereby studied their spatial distribution in three dimensions. He found that they seemed to be grouped about a particular direction in the Milky Way. Shapley suggested that both the globular clusters and the stars in our Galaxy were isotropically distributed about the centre of the Galaxy. This lay in the general direction of the constellation Sagittarius but at a far greater distance of about 17 kpc from the Solar System. At a single stroke, this placed the Sun far from the centre of the Galaxy. Following the displacement of the Earth from the centre of the cosmos by Copernicus, this discovery brought about the second revolution concerning our place in the Universe.

Why had star counting failed and Shapley succeeded? The culprit was the presence of large quantities of dust spread throughout the plane of the Galaxy. The dust absorbs visible light and prevents us from seeing parts of the Galaxy more distant than a few hundred parsecs in the plane of the Milky Way. Because the Solar System is located in the plane of the Milky Way it is easy to figure out why the stars are distributed approximately uniformly. Imagine being in a forest with a heavy ground fog. The nearby trees appear approximately uniform all around and the more distant trees gradually become invisible as they are enshrouded by the fog. These observations suggest we are at the centre of the trees. If the fog suddenly lifted, then we could see that the edge of the forest lay in a particular direction and we were not at the centre of the forest at all. A view from a helicopter above our position would also have shown us the true picture! Shapley's globular clusters lay outside the plane of the Milky Way, and were not troubled by dust absorption to a significant extent and were therefore easily visible over enormous distances.

In fact two globular clusters lay very close to the Milky Way and these had presented a puzzle to Shapley. He had determined that the diameters of globular clusters were linked to their apparent brightness, yet these two particular clusters appeared too faint for their size. They had both been dimmed by the dust absorption in the plane of the Galaxy and were a clear clue to the presence of dust in the plane of the Milky Way, something that was not picked up by Shapley. The presence of dust will be important for our study of active galaxies; however, in these cases we will not be considering the dust that is spread out thinly in the galactic planes, but rather heavy condensations of dust surrounding the central kiloparsec of a galaxy. As an aside it should be noted that in his studies of variable stars Shapley had also unearthed a clue to the fact that there were two different types of stellar population, later called population I and II. But the final proof of this discovery was a further thirty years in coming.

The true extent of dust in the Galaxy became apparent in the 1930s following the work of Trumpler who showed from studies of star clusters that dust must be prevalent in the plane of the Milky Way. We now know that the dust limits our view to about 2 kpc in the plane of the Galaxy even with the largest of optical telescopes. This is only about a quarter of the distance to the Galactic Centre, which remains totally invisible in the optical domain. We believe that the centre of the Galaxy is about 8 kpc distant. However, astronomers are not totally cut off from the secrets of the inner parts of the Galaxy or the cores of dusty regions in space; these can be observed by using wavelengths that are not significantly absorbed by dust. Here radio and infrared astronomy flourish. Plate 1 shows a picture composed of data from the COBE satellite and shows a composite image in three infrared colours. These penetrate the dust and we can see directly to the central regions. The bulge of the central regions of the Galaxy is clearly seen along with the very thin and flat spiral arms in which the Sun is located.

In the years following the discovery of carbon monoxide (CO) in space in 1970, we now know that our Galaxy contains a cold gaseous molecular medium. This cold contribution is a vital ingredient and the giant molecular clouds provide the source of material and the nurseries for star formation. This cold component will not figure highly in our discussion of active galaxies but has an important bearing on being a potential fuel supply for a central engine. In the past year or so, observations of the distribution of cold molecular gas along with data from the COBE satellite have persuaded many astronomers that the Milky Way galaxy is not just an ordinary spiral galaxy, but is in fact a barred spiral.

1.2.3 The Universe of galaxies

The story of the discovery of the Universe of galaxies starts with a rather unexpected character, a French comet-hunting astronomer called Charles Messier. In his quest for discovering new comets he became annoyed by the faint misty patches of light that could so easily be mistaken for a new comet. To overcome this frustration he made a catalogue of these 'nebulae' as they were called (from the Greek word for mist) so that he would not be misled in his comet hunting. Messier's original catalogue of nebulous objects was completed in 1784 and is probably the most famous of astronomical catalogues. Although it contains a miscellaneous assortment of 103 objects (gaseous nebulae, globular clusters, open star clusters and galaxies), these are the brightest nebulae in the sky and the list is still used today as a source of pleasure for countless amateur astronomers and as a godsend to teachers of astronomy on clear evenings.

Other astronomers studied these nebulae voraciously. With the advent of the spectrometer and new telescopes (see section 1.4) the Messier objects were better classified and subsequently grouped into star clusters, globular clusters, gaseous nebulae and the so-called spiral nebulae. This last category had been named by Lord Rosse due to his perception of their shape when first observed through his giant 72-inch reflecting telescope in 1845. The status of the spiral nebulae provoked a vigorous battle between those astronomers who believed that they were giant groups of stars associated with our Galaxy and those who believed they were 'island universes' of stars, just like our own Galaxy but at enormous distances from us. A young Swedish PhD student called Lundmark correctly stated that the Andromeda spiral nebula (M31 in Messier's catalogue) was a remote galaxy of stars, but his evidence was rejected by many of the older members of the astronomical establishment. Their reasoning was that if Andromeda was at the great distance claimed by Lundmark, then one of the flare stars (novae) that he had used to calculate the distance would be incredibly powerful, indeed vastly more luminous than any of the novae that had been observed in our Galaxy. This posed severe problems. In fact Lundmark turned out to be correct; he had discovered supernovae, the explosive end-points of massive stars.

The major breakthrough finally came from observations using the newly constructed 100-inch reflecting telescope on Mount Wilson. These showed conclusively that the spiral nebulae were indeed 'island universes' of stars, galaxies like our Milky Way, but very distant in space. This momentous discovery was made by Edwin Hubble in 1924. It solved the debate over the status of the spiral nebulae and opened up our view of the cosmos. We were merely one galaxy among tens of thousands, soon to be millions of millions of galaxies as observation techniques improved. The Universe was indeed vast. Hubble's crucial observation was a great discovery and ranks as one of the most important in astronomical science.

Although galaxies are intrinsically very luminous, many shining with the light of more than a billion suns, the vast majority are so distant that they remain incredibly faint objects as seen from Earth. Indeed, only one large galaxy can clearly be seen with the unaided eye, the Andromeda galaxy (M31). At 650 kpc this is one of our very closest neighbours. It is indeed awe-inspiring to realize that when we look up at the night sky and see this faint misty blur, the light left the galaxy over two million years ago, travelling unimpeded through space only to be converted into an electrochemical impulse by the retina of the eye.

Hubble obtained a distance to M31 of 250 kpc, which although far too small a value, was sufficient to prove conclusively that it was indeed a separate galaxy. With this discovery extragalactic astronomy was born. (Ironically, Hubble himself preferred the term extragalactic nebula rather than 'galaxy', which was favoured by most astronomers including Shapley.) Extragalactic astronomy exploded into prominence and has never looked back. Because of his pioneering work on galaxies and the redshift relation (see section 1.8), Edwin Hubble is reverently referred to as the father of extragalactic astronomy and the Hubble Space Telescope (HST) was named in his honour.

We close this overview by noting that although the Universe is made up of galaxies, these are not isolated phenomena. More often than not they are found in families, ranging from small groups of galaxies to clusters and enormous superclusters spanning many megaparsecs of space. Further studies of galaxies and clusters led to the modern study of cosmology. Previously this had been the province of philosophy, incorporating armchair theories and abstract mathematical modelling, far removed from the aspersion of observational science. Cosmology had been reborn, and it rapidly sprang forth into a youthful observational discipline, full of potential.

1.3 A CLASSIFICATION SCHEME FOR GALAXIES

With the discovery that spiral nebulae were galaxies of stars, the race was on to search for and identify more galaxies. Soon astronomers were finding galaxies of all shapes and sizes, and it became clear that some form of classification scheme was needed. The first years of this new discipline were spent attempting to measure the major aspects of galaxies, such as distance, size and luminosity. However, one of the easiest things to assign was the shape of the galaxy, and this formed the basis for the first classifications.

In any classification scheme, some basic parameters are grouped together according to established rules. These will doubtless change with time, but nevertheless, the basic scheme may be expected to reveal fundamental clues about the life story of the objects and hopefully how they fit in with other known phenomena. As far as galaxies were concerned, the most basic of the features used for a classification scheme was their shape. Again, Hubble was the prime architect of these studies. He identified galaxies as belonging to three main groups: spirals, ellipticals and irregulars. This was refined further by subdividing spirals into those which had a bar in their centre; these were called barred spirals. Spirals and barred spirals were divided into three main subtypes according to the size of the nuclear bulge, the degree of opening of the spiral arms, and

Fig. 1.2. Examples of galaxy types. Left to right, top: M87 (E0), NGC147 (dwarf E5); centre, M31 (Sb), NGC1365 (SBb–note the prominent bar); bottom, NGC2997 (Sc) and NGC4321 [M100] (Sc). The photographs are from the Anglo Australian Telescope apart from NGC147 and M31 which are from the Hale Observatories.

the fragmented nature and presence of star formation within the spiral arms. These were called Sa, Sb and Sc (SBa, SBb, SBc for the barred types) and are illustrated in fig. 1.2. Elliptical galaxies were subdivided according to their ellipticity, being classed as E0 (appearing spherical in shape) to E7 (highly elongated). Galaxies that fitted neither the elliptical nor spiral classification were lumped together and called irregular galaxies.

As the morphological studies of galaxies progressed and more features could be identified in photographic images, the classification became more complex. Spirals were subgrouped into numerous forms depending on whether they showed inner or outer ring features. We shall restrict ourselves to the basic Hubble types of galaxy, albeit modified. to include the intermediate spiral categories of Sab, Sbc and two new types called Sd and Scd. With better observational evidence, it became clear that the irregular galaxies could also be separated into two groups, termed irregular type I and II. We now know that the former are normal but on the whole relatively small, spiral galaxies seen edge-on while the latter are oddities and will figure prominently in this book. Indeed, these gave the first clues to the presence of active galaxies

1.4 THE USE OF GALAXIES AS COSMOLOGICAL PROBES

Before moving on to learn about the discovery of active galaxies, there is one more vital tool for extragalactic astronomy that we must describe. This came from another major observational discovery by Hubble and its importance cannot be over-emphasized. Hubble's discovery of the expanding Universe began the modern concept of cosmology as an observational science. This great discovery came about in the late 1920s.

Working with his close colleague Milton Humason, Hubble investigated the faint absorption lines in the spectra of galaxies. The lines (see section 2.6) came from absorption by elements in the outer atmospheres of the stars in the galaxies (like all stars, the Sun shows prominent absorption lines in its spectrum). Hubble and Humason found that the wavelengths of the absorption lines were shifted with respect to their measured laboratory values. If this was attributed to the Doppler effect (see below) then it showed that the galaxies were moving through space with respect to our Galaxy. Surprisingly, virtually all the galaxies seemed to be moving away from us, albeit with differing velocities. Hubble was fairly sure of the distances of the galaxies he had observed, and when he constructed a simple graph plotting the distance of the galaxies versus their velocity of recession, he obtained a remarkable correlation (shown in fig. 1.3).

Instead of the graph being a scatter diagram with points distributed randomly, he was convinced that he could see a clear relation between the two parameters. He made the further speculation that the correlation was a straight line. Inspection of Hubble's original plot showed that he made this statement more as an act of faith rather than from a strict statistical analysis of the goodness of the fit of a straight line, or even a reasonable correlation between the two quantities. There was considerable scatter on the points on the plot and the straight-line fit was far from obvious. But like many a great scientist, Hubble made a bold intuitive leap and obtained the correct answer that has been completely verified as distances were better determined. His farsighted conclusion was that the velocity at which a galaxy recedes from us (the recession velocity) is directly proportional to the distance of the galaxy. This is known as Hubble's Law of recession of galaxies (usually shortened to just Hubble's Law) and is one of the most fundamental and important discoveries of observational cosmology. Hubble's Law is written

$$v = H_0 D, \tag{1.3}$$

where v is the velocity of recession along the line-of-sight as measured by the Doppler shift of the spectral lines and expressed in km s^{-1}, D is the distance of the galaxy expressed in megaparsecs, and H_0 is called the Hubble constant. From eqn. 1.3, the Hubble constant has units of kilometres per second per megaparsec (km s^{-1} Mpc^{-1}). The subscript '0' refers to measurements made at the current epoch and is a standard notation found in cosmology.

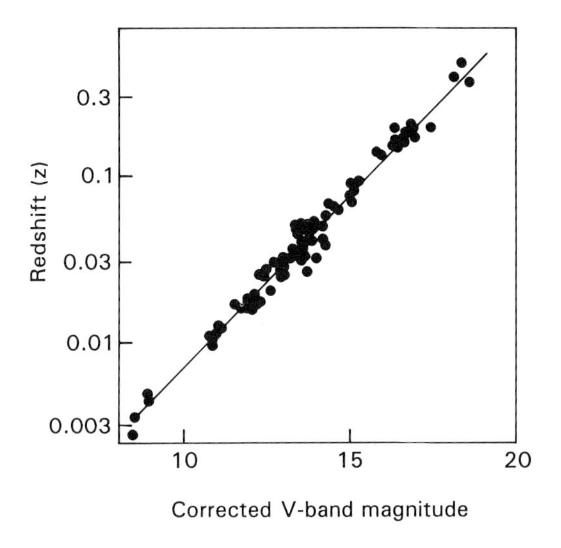

Fig. 1.3. Hubble's Law of the recession of galaxies. The ordinate (y-axis) is the redshift of the galaxy and the abscissa (x-axis) is the optical apparent magnitude. The dots are representative of various measurements. Note that Hubble only observed out to a redshift of $z < 0.005$ when he published his law of recession in 1929.

The determination of the value of this constant is critical for investigating the past and future history of the Universe, the realm of cosmology. Hubble obtained a value of 550 km s^{-1} Mpc^{-1}, but it turned out that his distance estimates were significantly too low and as these have been refined over the years the adopted value of H_0 has fallen and currently lies somewhere between 50 and 100 km s^{-1} Mpc^{-1}. This factor of two spread in uncertainty is directly attributable to the difficulty of determining distances to galaxies, even today. This has ramifications for our story of active galaxies; in order to determine their luminosities and their distribution in space, their distances must be known. The prime method for determining distances to all but the closest of active galaxies is to make use of Hubble's Law. The astronomer measures the redshift of a galaxy by identification of the lines in its spectrum (see next section) and so determines its velocity of recession. Equation 1.3 is then used with a 'preferred' value of the Hubble constant to obtain the distance of the galaxy. We shall see an example of this calculation in chapter 2 when we deduce the luminosity of the quasar 3C273.

1.5 REDSHIFTS AND COSMOLOGY

How exactly is the redshift of a galaxy measured? The displacement, $\Delta\lambda$, of the wavelength of a spectral line in the distant galaxy, from that of the same spectral line but measured in the laboratory is determined. This latter wavelength is termed the rest wavelength and is usually denoted by λ_0. The displacement in wavelength ($\Delta\lambda$) can be interpreted by the Doppler effect to give the velocity, v, along the line-of-sight of the galaxy with respect to the observer on the Earth. The Doppler effect is very simple but tremendously important in astrophysics. It states that if a gas which is either emitting or absorbing light is in motion with respect to the observer, then the resulting spectral line is moved from its rest wavelength (λ_0) to a different wavelength (λ_1) according to the relation

$$\frac{\lambda_1 - \lambda_0}{\lambda_0} = \frac{\Delta\lambda}{\lambda_0} = \frac{v}{c}, \tag{1.4}$$

where c is the velocity of light, and v is the line-of-sight velocity of the emitting material, which can be either toward or away from the observer. Gas moving toward the observer shifts the wavelength of emission to shorter wavelengths, i.e., to the blue, while gas moving away shifts it to longer wavelengths, i.e., to the red. In terms of equation 1.4, convention has allocated redshifts to positive velocities and blueshifts to negative velocities. Hubble's observations of galaxies showed that virtually all were moving away from us and therefore had redshifts (M31 was a notable exception and it is moving towards us). The term redshift is often written

$$z = \frac{v}{c}\ . \tag{1.5}$$

Although this is frequently quoted for the redshift equation, it is an approximation, valid only for small values of v/c. We realize that equation 1.5 cannot be strictly correct since many observed quasars have redshifts much greater than 1, which would imply that they are receding from us at a velocity exceeding the speed of light. Einstein's special theory of relativity states that an object cannot accelerate to, or exceed the speed of light, and that information cannot be transmitted at a velocity greater than that of light in vacuum. Equation (1.5) is valid only for the small values of v/c that we are used to dealing with in the normal everyday world. It breaks down for v/c > 0.3 and for these situations one must use the correct relativistic equation for the redshift, z, which is given by

$$(1+z)^2 = \frac{(c+v)}{(c-v)} \tag{1.6}$$

and is shown in fig. 1.4.

Using the full relativistic redshift expression (eqn. 1.6) we see that in the case of a quasar at a redshift of 1.0, we obtain a value for v/c of 0.6. The velocity of recession is therefore 60% of the velocity of light and the quasar is receding from us with a velocity of 1.8×10^5 km s^{-1}. Substituting this value into the Hubble Law distance equation (equation 1.3) and using a value of the Hubble constant of 50 km s^{-1} Mpc^{-1}, we find the quasar is at a distance of 3.6×10^3 Mpc. (In fact this is an overestimate for two reasons: we have neglected to take account of the fact that when considering cosmological distances and times the Hubble constant changes with cosmic epoch, and that the curvature of space effect can also be important.) However, for this simple example and unless absolutely essential, we will generally ignore cosmological effects such as the curvature of space.

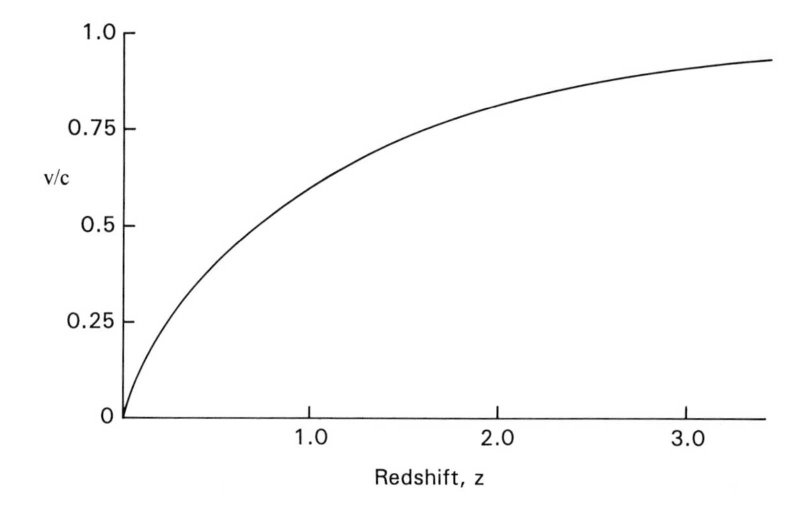

Fig 1.4. Relativistic relation of recession velocity v/c and redshift z.

A distance of 3,600 Mpc is enormous and represents a very much earlier epoch of the Universe than we see in our local neighbourhood. How far back in time is this? We can take a very simplistic approach and calculate how long the light has taken to reach us. We do this by converting the distance of 3,600 Mpc into light years, which gives $3.6 \times 10^9 \times 3.262 = 11.7 \times 10^9$ light years. This simple treatment shows us that the quasar existed almost 10^{10} years in the past. We will give a more refined treatment of this same question in section 1.7, after we have been introduced to some cosmology. Nevertheless, our current belief is that the Universe is about 1.3×10^{10} years old. Therefore our simplistic treatment is telling us that the quasar is presenting us with information about conditions in the early Universe.

1.6 THE BIG BANG COSMOLOGICAL PICTURE OF THE UNIVERSE

The above discussion has been firmly based on the big bang cosmological picture. As we shall see, considerations of active galaxies in terms of a 'quasar epoch' and potential collisional-triggering mechanisms are closely linked to cosmology. We will now present a very brief overview of a simplified picture of the big bang and subsequent evolution of the Universe. For a more detailed discussion see Roger Tayler's *'The Hidden Universe'* in this series.

In the beginning there were quantum fluctuations in the void that spontaneously erupted and exploded into the seed of the Universe we know today. At a time very much less than one second after the big bang there was a sudden period of inflation and exotic particle reactions took place. At an age of about 1 second, the Universe had a thermodynamic temperature of about 10^{10} K and was composed of photons (electromagnetic radiation), neutrinos, neutrons, protons and electrons. The particles and radiation were closely coupled and were in thermodynamic equilibrium. Because the Universe was expanding rapidly, the density and the temperature of the radiation field were also rapidly falling and in the next few seconds the neutrinos de-coupled from the plasma. (Decoupling is a technical term meaning that the particles stopped interacting with the other constituents, especially radiation.) The neutrinos cease interacting when the plasma density falls to a sufficiently low value that the neutrinos effectively see a transparent medium. Stated another way, the mean free path for neutrino interactions suddenly becomes extremely large (see chapter 4 for mean free path arguments). From this time onwards neutrinos are consigned to continue their ghostly paths through the Universe, without significant interaction with matter or radiation, until the end of time. The importance for us is that once the neutrinos decoupled the ratio of neutrons to protons is frozen.

However, we know that this ratio cannot remain fixed for long because the free neutron is an unstable particle, decaying with a timescale of around 900 seconds into a proton, electron and anti-neutrino. But before this decay could really get going, another process took over, representing another form of freezing-in of interactions. In those first few seconds, one of the nuclear reactions that was taking place was the combination of neutrons and protons to form deuterium nuclei. Deuterium is an isotope of hydrogen, the nucleus of which contains a proton and a neutron and is usually written as ^{2}H. Once formed in the very early Universe, the nuclei are immediately split apart again by the high-energy photons present. But, after about 100 seconds, when the temperature in the Universe had fallen to about 10^9 K; there were insufficient high-energy photons to continue the break-up of deuterium and at this point its abundance began to increase dramatically. The deuterium nuclei, however, had a brief moment of glory, because almost as soon they were created, they combined in pairs to form something even more important. This was the nucleus of the element helium, ^{4}He, often referred to as an alpha particle for historical reasons relating to its discovery.

These dramatic events happened during the first four minutes of the Universe and as a process, the production of helium is now relatively well understood. Following this phase of helium production by big bang nucleosynthesis, the building up of heavier elements ceased. It did not resume until a very much later epoch, when nucleosynthesis in stellar interiors and stellar explosions (supernovae) produced all the heavy elements we see today. The reason the epoch of big bang nucleosynthesis was so brief and did not progress beyond helium is because of the extreme difficulty in forming stable atomic nuclei with mass numbers 5 or 8. The mass number is the number of nucleons (protons and neutrons) in the nucleus; hydrogen has a mass number of 1, deuterium 2, and helium 4, etc.

This same difficulty faces nucleosynthesis inside stars. However, in this case there are ways by which the mass numbers of 5 and 8 can be skipped over to get to elements such as carbon, nitrogen, oxygen and beyond. Because these reactions were not possible in the early Universe we are left with the amazing conclusion that virtually all the helium we see around us today is the direct product of the first few minutes of the life of the Universe. Indeed, the observed ratios of the hydrogen to helium to deuterium mass abundances in local regions of the Universe ($0.75 : 0.25 : \sim 10^{-4}$) provide a very strict test of the conditions in the first few minutes after the big bang. The excellent agreement between theory and observation is one of the major platforms on which the success of the hot big bang cosmological model stands.

Following this helium generation epoch, all the remaining free neutrons in the rapidly expanding Universe decayed by the time of ~ 20 minutes. Being electrically charged, the protons and electrons continued to interact strongly with the high-energy photons of the electromagnetic radiation field. In this process, energy was exchanged between the particles and the photons, ensuring that the matter and radiation fields were closely linked. The two energy fields were in thermodynamic equilibrium and they remained in this state until the next milestone in the cosmic evolution, the so-called recombination epoch, when radiation and matter became decoupled from each other.

This happened when the temperature of the radiation field fell to about 4,000 K. At this time the electrons and protons began to combine to form neutral hydrogen atoms and the Universe passed from a state of being an ionized plasma to one comprising a neutral gas. At earlier epochs, any hydrogen atoms that happened to form were immediately re-ionized due to the high energy and number density of the photons of the radiation field. But when the temperature fell to around 4,000 K, the photons no longer possessed sufficient energy to maintain the hydrogen in an ionized state. Hence, over a very short time, the Universe rapidly changed from a predominantly ionized state to a neutral state. This is what is all too often termed the recombination epoch.

We should note that this name is a real misnomer, designed to confuse the student. Usually one associates recombination with a concept whereby matter, which was once neutral, is ionized and then the nuclei and electrons recombine to become neutral again. However, this is not what happened in the early Universe. At all times before the recombination epoch, the hydrogen remained a fully ionized plasma; it became neutral only when the radiation and matter decoupled. However, like so many oddities in life, we seem to be stuck with the name, but I will do my best to assist the desired change by referring to this epoch as the epoch of radiation and matter decoupling. It is relatively easy to calculate that this occurred at a time of around a few hundred thousand years after the big bang and at a redshift, z, of about 1,500, measured from the current epoch.

The decoupling epoch represents a crucial landmark in the evolution of the Universe. Before this time, both the free protons and electrons exchanged energy with the radiation field by a process termed scattering. This ensured that all information regarding spatial structures in the very early Universe was smeared out. The light was scattered as in a dense fog and any imprints of the very early Universe vanished in the mists of time. However, once the protons and electrons combined to form neutral atoms of hydrogen and helium, the neutral gas no longer interacted with the radiation field and the matter and photon fields were free to evolve on their own. They have decoupled.

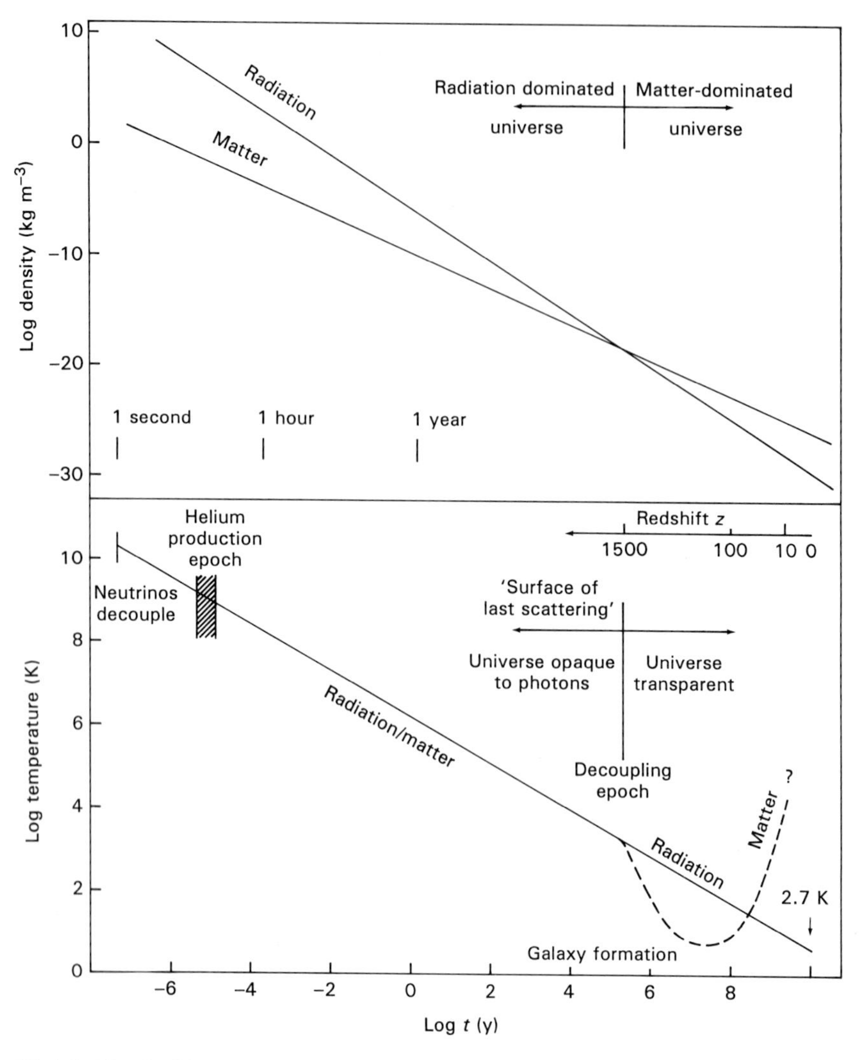

Fig 1.5. Sketch of the evolution of the Universe from the big bang to the current epoch showing a number of important epochs. Note the very short lookback time for z = 5.

This fact is tremendously important because the photons from this epoch onwards are no longer scattered by the matter, travel in straight lines (called geodesics), and are observable today. This epoch of decoupling is the earliest epoch of the Universe that we can directly observe by means of electromagnetic radiation. What is more, we actually detect photons from this precise epoch. They are all around us in the form of the celebrated 2.7 K cosmic background radiation field discovered by Arno Penzias and Robert Wilson in 1964. From a temperature of 10^{10} K at a time of 1 second, the Universe has expanded and the all-encompassing radiation field has cooled dramatically, to a value of about 2.7 K at the present epoch.

Another way of looking at these cosmic background photons is to realize that they originated at the surface of last scattering in the Universe. They therefore provide clues to the inhomogeneities of the density of matter at that epoch, the precise inhomogeneities that eventually formed galaxies and clusters of galaxies, a topic of great current debate and research. The spatial distribution of the small scale structure of the cosmic background radiation gives us direct clues of the clumpiness of the density of material at the epoch of decoupling of radiation and matter. In fact the Universe appears to be incredibly smooth on the small scale at this epoch, posing severe problems for the theories of galaxy formation.

In passing, we should note that on the large scale the cosmic background radiation shows a strong dipole characteristic; the radiation field appearing very slightly warmer in one direction and very slightly cooler 180° away. This is the dipole anisotropy and presents us with clues to the motion of the local volume of space with respect to the cosmic background radiation field.

After the decoupling epoch, the Universe continued to cool and eventually the first galaxies of stars formed. Although we do not know the precise epoch of galaxy formation, we can pose some limits. Galaxies have been clearly identified at redshifts of greater than 2 and quasars are now known with redshifts of order 5. Given that we firmly believe quasars are galaxies with active nuclei, this puts a lower limit on how recently galaxies have formed. Current belief is that galaxy formation occurred between redshifts, z, of 10 to 100. This is an area of intense astronomical research and speculation. It includes scenarios with 'cold dark matter' (CDM) in the Universe, something we will not touch on. Once galaxies of stars formed, synthesis of the heavier elements commenced. The massive stars are especially important because after a brief life of around ten million years, they explode as supernovae. In so doing they scatter the newly synthesized heavier elements into the interstellar medium. New stars then form from this enriched interstellar debris, and as these stars evolve and explode, the interstellar medium of the young galaxies becomes richer in elements such as carbon, etc. Some, and perhaps many stars formed with planetary systems and in at least one such system, one planet formed on which life began and evolved. And here we are today.

The above picture seems simple enough but many details remain to be clarified. It should be made clear that this is just a scenario, science's best estimate of the evolution of the Universe. We do not understand how the Universe came into being and we do not have a precise theoretical understanding of how galaxies or stars formed. Nevertheless, an obvious question for us to ask is how active galaxies come into the picture. Why does a galaxy become active? Do all galaxies undergo some epoch of activity? It turns out that these are two of the most fundamental questions to be answered concerning active galaxies and will form much of the discussion in the later chapters of this book. If we believe our premise that active galaxies are powered by supermassive black holes, then we also need to ask how these black holes formed.

Active galaxies may be triggered by close interactions and mergers, something we will follow up in more detail later. However, we can already speculate that close encounters are more likely to occur at earlier epochs in the Universe in rich clusters of many thousands of galaxies. Hence, the scenario paints a picture that suggests that AGNs should be more numerous in rich clusters and at earlier epochs. This is a prediction that can be tested observationally and searches have been made to determine whether quasars are found preferentially in clusters of galaxies. Although these observations are extremely difficult the best evidence seems to favour quasars being found in rich clusters of galaxies at moderate redshifts. Therefore we believe we have at least one clue as to why some galaxies are active.

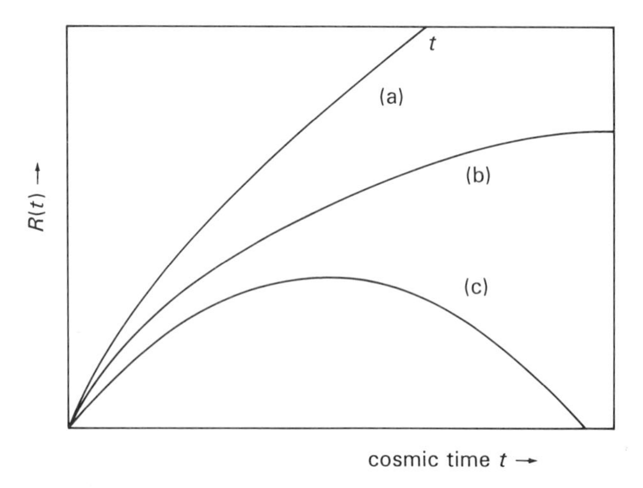

Fig. 1.6. The three standard cosmological models for the evolution of the Universe from the big bang. Model(a) is for an open Universe, model(b) is the critical density Universe (Einstein–de Sitter case), while model(c) is a high density closed Universe. The scale factor R(t) is linked to the radius of the Universe.

Another very important consequence of the big bang cosmology is the behaviour of the expansion of the Universe from a time of $t = 0$. For a hot big bang Universe, three possible models are shown in fig. 1.6. Model A represents an open Universe, in which the Universe continues to expand at all times. Model B is a Universe at the critical density; one of the consequences of this model is that the expansion of the Universe comes to a halt at extremely large values of cosmic time (technically infinity). Model C is the closed Universe, where the expansion decelerates from the big bang and comes to a halt at some finite cosmic time, beyond which the Universe begins to contract, eventually resulting in a cosmic crunch at some very distant time in the future. Which model we live in depends on the average density of the Universe. The critical density case has just enough matter to barely close the Universe, the open model having a lower density, whereas the closed Universe has a higher density, sufficient to close the Universe and cause it to collapse back on itself.

Inflation theories of the evolution of the very early Universe, at epochs many millions of times less than 1 second, require that we live in model B. On the other hand, observations fail to find enough matter to satisfy this prediction, being short by a factor of about ten. This has driven large searches for an 'invisible' ingredient. Cold dark matter is the current best favourite and is believed by many cosmologists to be the major constituent of the Universe, providing just sufficient mass to achieve the critical density. This material is nothing like the everyday matter made up of elementary particles such as neutrons, protons, electrons, etc. (called baryons). Instead it is termed non-baryonic, being composed of as yet undiscovered exotic particles.

Determining the model that best describes the Universe in which we live is of particular importance for cosmology and the epoch of galaxy and quasar formation. Calculations of the properties of objects at high redshift depend on which of these models pertains. This boils down to a dependence on two parameters, the Hubble constant and the deceleration parameter (see later). For those objects where the particular value of the Hubble constant is important, usually one sees expressions written that include for the Hubble constant the parameter, *h*, defined by $H = 100h$ km s^{-1} Mpc^{-1}. So if one preferred a value of 50 instead of 100, *h* would be 0.5 and could be substituted in the calculations to give the desired result.

We can see from fig. 1.6 that the expansion of the Universe decelerates with time in all the big bang models (apart from one in which the Universe is completely empty). In Model B, there is a precise and simple relation between the deceleration parameter, q, and the density of the Universe. In this case $q = 0.5$. In the open Universe, the deceleration parameter is smaller than 0.5 (which is obvious from studying the change of the gradient of fig. 1.6), whereas in the closed Universe, $q > 0.5$. In this book we shall consider only a Universe that is either open or at the critical density, and indeed we shall only need to revert to cosmological aspects in chapters 7 and 9 when we discuss the formation and evolution of quasars.

Calculations turn out to be relatively simple for Model B, which is based on the Einstein–de Sitter model, because the geometry of space-time is flat. The Universe expands according to the simple relation between the scale factor R(t) and cosmic time, t, by

$$R(t) \propto t^{2/3} . \tag{1.7}$$

For our purposes the scale factor, R(t), can be thought of as the factor by how much the Universe scales up or down in size (expands or contracts) and simplistically we can think of it as referring to the size of the Universe. The rate of change of the scale factor with time is therefore

$$\frac{dR}{dt} \propto \frac{2}{3} t^{-1/3} . \tag{1.8}$$

This is important as it allows us to calculate the Hubble constant at various epochs. The Hubble constant is also defined by

$$H = \frac{1}{R}\frac{dR}{dt} . \tag{1.9}$$

From $t = 0$, R increases and dR/dt decreases so the Hubble constant, H, decreases with time. For the critical Universe, it approaches zero at extremely large values of cosmic time, when the Universe becomes static and the galaxies nearly come to rest with respect to each other. The Hubble constant is related to the critical density, ρ_{crit} by the expression

$$\rho_{crit} = \frac{3H_0^2}{8\pi G} . \tag{1.10}$$

For a value of $H_0 = 50$ km s^{-1} Mpc^{-1}, $\rho_{crit} = 4.7 \times 10^{-27}$ kg m^{-3} ($\sim 7 \times 10^{10}$ $M_{\odot}$ Mpc^{-3}). It is also important to be able to calculate the cosmic time at which events happen and to relate this to the redshift, the observational parameter we measure. For the Einstein-de Sitter model, the intersection of the gradient of the R(t) versus t curve with the R = 0 axis in fig. 1.6 occurs at a time τ_0 from the current epoch, t_0. i.e.

$$\tau_0 = \frac{R_0}{dR_0/dt} = \frac{1}{H_0} \quad (1.11)$$

and substituting for R(t) from equation 1.7 we obtain

$$\tau_0 = \frac{3}{2} t_0 \ , \quad (1.12)$$

so

$$t_0 = \frac{2}{3}\frac{1}{H_0} \ . \quad (1.13)$$

In *principle* we can measure the Hubble constant and hence determine the age of the Universe, t_0, the time to the present epoch since the big bang. For a value of $H_0 = 100$ km s^{-1} Mpc^{-1}, $t_0 = 6.5 \times 10^9$ y, whereas for a value of $H_0 = 50$ km s^{-1} Mpc^{-1}, $t_0 = 1.3 \times 10^{10}$ y. For an open Universe, the expansion has been faster and so the gradient of the R(t) versus t curve of fig. 1.6 is steeper at the current epoch. Hence the Universe is actually older and we have

$$t_0 > \frac{2}{3}\frac{1}{H_0} \ . \quad (1.14)$$

We can now extend this to incorporate the redshift, z. In general we have

$$(1 + z) = \frac{R_0}{R(t)} \ , \quad (1.15)$$

and for the Einstein-de Sitter case

$$(1 + z) = \frac{R_0}{R(t)} = \left(\frac{t_0}{t}\right)^{2/3} \ . \quad (1.16)$$

This will be very useful. To give some feel for what we have just accomplished, let us look at some examples. A galaxy at a distance of 100 Mpc has a redshift ($H_0 = 50$) of $z = 0.017$, so $t/t_0 = 0.976$ and we are only looking back about 2.4% of the age of the Universe. On the other hand, a redshift 5 quasar has $t/t_0 = 0.068$. This means that we are seeing this quasar when the Universe was only about 7% of its current age and we have a look-back time of 93%. When we view the cosmic background radiation from the epoch of decoupling at $z \approx 1500$, this gets us to a staggering look-back time of around 99.998% of the age of the Universe. This dramatically illustrates the logarithmic nature of the abscissa on fig. 1.5. The implications for the look-back time and how young the Universe was at redshifts of only around 5 is something that is not always appreciated.

To conclude this section we have seen that there are two key parameters required to describe our cosmological evolution, the Hubble constant and the deceleration parameter. Unfortunately, neither of these are known to a good accuracy. Selecting $H_0 = 75$ km s^{-1} Mpc^{-1} would give a value somewhere in the middle of the current favoured range of 50 to 100. For an Einstein–de Sitter model (favoured by inflation) the lower values make the Universe proportionally older (and more consistent with the ages of the oldest stars in the Galaxy and globular clusters) and somewhat aids the problem of galaxy formation, but it exacerbates the difference between the critical density required for inflation theories and the actual baryonic density measured—hence requiring more of a drive for CDM solutions.

1.7 THE DISCOVERY OF ACTIVE GALAXIES

There are over a ten billion galaxies visible in the Universe using modern telescopes and the vast majority are termed 'normal' galaxies. These radiate by the combined output of their billions of stars and as we saw in section 1.3 most fall into the category of spiral, elliptical or irregular galaxies. Interspersed among this universe of normal galaxies are those which do not conform to this simple picture. Many of these galaxies have the same morphological shape as their normal counterparts but their claim to fame lies in their much greater power output (luminosity). Here they outshine their compatriots by up to a thousand-fold. These are the active galaxies and many of them are variable in output, something very rare in such giant structures. As we saw earlier, this variability can be used to place constraints on the size of the power source. Incredibly, this has been shown to be confined to a region of space much less than 1 pc! The luminosity of a thousand galaxies produced in a volume of space less than the distance to our nearest star is an awe-inspiring concept and requires explanation.

Somewhat surprisingly there is no unanimous agreement amongst astronomers as to the precise definition of an active galaxy. However as our story unfolds, this apparent difficulty will become clear and will not cause us concern. Indeed, we shall use this very fact to drive home the idea of unified models. For the time being let us assume that a reasonable working definition for an active galaxy goes along the following lines: an active galaxy is one in which a significant fraction of its total luminosity is radiation not ultimately attributed to stellar photospheres. In other words, we are seeing something other than the starlight that typifies the dominant electromagnetic output of a normal galaxy.

It is these very challenging aspects of high luminosity and non-stellar emission that make active galaxies such an exciting field of astrophysical study. AGNs are the most powerful bodies in the Universe and we shall see that this activity is manifest throughout the entire electromagnetic spectrum, so providing a rich harvest of data for observational astronomers. (Technically speaking, Type II supernovae are actually more powerful, emitting up to 10^{46} J, mostly in the form of neutrinos, but over a timescale of only a few seconds.) Nevertheless, even with the most modern of observational facilities, active galaxies still pose many unsolved puzzles. Indeed, the quest to understand these exotic objects has provided the driving force for many of the major current astronomical telescopes and satellites. Although the puzzle has not yet been solved and many pieces from the jigsaw remain to be placed, astronomers now believe they have a reasonably clear picture of the overall properties of AGNs and a working understanding of how they are powered. As we noted at the beginning of this book, nearly all astronomers are now convinced that the primary source of energy powering an AGN is the gravitational field of a supermassive black hole lying at, or very close to, the centre of the galaxy. This consensus has been built up due to the accumulation of observational data that give support to the above hypothesis. Whether this scheme is the whole truth and nothing but the truth will only be verified by further observations in the years to come. But for the moment, it is the best we have.

Fig. 1.7. The Seyfert galaxy NGC4151. The very bright nucleus is readily apparent and almost dominates the starlight from the underlying spiral galaxy. If NGC4151 were much further away, it would only be with great difficulty that we could see the underlying galaxy. (Caltech/Hale Observatories photograph.)

Although some of the early (short exposure) photographs revealed the jet-like structure in the galaxy M87 (not shown in fig. 1.2), it is reasonable to claim that active galaxies, as a class distinct from normal galaxies, were discovered during the 1940s by the astronomer Carl Seyfert. He found that a handful of spiral galaxies possessed extraordinary bright and point-like nuclei. On short exposure photographs, these nuclei looked like stars but spectroscopic studies revealed they possessed strong and broad emission lines. It had been known since the time of Hubble and Humason that galaxies had absorption lines due to the stars, but to see emission lines was very rare and was quite unexpected. Furthermore, these emission lines were broad, very much broader than the widths of the absorption lines in galaxies. The observations of broad emission lines, often from highly excited or ionized atomic species, is now recognized as one of the key features exhibited by all classes of active galaxy. The interpretation of the position, strength and width of these emission lines formed one of the main tools of active galaxy research for many years, and is the principal tool for probing the conditions of the emitting gas in the nuclei of these galaxies. Another major bonus resulting from the presence of strong emission lines that should not escape our attention, is enabling the distance of the AGN to be determined through the use of the Hubble Law (equation 1.3).

With the discovery of active galaxies, astronomers were back at stage one in terms of a classification picture. They needed to make objective decisions concerning very subjective pieces of observational evidence, initially collected in only a very small part of the electromagnetic spectrum, that of visible light. Unlike the picture for normal galaxies, active galaxies have not succumbed to a completely satisfactory classification scheme. However, as we progress through this book we will be reassured to find this is not a problem. In fact it can be see as a positive advantage, freeing us from the mire of the classification scheme that has been built up (see chapter 3) and allowing us to look at the physics of the luminosity generation. This has led to the unification and grand unification scenarios which we will shamelessly exploit as a working hypothesis.

Being a very new topic, the classification of active galaxies went hand in hand with new discoveries, but unfortunately, the naming of the new types was neither systematic nor unique. This aspect is a cause of much of the confusion surrounding the topic of active galaxies for the student who is new to the field. On the other hand, this confusion over classification is common to most new areas of science. History has revealed that better understanding eventually brings about a re-ordering, and often unification, whereby everything fits neatly into place based on sound physical principles and predictions. The good news is that active galaxy research has just about arrived at this juncture. One of the highlights of the classification scheme was the successful spectroscopic subdivision of the Seyfert class; this ultimately paved the way for unification.

Let us look at some of the differences in classification schemes. The galaxies discovered by Carl Seyfert were called 'Seyfert' galaxies for obvious reasons, but naming a class of object after the discoverer became the exception rather than the norm. The only other example of this is Markarian galaxies. Classifications of active galaxies took many forms: (1) after the name of the astronomer who undertook a search looking for specific characteristics; (2) from the identification of unusual continuum emission in some part of the electromagnetic spectrum; (3) from unusual spectral properties; and (4) possessing unusual morphological characteristics.

It is immediately obvious that these criteria are by no means mutually exclusive and many active galaxy classes possess at least three of the above properties – hence the problem. Seyfert and Markarian galaxies are representative of category (1), but Seyfert and Markarian used criteria (2) and (3) in their initial searches for these new types of galaxy. When wavelength selection comes into play the picture becomes distressingly more complex. Radio galaxies fall into category (2), but quasi-stellar objects (quasars) fall into (3), although historically all the first examples of quasars were discovered through the use of criterion (2) as they all happened to be radio-loud objects! Once we add LINERs (low ionization emission line nuclei), BL Lac objects, OVVs, blazars and broad-line radio galaxies, the picture begins to take on a bewildering confusion.

Not to worry; in this book we will draw order out of this apparent chaos and present active galaxies in terms of a general luminosity hierarchy. Indeed we will end up with a unified theory of what makes an AGN and why we see such apparently different manifestations of the same basic phenomenon.

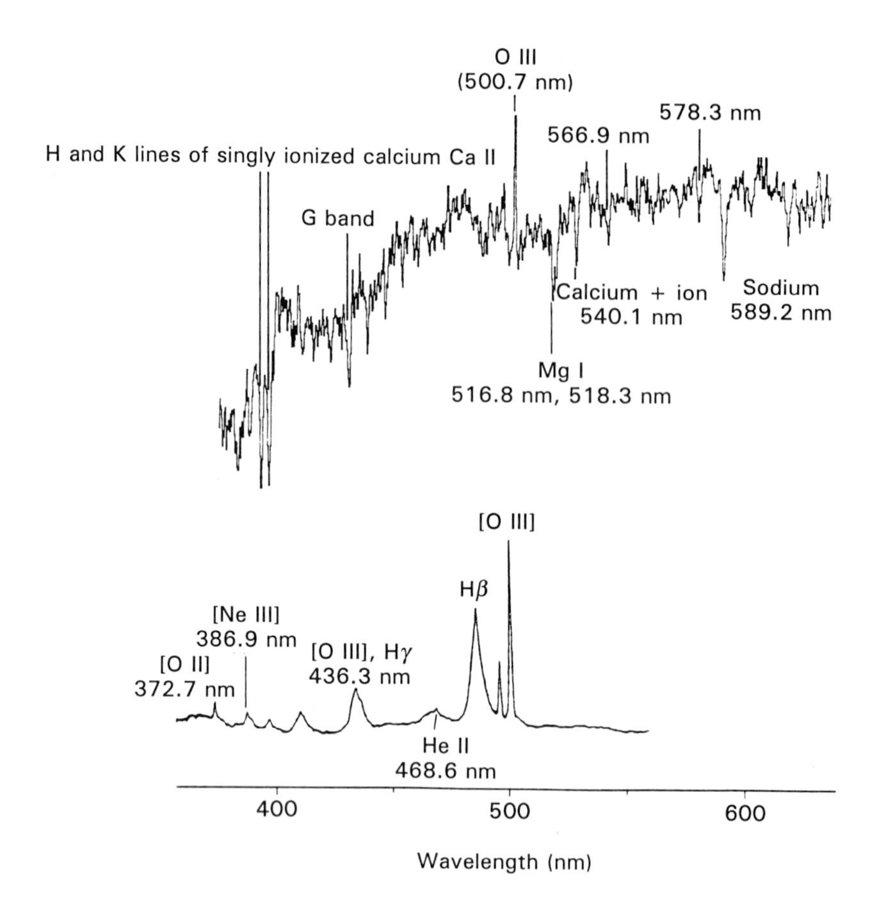

Fig 1.8. Optical spectrum of a spiral galaxy (top) showing absorption lines due to the stars of the galaxy and narrow emission line of doubly ionized oxygen (OIII) due to ionized hydrogen regions around star formation sites. Below is an active galaxy (Mkn509, a Seyfert) whose spectrum is dominated by the very strong and broad emission lines of hydrogen. Note the narrow emission lines of OIII. (from Audouze & Israel, *Cambridge Atlas of Astronomy*, Cambridge University Press, 1985)

We should pause for a moment to consider bias, and the author should come clean and say where he stands. Scientists have biases and astronomers are no exception. These biases are formed from a study of the observational evidence, or the beauty and simplicity of a new theory. Sometimes they occur because of the individual protagonists, the persuasive nature of their arguments and their reputations. We have already met a famous example in that Hubble deeply believed in the law of recession of galaxies and history proved him correct. Along with many other astronomers today, my bias is to believe that the luminous AGNs are all related to one specific phenomenon, the presence of a large gravitational mass at the centre of the galaxy. Such parameters as the size of this mass, the local surrounding material, the magnetic field in the host galaxy, the orientation of the central engine and galaxy with respect to us, as well as the presence of nearby interacting companion galaxies, all probably contribute to presenting the astronomer with pieces of observational evidence that he or she uses in order to place the objects into the slots of a classification scheme. Whether there are discrete pigeonholes, rather than a long sorting counter, is a point at the crux of current active galaxy research. My bias is to come down on the side of the long sorting counter, and much of the underlying theme throughout this book will centre on this crucial aspect of unification.

1.8 AN OBSERVATIONAL SCIENCE

Active galaxy research brings together theory and observation. Chapter 2 will discuss some of the fundamental aspects of observational astronomy. These include measurement units, instrumentation, and observing techniques, all of which have improved dramatically over the years since Hubble determined M31 was a nearby galaxy. For the theorist, the advance of computer technology has allowed detailed numerical solutions of highly complex physical processes to be undertaken. Calculations, which years ago could only be dreamed about or which could only be crudely estimated, can now be treated in great detail through numerical simulations. These can even provide spectacular video sequences of galaxy interactions showing mergers, tidal tails and subsequent bursts of star formation. Another amazing example of numerical modelling is the dynamics of spectacular jet outflows of radio galaxies (figs. 8.2 and 8.3). Computers have made huge inroads into the efficiency of the observing process and for the observational astronomer the times have greatly changed. Computers have provided remote control of instrumentation, precision pointing and tracking of telescopes along with major advances in data analysis, including on-line pre-processing of digital images and spectra at the telescope. Powerful techniques with exotic names such as 'maximum entropy' the 'AIPS cookbook' are available for image restoration and the suppression of noise (which should come with a warning label 'to be used with care'). A number of the images shown in this book (especially those demonstrating the highest spatial resolutions) have been produced using such techniques.

Even in the actual technique of observing, progress has been dramatic. In the early years of active galaxy research, and by that we mean the first half of this century, photographic plates were the norm for Edwin Hubble and his followers. Exposures of many hours were required to obtain sufficient light to register an image on the plate. The pioneering astronomers needed to be very proficient in manual guiding of the telescope during these long exposures. Indeed, until as late as 1960, truly great extragalactic observational astronomers needed to be hardy individuals, guiding a telescope to pinpoint accuracy for many hours in the freezing cold of a telescope dome.

Today all that has changed. As optical and infrared observers we have become wimps but this represents a great step forward. We are completely banished from the cold of the observing floor and telescope because the human body is an unwanted source of heat in the carefully cooled airflow of the telescope dome. Technology has allowed the astronomer to move into the comfort of a warm control room, surrounded by computers, visual display screens and keyboards. Modern optical and infrared telescopes are equipped with ultra-sensitive TV cameras which project the field of view of the telescope into a cosy control room where the astronomer (or more usually a highly trained night assistant who is trusted with the care of the multi-million dollar telescope) can position and guide the telescope by means of a computer keyboard. Even better, the telescope can also be 'autoguided' on an object for many hours without the telescope operator even needing to touch the controls. Computer telescope control and computer data acquisition are standard on all large telescopes and there is no doubt that in many respects the 'art' has gone out of observational astronomy.

The giant 4–10-m diameter ground-based optical/infrared telescopes are expensive facilities and as such are designed to ensure that they can be operated at the highest possible efficiency. Telescope time is highly prized and the opportunity to gain time on a large telescope is fought over with grim determination amongst the astronomical community. Automation and computer control are standard and the improvement of electronic communication networks (the information superhighway and the Internet) augurs for dramatic change. There is the distinct possibility that in the near future astronomers will no longer routinely travel to distant telescopes but instead will be able to sit in their institutions, or even homes, and 'observe' remotely, the information being fed to them in real time over the ever–improving data communication lines. These, and the World Wide Web are a fundamental part of modern astronomy.

The general reader might be surprised to learn that the pressure to obtain results from the telescope is so great that during a good night when things are really going well and the pace is frenetic, from conversations with most of my colleagues, it seems that going outside to look at the night sky is a very rare occurrence. A tiny spot of light on a display TV screen is the nearest most professional ground-based optical–IR astronomers now see of the real heavens! Somewhere the wonder has gone out of just looking at the night sky and marvelling. Such is technological progress. The public or cartoon image of an ancient bespectacled and bearded astronomer, wearing a white lab coat and peering through the eyepiece of a giant refracting telescope, is definitely from a long-gone age.

In the next chapter we will take a detailed look at observing techniques and facilities across the entire electromagnetic spectrum. This will enable us to appreciate the power of observations and understand the limitations of the current technology and potential selection effects caused by instruments and observational programmes. In chapter 3 we will undertake a detailed study of the classification scheme for active galaxies. A wealth of observational data will be used to formulate a coherent overview of some basic parameters of active galaxies. These will be used in presenting scenarios of unification themes, culminating in grand unification presented in chapter 9. However, to appreciate fully the study of the physical processes at work in active galaxies, it is important that we understand the basic concepts concerning line emission and absorption and a number of other detailed physical radiation processes. These are fundamental in the study of active galaxies and are presented in chapter 4 in sufficient detail for our requirements but with consideration for the general reader. By necessity these processes are described by mathematical equations. So that the general reader can obtain the fullness of the picture without recourse to a thorough understanding of the physics and mathematics, summary sections are provided. This should also help to ensure that the student can fully appreciate how the topic fits into the big picture. This is all part of the challenge of the study of AGNs. It is relatively easy to understand the details; the difficulty comes in fitting all of these details into the big picture. That is the challenge for this book. We intend to provide a comprehensive tour of the field of active galaxies.

1.9 CONTROVERSIES

In keeping with virtually the entire astronomical community we shall accept the cosmological explanation of the redshift of a galaxy. This means that we interpret the redshift of a galaxy or quasar as being due to its distance from us and the general expansion of the Universe. We can therefore use Hubble's Law (equation 1.3) to determine distance. Because this acceptance of the cosmological interpretation of the Hubble Law is crucial to our study of active galaxies, it is useful to pause and consider why there was any doubt about its use in the early days of AGN study. After all, the Hubble Law was determined from normal galaxies and there has never been any serious challenge to its validity, so why were active galaxies different?

Doubts set in amongst a small number of astronomers with the discovery of the first quasars. The problem was, and remains today, the quasar luminosity and the size of the emitting region. Taking quasars to be at their cosmological redshift distance, the first quasars were found to be the most distant and most luminous objects in the Universe. This tremendous luminosity, sometimes amounting to that of 1,000 powerful normal galaxies, must, because of variability arguments, be emanating from a volume of space only a few times the size of the Solar System. Such a huge power from an incredibly small volume of space could not be explained by the normal physical processes with which astronomers were used to dealing. The most efficient of these was nuclear fusion, the power source for stars, and the failure of nuclear fusion to provide a satisfactory explanation for the luminosity of quasars caused some astronomers to question the validity of the quasar distances. They suggested that an alternative explanation might be that the observed redshift was not due to the great distance of the object, but to some other effect such as a gravitational redshift, or a very local high velocity due perhaps to the objects being expelled from our Galaxy. In either case, quasars would be much closer and the energy problem would then disappear.

However, these arguments are certainly invalid. The introduction of the energy release from the gravitational potential well of a supermassive black hole solves the luminosity problem and this concept sits comfortably with the current generation of astronomers. Another crucial reason for rejecting the non-cosmological interpretation of quasars is that in some of the more nearby quasars and other powerful active galaxies we can clearly see the underlying galaxy of stars. Although the redshift determined from the absorption lines of the stars in the host galaxy has so far been deduced for only a relatively small number of these cases, they are all in total agreement with the redshift as deduced from the broad emission lines of the active nucleus. Because there is no suggestion of problems in deducing distances from the redshifts of stellar absorption lines, the same should apply to the emission lines. Furthermore, the drive for better and more sensitive facilities has enabled the 'fuzz' surrounding many quasars to be observed, and the weight of evidence suggesting that the fuzz is starlight from the underlying host galaxy is very strong in my opinion. To show how far this subject has progressed recent images from the HST and the 10-m diameter Keck telescope on Mauna Kea have revealed the underlying galaxy associated with a radio galaxy (4C 41.17) at a redshift of $z = 3.8$.

A further area that caused consternation was the discovery of blobs of radio emission blasting out from some radio galaxies and quasars with apparent velocities exceeding the speed of light. This phenomenon is referred to as 'superluminal motion' and is seen in spectacular radio jets. Although we now have a satisfactory explanation (which is presented in detail in chapter 8), it should be noted that this still requires bulk motion of material accelerated to velocities close to the speed of light, perhaps as high as 80–90%. This concept has never fitted comfortably with some eminent astrophysicists and it still far from being clear how such high speed bulk motion is produced. Nevertheless, theorists have been hard at work to explain such events, and we should pay tribute to the fact that the concept of superluminal motion was first proposed by an eminent member of their group. As a consequence of his work on jets in radio sources, Martin Rees proposed the existence of superluminal motion years before it was observed by radio astronomers who had to wait until the technique of VLBI was developed in the 1960s (section 2.4.4). The name of Martin Rees figures prominently in theories of AGNs. Indeed, Rees and his students and even their students have made tremendous advances in our understanding of the workings of the central engine.

In terms of the non-cosmological redshift controversy, there are still a small number of astronomers who persist in claiming that at least some part of the observed redshift comes from some other property that is intrinsic to the active galaxy in question. The upshot is that the galaxy is nearer than we assume when applying the Hubble Law. Although this view is totally rejected by the vast majority of astronomers, it is important that the reader understand that this idea *must* still be thoroughly tested by comparison with observational data. This is the way by which science makes progress. Frequently it is the maverick, who, although at odds with the established view, makes the great breakthroughs and is eventually shown to be correct. To my mind Fred Hoyle stands out in this regard as a great astrophysicist. All ideas must be subject to the same degree of scrutiny and observational test, no matter how outlandish they may seem at first sight.

1.10 SUMMARY

We have now set the scene for the study of active galaxies. We have reviewed our place in the cosmos and at the same time obtained an overview of the current picture of the origin and evolution of the Universe. We have been introduced to the scales of astronomical distances and sizes and have met the concept of the redshift as a measure of distance. We have seen that even though galaxies were only identified in the 1920s vast progress has been made in their study.

The discovery of a highly luminous subset of galaxies, dubbed active galaxies, led to a plethora of names and a rudimentary classification scheme. The nearby examples clearly showed that this activity originated in the central regimes of the galaxy and so the active galactic nucleus (AGN) became the precise object of study. Observations of very broad emission lines from highly ionized atomic species showed that there must be an intense source of energy present in these AGNs. Variability of a significant fraction of the luminosity of the AGN with timescales measured in weeks to days provides critical evidence of the very compact nature of the central engine. The failure of alternative forms of energy generation to explain this size-constrained luminosity led to the concept of the gravitational energy released by material accreting onto a supermassive black hole being the source of power for the AGN. This is the so-called 'central engine'.

This progress came with continually improving observing facilities: new detectors, new telescopes, new observing platforms. Satellites and telescopes in space opened up the electromagnetic spectrum. Hand-in-hand came new theoretical models, explaining the observations and sometimes providing theories and predictions that were way ahead of the observations. The study of active galaxies is now at a watershed, many pieces of the puzzle are now in place, and a concept of unification of the various observations into an underlying theme is now upon us. The following chapters will expand on the story of active galaxies, provide a firm basis for the physical processes taking place therein, and illustrate why the study of AGNs remains at the forefront of modern astronomical research in terms of challenge and excitement.

1.11 FURTHER READING

General and reviews

Arp,H.A., *Quasars, redshifts and controversies*, Interstellar Media, Berkeley, 1985.(An interesting book about why much of what is stated in this book is incorrect—this is the 'maverick' point of view.)

Cowley,A.P., 'Evidence for black holes in stellar binary systems', *Ann.Rev.Astron. Astrophys.*, **30**, 287-310, 1992.

Freedman,W.L., 'The expansion rate and size of the Universe', *Scientific American*, p54, November, 1992.

Hubble,E.P., *The Realm of the Nebula*, Yale University Press, 1936. (A very rare book but well worth reading; a 1935 lecture series given by Hubble, with an excellent foreword setting the scene of extragalactic research.)

Kaufmann,W.J., *Universe*, Freeman. 1995. (An excellent, lavishly illustrated student guide to astronomy, suitable for all readers. Updated annually and excellent value.)

Maran,S.P., *The Astronomy and Astrophysics Encyclopedia*, Cambridge University Press, 1992 (you name it, its in there somewhere—relatively advanced).

Osterbrock,D.E., Gwinn,J.A. & Brashaer,R.S. 'Edwin Hubble and the Expanding Universe,' *Scientific American*, p84, July, 1993.

Rowan-Robinson,M., *Cosmology*, Oxford Science Series, Oxford University Press, 1986. (An excellent introduction to cosmology and other aspects of astronomy.).

Rowan-Robinson.M., *The cosmological distance ladder: distance and time in the Universe*, Freeman, 1985. (A superb treatment of the distance scale and cosmology.)

Sandage,A., *The Hubble Atlas of galaxies*, Carnegie Institution Washington D.C., 1961. (Spectacular pictures of galaxies and good discussion of classification.)

Sandage,A. & Tammann,G.A., *A revised Shapley-Ames catalogue of bright galaxies*, Carnegie Institution Washington D.C., 1981. (Beautiful photographs of galaxies.)

Sciama,D., *Modern cosmology and the dark matter problem*, Cambridge University Press, 1993. (A very clear treatment of cosmological models, especially cold dark matter which is well explained for the uninitiated.)

Shapley,H., *Galaxies*, Harvard University Press, 1975 (Another rare book, but one which gives a historical perspective of Shapley's work.)

Shu,F.H., *The Physical Universe, an introduction to astronomy*, University Science Books 1989. (A brilliant book revealing the physical principles of astrophysics, includes techniques and instruments: excellent student material.)

Silk,J., *The big bang*, Freeman, 1980. (Another excellent book on hot big bang cosmology.)

Snow,T.P., *The Dynamic Universe*, West Publishing 1991, (Similar to Kaufmann.)

Tayler,R.J., *The Hidden Universe'* Wiley–Praxis, 1995. (An excellent and comprehensive view of astronomy today including cosmology.)

Tayler,R.J.,*Galaxies:structure and evolution,* Cambridge University Press, 1993. (An excellent review of the properties of galaxies.)

Weinberg,S., *The first three minutes*, Andre Deutsch, 1977. (A brilliant review of hot big bang cosmology, written for everyone—a masterpiece.)

Wynn-Williams,C.G., *The fullness of space,* Cambridge University Press 1992. (A superb discussion of the interstellar medium and modern observing techniques.)

Zuckerman,B.M., & Malkan,M.A., Eds., *The origin and evolution of the Universe*, Boston University Press, in press, 1995. (A symposium of astronomy for the general public—a must to read and enjoy.)

Specialized

Shabaz,T., *et al.,* 'The mass of the black hole in V404 Cygni', *Mon.Not.R.Astron.Soc.*, **271**, L10, 1994.

2

Observational progress

2.1 INTRODUCTION

Many of the major advances in the study of active galaxies have been intimately linked to improvements in observational capability. This chapter is devoted to studying the themes of observational techniques. We will begin by considering some of the basic measurement units used in astronomy. We will define the physical measurement of brightness (flux and luminosity), and relate these to the peculiar units of magnitudes used by optical astronomers. This leads directly to a crucial expression known as the *distance modulus*, an incredibly useful relationship between the apparent brightness, absolute brightness and distance of an object. Before tackling the development of telescopes, instruments and detectors, we will give a brief overview of potential biases and selection effects in the observational process. These are important and need to be understood and taken into account before we can obtain a true handle on the 'big picture' of active galaxies.

Next we will present an overview of optical instrumentation and consider the wealth of information obtained from sky surveys and their importance in opening up new fields of research. This leads us into the first steps of classification and the quest for understanding the physical processes of emission powering AGNs. We note again that classification is akin to botany: look, gather data, classify. To make further progress, more specific observations must be targeted to test predictions of the scheme or theory. These will often require advances in technology before they are possible, and this is the major force for the continual drive for new and more powerful facilities.

Obtaining a comprehensive picture of the physical processes at work in active galaxies requires an understanding of the processes themselves. Much of astrophysics entails the interaction of electromagnetic radiation with matter, and a useful concept is to consider interactions in terms of sources and sinks of radiation energy. As an example, stars are sources of radiation. On the other hand, dust in the plane of the Milky Way Galaxy is a sink in terms of the visible light from stars; it absorbs and scatters these wavelengths, and removes energy from the radiation field in the process. But even here the picture is more complex. In absorbing the visible light the dust is warmed and re-radiates into the surrounding space as longer wavelength radiation. The dust is therefore an optical absorber but an infrared emitter, i.e., it is a sink at visible wavelengths and a source at infrared wavelengths. In equilibrium, unless an internal heat source is present (such as in stars) any body will be both a source and sink, usually depending on the wavelength of observation.

These ideas are treated in detail in chapter 4, but here we give a summary of the basic concepts of radiating bodies. Although the continuum emission from radiating bodies (their brightness) is relatively simple to measure and basically is what is measured by the eye (albeit with some wavelength selectivity which we call colour), we will see that a quantum leap in information content can be gained from a detailed study of the line spectrum from the object. The field of spectroscopy lies at the heart of the observational study of active galaxies. To understand the wealth of information contained in the emission and absorption lines observed in active galaxies requires a review of some basic ideas of atomic physics. This will put us in good stead for our detailed study of AGNs beginning in chapter 3.

2.2 MEASUREMENT OF BRIGHTNESS

2.2.1 Flux, flux density and luminosity

The observational parameters relating to brightness, flux and luminosity are fundamental and are used prominently throughout this book. Readers who are well versed and confident in the meaning and use of these quantities can proceed to section 2.3. But before doing so ask: am I familiar with flux and flux density and do I understand what a bandwidth of detection expressed in hertz (Hz) means? If not, read on.

Before getting into the details of the definitions, we will review the fundamental units of measurement used in observations across the electromagnetic (e-m) spectrum. These differences have arisen due to the wide-ranging detection techniques used. In the ultraviolet through submillimetre regions, wavelength units are used. Optical and UV units are nanometres (nm, where $1\ \text{nm} = 10^{-9}\ \text{m}$); infrared, far-infrared and submillimetre units are micrometres (microns or μm, where $1\ \mu\text{m} = 10^{-6}\ \text{m}$). However, also in the submillimetre, and to all longer wavelengths, the radio astronomy units of frequency (MHz, GHz) are used due to the specific (heterodyne) detection techniques employed. At the shorter wavelengths of the X-ray and γ-ray domains, energy units (based on the energy of the individual photons expressed in electron volts, eV) are employed. Remember, the relationship between a photon energy and frequency is given by Planck's relation of $E = h\nu$, where E is the energy of the photon in joules (J), ν is the frequency of the photon (Hz or cycles per second) and h is Planck's constant of $6.63 \times 10^{-34}\ \text{J Hz}^{-1}$. The conversion from joules to electron volts is $1\ \text{J} = 1.6 \times 10^{-19}\ \text{eV}$. Fig. 2.4 shows the link between these different parts of the e-m spectrum.

Luminosity is the single most important quantity to determine about a radiating body. The luminosity is defined as the total flow of energy outward from a body per unit time over all wavelengths. It is the power output of the body. The luminosity, L, is therefore J s^{-1}, which is also the unit of power expressed in watts (W). The luminosity is both the rate at which a body is losing energy and the power output of the body. The luminosity is an *intrinsic* property of all radiating bodies and depends on the temperature and size of the body, but *not* the distance from which it is viewed.

Although the luminosity is an intrinsic parameter of a radiating body, as observers we measure the apparent brightness, or *flux*. This is the luminosity diluted by the distance of the body from the observer. The flux of energy arriving at an observer from a radiating body is defined to be the amount of energy, measured over all wavelengths, collected per unit time crossing unit surface area of a detector that is normal to the direction of the radiation. The units of flux are therefore joules per second per square metre (J s^{-1} m^{-2}), or watts per square metre (W m^{-2}). Fig. 2.1 shows a body radiating a luminosity, L, outwards into all space. As the distance, d, from the radiating body increases, the unit area of the detector collects less energy per unit time. How much less is just the difference in the ratio of the collecting area of the detector, say 1 square metre, to the total surface area of the sphere of radius, d, centred on the radiating body. The surface area of the sphere is $4\pi d^2$ and as the radius of the sphere increases, the surface area increases proportional to d^2. Therefore the flux measured by the detector falls by $1/d^2$, which is the famous inverse square law of radiation. The flux therefore depends on the luminosity of the body and inversely as the square of the distance from which it is measured.

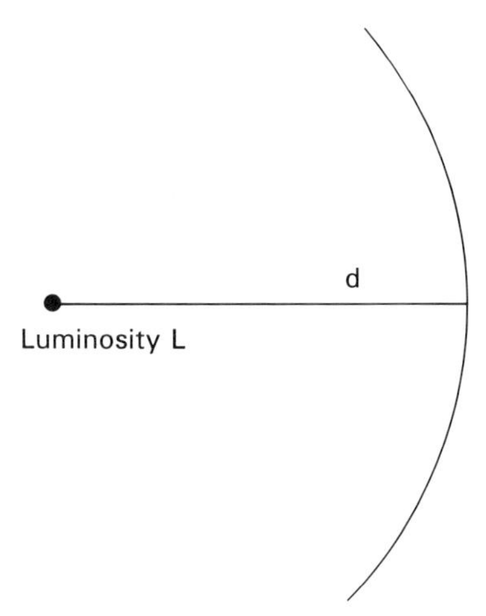

Fig. 2.1. The inverse square law of radiation.

So far so good, but there is yet another complicating factor. We are used to measuring brightness in everyday life, and we know that one of the problems in comparing the brightness of objects is how to handle the fact that they may have differing colours. It turns out that the eye-brain combination is not too good at comparing the brightness of bodies of very differing colours, such as red and blue. For astronomers working across a wide range of the electromagnetic spectrum, it is crucial to be able to compare the brightness in different parts of the spectrum. This requires care in calibration of detectors and instruments, moreover agreement over the precise output of a calibration source over the entire e-m spectrum. To assist in this process, the brightness (flux) can be measured and defined at a particular wavelength, or a very small range of wavelengths, rather than over a very large range of wavelengths.

To obtain a measurement of the flux at a particular wavelength a new unit of measurement is used. This is called the *flux density,* which is defined as the flux per unit bandwidth of observation. The bandwidth is a unit of frequency, the hertz (Hz), and so the units of flux density are watts per square metre per hertz ($\mathrm{W\,m^{-2}\,Hz^{-1}}$). We shall manage to discuss the physical processes in AGNs without much recourse to the nitty–gritty details of flux density. Nevertheless, it is useful to expand on this concept in a little more detail as it is frequently found to be a source of confusion for students. We begin by remembering that the wavelength, λ, and frequency of electromagnetic radiation, ν, are related to the velocity of light, c, by

$$c = \nu \lambda \,. \qquad (2.1)$$

A detection system, like the eyeball, has a *bandwidth*, or spread of wavelengths, over which radiation can be detected. The response of the human eye is shown in fig. 2.3. Although the eye can detect light from a wavelength of around 400 nm to 680 nm, the bandwidth is usually taken to be the range of wavelengths over which the response of the system is about 50% of the maximum response. This is called the full width at half maximum (FWHM). For the eye the bandwidth ranges from 510 nm to 610 nm, giving a bandwidth, $\Delta\lambda$, of 100 nm. It is convenient to be able to express the bandwidth in terms of the frequency, $\Delta\nu$. To do this we must first re-arrange eqn. 2.1 and differentiate it to obtain

$$d\nu = -\frac{c}{\lambda^2}\, d\lambda \,. \qquad (2.2)$$

Here, $d\nu$ and $d\lambda$ translate to the bandwidth terms of $\Delta\nu$ and $\Delta\lambda$, but the accuracy of the approximation decreases as the size of the bandwidth increases. To convert a wavelength bandwidth ($\Delta\lambda$) to a frequency bandwidth ($\Delta\nu$), we can use eqn. 2.2, with λ being the centre wavelength. For instance, in the example above the bandwidth of the eyeball can be expressed as 9.6×10^{13} Hz.

Hence flux density is a form of normalised measurement unit, allowing comparison of the measure of the apparent brightness of objects at a particular wavelength (or frequency) when measured through an idealised detector that has a detection bandwidth of 1 Hz. The flux density, often written as S_ν, is properly defined as the flux per unit frequency interval. Because these units were first used by radio astronomers, the unit of flux density is now called the jansky (Jy), in honour of the founder of observational radio astronomy. In terms of definitions, $1\ \mathrm{Jy} = 10^{-26}\ \mathrm{W\,m^{-2}\,Hz^{-1}}$. For fainter sources the milli-jansky (mJy) and microjansky are used (μJy).

2.2.2 Magnitudes and the distance modulus

We now move on to consider the optical astronomers' units of brightness, called magnitudes. In general we shall avoid using magnitudes in this book and will prefer the units of flux and luminosity, but because magnitudes are so widely used in books and articles these are summarized below. In terms of the inverse square law, the flux (F) or flux density (S_ν) are clearly related to the luminosity (L) by the distance of the object. For simplicity we will continue by using wavelength scales and flux rather than frequency and flux density. Consider again fig. 2.1. An observer located on the surface of the sphere of radius d measures the flux, F, to be

$$F = \frac{L}{4\pi d^2} . \qquad (2.3)$$

The Greek astronomer, Hipparchus, decided that the naked eye stars could be ranked in six levels of brightness, most of the brightest being termed the first magnitude and the faintest being the sixth magnitude. It turns out that the response of the human eye is logarithmic in nature, as opposed to linear, and the English astronomer Pogson put everything on a firmer footing with the introduction of the modern magnitude scale. This is defined so that a difference of five magnitudes corresponds to a brightness ratio of exactly 100, which naturally reflects the logarithmic response of the eye. So we have that the magnitude difference, $m_1 - m_2 =$ *a function* (b_1/b_2), and with $m_1 - m_2 = 5$ and $b_1/b_2 = 100$, then we can write

$$m_1 - m_2 = -2.5 \log \left(\frac{b_1}{b_2}\right), \qquad (2.4)$$

where m_1 and m_2 are the magnitudes of two bodies, 1 and 2, which have respective brightness b_1 and b_2. But we now know that we should replace the term for brightness by the flux, F. Therefore we can write that for two bodies of brightness F_1 and F_2, the magnitude difference is given by

$$m_1 - m_2 = -2.5 \log \left(\frac{F_1}{F_2}\right) . \qquad (2.5)$$

The magnitude in eqn. 2.4 is the *apparent magnitude*, the apparent brightness of an object measured by an observer at some arbitrary distance. So eqn. 2.5 gives only a relative measure between the two objects. To incorporate this into a scale against which all other objects are referred, a zero point is introduced. This is the same idea as the zero point in temperature measurement, such as 0° C or 32° F. Historically the zero point for magnitudes was chosen to be the very bright northern star Vega (α Lyra), which was assigned an apparent magnitude of 0. The apparent magnitude scale is measured with respect to this and hence the apparent brightnesses of all objects are measured with respect to the apparent brightness of Vega. For any star of apparent magnitude, m, we can thus write

$$m - m(0) = -2.5 \log \left(\frac{F}{F(0)}\right), \qquad (2.6)$$

where m(0) is the magnitude of the zero point calibrator (Vega), which for convenience we equate to zero. Therefore we choose to make m(0) = 0. Now F(0) is the flux from Vega and hence

$$m = -2.5 \log \left(\frac{F}{F(0)}\right) . \qquad (2.7)$$

We have now determined a single scale for comparing the apparent magnitudes of all radiating bodies. However, we are much more interested in the luminosity (L) of bodies, rather than the accident of nature that decides their apparent magnitudes due to their distances from us. The luminosity can be expressed in terms of an *absolute magnitude* (M). This is always expressed by a capital M to distinguish it from the apparent magnitude, which by convention is written with a lower case m. The absolute magnitude is defined to be numerically equal to the apparent magnitude when measured at a distance of 10 pc from the source. For the absolute magnitudes, M, of two bodies, 1 and 2, of luminosity L_1 and L_2 , we have

$$M_1 - M_2 = -2.5 \log \left(\frac{L_1}{L_2}\right). \qquad (2.8)$$

As before we need to rationalize the overall scale by defining a zero point by setting $M(0) = 0$, the corresponding luminosity being $L(0)$. This then gives us

$$M = -2.5 \log \left(\frac{L}{L(0)}\right). \qquad (2.9)$$

But usually we do not know the luminosity of a body, we only measure the apparent brightness, F. We can now use the inverse square law between luminosity and flux (eqn. 2.3), and substitute for L to obtain

$$M = -2.5 \log \left(\frac{F\, 4\pi d^2}{F(0)\, 4\pi d(0)^2}\right). \qquad (2.10)$$

This can be re-arranged for simplicity to give

$$M = -2.5 \log \left(\frac{F}{F(0)}\right) - 2.5 \log \left(\frac{d^2}{d(0)^2}\right). \qquad (2.11)$$

We note that the first part of this equation is just that of the apparent magnitude, hence we can re-write eqn. 2.11 by

$$M = m - 2.5 \log d^2 - 2.5 \log d(0)^{-2}, \qquad (2.12)$$

or,

$$M = m - 5 \log d + 5 \log d(0) \quad . \qquad (2.13)$$

We saw above that the absolute magnitude is equal to the apparent magnitude at the standard distance of 10 pc, so we can now substitute 10 pc for d(0) in eqn. 2.13 to give

$$M = m - 5 \log d + 5 \qquad (2.14)$$

which is more usually written as

$$m - M = 5 \log d - 5, \qquad (2.15)$$

with the units of the distance, d, being in parsecs.

Eqn. 2.15 is known as the distance modulus equation. It links the apparent and absolute magnitudes with the distance of the object. It is an extremely useful concept. If the distance can be determined, then a measurement of the apparent magnitude will enable the absolute magnitude (the luminosity of the body) to be determined. On the other hand, if the luminosity is somehow known then the equation can be used to derive the distance of the object. Use of eqn. 2.15 is one of the key stages in leading up to the magic of the diagram that provides the foundation for stellar evolution, the Hertzsprung-

Russell diagram. Although we shall not use magnitudes, the usefulness of the distance modulus equation is in the relation between the flux, the luminosity and the distance.

2.2.3 Intensity and solid angles

The *intensity* is a property that is intrinsic to the source and like luminosity is independent of the distance to the observer. The intensity is defined as the flux per unit solid angle, and now is an appropriate time to review angular measure. We will begin with the *radian*. We can start from a definition that a full 360° is equivalent to 2π radians. Consider the segment of a circle shown in fig. 2.2(a). The angle θ is related to 360° by the same ratio as the length of the arc, a, to the entire circumference. Therefore, in degrees we have

$$\frac{\theta}{360} = \frac{a}{2\pi r} \quad \text{or} \quad \theta = 360\frac{a}{2\pi r}\,, \tag{2.16}$$

but we know that there are 2π radians in 360°, so we can convert θ to radians by substitution in eqn. 2.16 to give

$$\theta = \frac{a}{r} \quad \text{radians.} \tag{2.17}$$

For small angles, the arc of the circle, a, can be approximated by the straight line, x. This then gives us

$$\theta = \frac{x}{r} \tag{2.18}$$

and for small angles this is the simple relation that we shall use. It is easy to calculate that there are 206,265 arcseconds in 1 radian (note the link with the definition of the parsec in section 1.2).

We can now extend the above argument to a three–dimensional, or solid angle and consider fig. 2.2(b). The solid angle, Ω, is measured in a unit called a *steradian*. The radius of the sphere is r and A is the area of the shaded surface. The solid angle is defined by

$$\Omega = \frac{A}{r^2} \quad \text{steradians.} \tag{2.19}$$

We can compute the number of steradians in a sphere in an analogous way to the above treatment for radians. For a complete sphere, the surface area $A = 4\pi r^2$, resulting in there being 4π steradians in a sphere. The relation between the solid angle, Ω, in steradians and the linear angle, θ, is given by

$$\Omega = 2\pi(1 - \cos\frac{\theta}{2})\,. \tag{2.20}$$

For small angles, eqn 2.20 simplifies to $\Omega = \pi\theta^2/4$, where the linear cone opening angle θ is expressed in radians.

Returning to the intensity, I, this is the flux per unit solid angle, hence the units are $W\ m^{-2}\ sr^{-1}$ and it can be written as

$$I = \frac{F}{\Omega} \, . \qquad (2.21)$$

We can illustrate the difference between flux and intensity by considering eqn. 2.3. For a given distance from a radiating object subtending a solid angle Ω, we can measure the flux, F, and the intensity I (the flux per unit solid angle). If we halve the distance to the object, then we see that the flux increases by a factor of four because of the inverse square law. On the other hand, we see from eqn. 2.19 that the solid angle also increases by the same factor. Therefore the intensity does not change with the distance from the body.

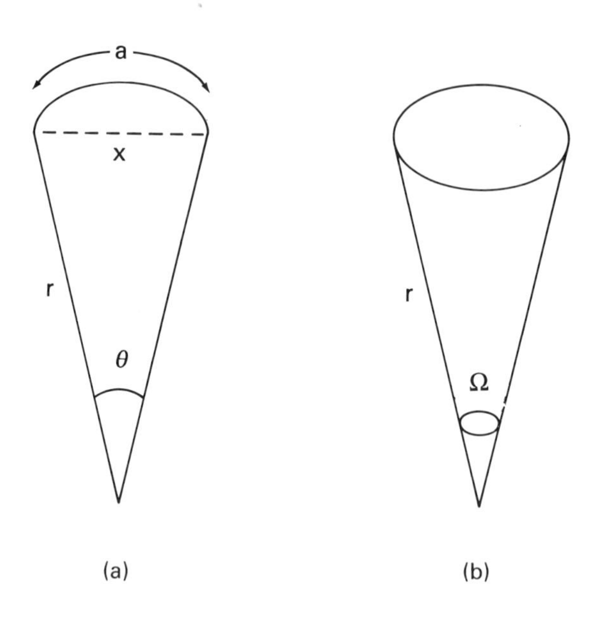

Fig. 2.2. Diagram showing (a) Linear angle, (b) solid angle.

To give an example of angular measure, flux and intensity we can take a familiar example. The Sun lies at a distance of 1.496×10^8 km and has a radius of 6.96×10^5 km. The angular diameter is then 9.305×10^{-3} rad. Multiplying this by 206,265 gives the answer in arcseconds, which is 1.92×10^3, or 32 arcminutes. The solid angle can be found from eqn. 2.19, and is 6.8×10^{-5} sr. We measure a total flux of 1,360 W m^{-2} and hence the luminosity of the Sun is (from eqn. 2.3) 3.83×10^{26} W and the intensity is 2.0×10^7 W m^{-2} sr^{-1}.

2.2.4 Photometry

We saw that flux density is the preferred quantity of measurement because it defines precisely the wavelength of the measurement. However, this requires either detectors with precisely known bandwidths or some means of isolating a small range of wavelengths. The usual method of achieving this is to use bandpass filters in the optical train leading to the detector. Bandpass filters only allow a certain range of wavelengths to pass through; all other wavelengths are blocked. The observational measurement of brightnesses using such techniques is called photometry.

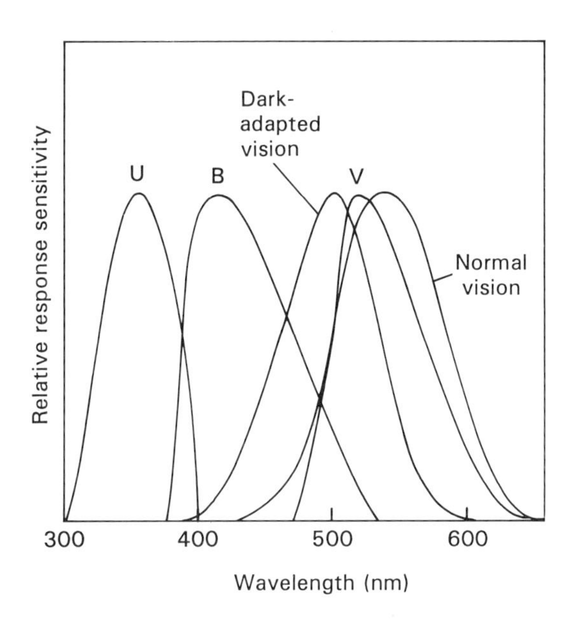

Fig. 2.3. The spectral response of the dark-adapted eye and the U, B, V passband filters

In the optical, the standard bandpass filters are known as U, B, V, R, I, standing for ultraviolet, blue, visible (yellow-green), red, and infrared, although in reality the U and I filters are better called 'near ultraviolet' and 'far red'. The transmission profiles of the UBV filters are shown in fig. 2.3 along with the spectral response of the human eye. Apparent magnitudes measured through these filters are denoted for convenience (and historical use) by U, B, V, R, I, which is a shorthand for m_U, m_B, m_V, m_R, m_I. Note that these are all *apparent* magnitudes, the magnitudes measured by a detector and a telescope, and must not be confused with the absolute magnitude even though U, B, V, R, I are denoted by upper case (just to confuse students). The absolute magnitudes corresponding to the apparent magnitudes of U, B, V, R, I are written as M_U, M_B, M_V, M_R, M_I. Infrared astronomers use bandpass filters referred to as J, H, K, L, M, N, Q, ranging from 1 μm to 20 μm.

Another important and useful relation in observational astronomy is the *colour index*. This is the difference between apparent magnitudes measured through two different bandpass filters and it is directly related to the temperature of the body. The commonest colour-indexes are those deriving from the U,B,V filters and we frequently find references to (U–B) and (B–V). The zero points of the UBVRI filters were selected to be that of an A0 star, and we find that (U–B) = (B–V) = 0 for such a star. Vega is an A0 star and has a photospheric temperature of around 10,000 K, and so stars having (B –V) ~ 0 have temperatures around 10,000 K. Stars with negative values of (B–V) are hotter than this, while positive values of (B–V) correspond to cooler temperatures. The Sun has a photospheric temperature of 5,800 K and a (B–V) of +0.65. The colour index is extremely useful because from two very simple measurements a reasonable estimate of the temperature of the emitting body can be made.

A final comment on the measurement of brightness is to note that for bodies which radiate across a wide range of the electromagnetic spectrum (such as active galaxies), to obtain the total emitted flux or luminosity we must ensure that it is measured over *all wavelengths* because there is no guarantee that most of the luminosity is emitted in the visible. Indeed, it is more often the case for AGNs that other regions dominate. A measurement over all wavelengths or frequencies is termed the *bolometric* magnitude and is a true measure of the total output from a body. Unfortunately, it is a quantity that is not readily measurable for two reasons: no single detector works over the entire electromagnetic spectrum and the Earth's atmosphere is only transparent to some of these regions (fig. 2.4). However, the bolometric magnitude can be derived from the compilation of measurements made in different parts of the electromagnetic spectrum. This then gives the total flux from the radiating body, and if its distance is known, we can immediately obtain the absolute magnitude from the distance modulus eqn. 2.15. The luminosity expressed in watts can subsequently be obtained by a suitable conversion factor.

2.3 OBSERVATIONAL BIASES: A WARNING

The study of active galaxies relies on observational data. As for any observational science, potential selection effects that might seriously distort conclusions derived from the data must always be borne in mind. Many areas of observational astronomy suffer from selection effects to quite a high degree and the classification of active galaxies is one of them. A simple demonstration of a selection effect is to view the night sky with your eyes. What we see with the naked eye represents a wavelength–selected, flux–limited sample. The naked eye can only see down to a certain level of brightness, about magnitude 6. Fainter stars, of which there are a vast number, are invisible. Binoculars and telescopes reveal these fainter stars and so the flux–limited sample of the naked eye is a bias but can be readily understood. A careful scientist will endeavour to investigate potential selection effects contained within the data and make suitable allowances in drawing conclusions.

One potential problem in terms of bias and classification of active galaxies is knowing whether the brighter members of a class, usually the first members discovered and subsequently studied in greatest detail, are truly representative of the entire class as a whole. As more objects are discovered, then statistically speaking the average properties of the class can be better determined. However, even here, it will often be the extreme examples of the class that attract the most study as they usually pose the most challenging questions. In terms of obtaining sufficiently large samples, the difficulty comes about by (a) how to obtain a large sky coverage to a required flux limit and (b) how to recognize the objects under study. The technique of obtaining large samples of objects is frequently tackled by sky surveys. These are crucial observational programmes and later we will take a brief look at sky surveys and to see how they played a major part in the development of the study of active galaxies.

Classification schemes can clearly be fraught with danger because of observational biases. We saw in chapter 1 that the classification of active galaxies is based on numerous factors including morphology, emission lines, or strong emission in other regions of the electromagnetic spectrum apart from the visible. Morphology will obviously be a highly distance–dependent factor because as the distance increases, the image is smaller and the underlying galaxy becomes harder to recognize. The critical aspect for AGNs is the high contrast between the bright point-like nucleus and its underlying galaxy. At even farther distances there will come a point when the galaxy is merely an unrecognizable smudge, totally dominated by the active nucleus. For higher contrast objects at greater distances we only see the nucleus, so what do we call this based on morphology?

We will see in the next chapter that the original definition of a quasi-stellar object was for a point-like emitter showing broad, highly redshifted emission lines, but excluded an underlying galaxy. The original definition of Seyfert galaxies was for a bright, point-like nucleus showing strong emission lines and residing in (usually) a spiral galaxy. Seyferts are now classified solely on their spectral characteristics, but we can still ask at what point quasars and Seyferts differ. We will see that there is a gradual merging of Seyfert 1 and quasar classes and the question of how we differentiate between a low–luminosity quasar and a distant high–luminosity Seyfert lies at the heart of what we understand by an active galaxy class. We will return to this very important point at numerous places in this book. One must always remember that how objects appear depends critically on how they are observed, and with what apparatus.

Before leaving this section we should note that an obvious candidate for introducing a misleading bias is to use a selection criterion that is based solely on a particular wavelength. Radio galaxies are an example. The first radio surveys were of tremendous importance and as far as extragalactic astronomy was concerned they brought about their own revolution and introduced a flourishing field of study that is still very active today. The radio surveys naturally selected those galaxies that had powerful radio emissions; however, as soon as radio mapping was undertaken it was realized that the source of this emission was not the same for all the galaxies. Some galaxies emitted radio emission only from their nuclear regions, some from the general body of the galaxy, while others produced gigantic lobes of radio emission on each side of the galaxy. So we can sympathize with the historical reason for such a selection. As any new wavelength of study is opened up, classification based on that wavelength is bound to happen. It is only later that these original classifications need to be reorganized according to physical processes rather than a wavelength category. The term radio galaxy covers a range of objects, including narrow-lined radio galaxies and quasars. The narrow-lined radio galaxies will feature in chapter 9 when we discuss unification ideas, and in many respects these objects hold the key to unification of radio loud and radio quiet quasars in terms of the fundamental central engine.

2.4 MAJOR TECHNOLOGICAL ADVANCES

2.4.1 Astronomical science

Optical astronomy has been established for hundreds, if not thousands of years, while all other wavelengths of study are products of this century. In historical order, radio astronomy started in the late 1930s, followed by ultraviolet, infrared and X-ray astronomy in the 1960s. Although still very much a new field, gamma-ray astronomy is making excellent progress.

Major discoveries have usually gone hand-in-hand with technological advances: either larger and more powerful telescopes or major advances in detection equipment. The telescope is a flux collector, and even optical telescopes have changed over the years. The first telescopes were built with glass lenses and were called refractors because the light was brought to a focus by refraction through the glass. However, these eventually reached the limit of practicality, both in terms of being able to construct a large glass lens and because these huge lenses were very heavy. This latter aspect meant that because the lenses were mounted at the end of a long tube, this had to be very stiff to eliminate, or at least minimize, gravitational bending. Also, because they were supported only around their edge, the very large lenses tended to deform under their own weight. These two factors limited the size of refracting telescopes to about 1 m diameter. The demise of the giant refracting telescope was driven by the discovery by Foucault in the 1850s of how to make a mirror by depositing silver on glass. This put the writing on the wall for the refractor as, in principle, mirrors could be made much larger than the world's largest telescope lenses. By the turn of the century the age of the giant refracting telescope was over and the age of the reflecting telescope began, culminating in the 8 m and 10 m diameter optical telescopes of today.

Another interesting aspect of the application of computer technology is the mounting and subsequent steering of ground-based telescopes. The enormous size of radio telescopes resulted in nearly all of them being mounted using what is called the *alt-azimuth* system. This means that they rotate east–west about an axis vertical to the ground and are tipped north–south about an axis horizontal to the ground. In other words they have the normal up–down and left–right movements associated with everyday life. The names derive from the coordinate system of altitude (the elevation above the horizon) and azimuth (the bearing measured eastwards from north). To track a celestial object across the sky, alt-azimuth mounted telescopes must drive in two directions simultaneously. This was acceptable in the radio regime because the errors introduced by movements in the tracking were small compared to the poor angular resolution of the telescope. Furthermore, as an object is tracked across the sky, the image rotates in the field of view, again something that was tolerated in low spatial resolution radio astronomy. Neither of these were acceptable for optical astronomy and until the 1980s, all large optical telescopes were mounted on an equatorial mount.

In this arrangement the rotation axis of telescopes left–right movement is no longer vertical but instead is parallel to the spin axis of the Earth. The telescope then tracks from east-to-west parallel to the Earth's equator and so follows stars and celestial objects perfectly without the need for additional up–down motions. The telescope is set to the north–south position (corresponding to the astronomical coordinate system called declination) and then driven in a single east–west direction (right ascension) to compensate for the rotating Earth. Furthermore, there is no image rotation. This was the situation until the very early 1980s, when the improvement and relative cheapness of computer technology enabled optical telescopes to be built using alt-az mounts. These are much cheaper to construct, an important consideration in financing a multi-million dollar telescope and all of the recent very large optical/infrared telescopes built or being built use this mounting arrangement. However, to preserve image coherence, either the detector or the image must be counter-rotated as the telescope tracks. This is also easily possible with modern technology.

2.4.2 Observational astronomy: from radio waves to gamma-rays

Astronomers need to be able to observe over the entire electromagnetic spectrum. However, the Earth's atmosphere is not transparent at all wavelengths, which is just as well for us. It is mostly opaque at ultraviolet (UV) wavelengths, protecting animal life from the ravaging solar UV radiation. It also blocks out other significant regions of the electromagnetic spectrum in which astronomers would very much like to observe. Therefore specialized observing platforms that carry telescopes above the atmosphere are required. Telescopes themselves serve to collect the energy from space and to bring it to a focus on a detector, which converts the incoming energy into a recognizable signal. This section will look at telescopes and detectors from radio waves to gamma–rays.

Fig. 2.4 shows the transmission of the Earth's atmosphere over the electromagnetic spectrum. At sea level, the atmosphere is transparent in only two wavelength regimes: the optical and radio. Far infrared, UV, X and γ-ray telescopes are space-based, while infrared and submillimetre observations are made from high mountain-top sites. It also turns out that secondary events emanating from the impact of the highest energy γ-rays from space onto the atoms of the atmosphere can be observed from the ground.

To put more meat on the bones of fig. 2.4, the traditional radio regime extends from wavelengths of metres down to millimetres and the radio telescope can be built more or less anywhere. As the wavelength decreases further, the location of the telescope becomes more critical. Purpose-built submillimetre facilities, such as the James Clerk Maxwell Telescope, are constructed on the highest and driest of sites, the best example being the 14,000 ft (4.2 km) summit of Mauna Kea in Hawaii. Such high, dry sites are also the domain of infrared telescopes making observations from wavelengths of about 20 μm to ~1 μm. The atmosphere is essentially transparent from about 1 μm to 320 nm and as far as transmission is concerned, optical astronomy can be performed from sea-level locations. Nevertheless, for optical astronomy, high sites are still the order of the day to make best use of the stability of the atmosphere (as we will see below), so that the very best spatial resolutions can be obtained from the expensive telescopes.

A range of satellites has enabled telescopes in space to cover those regions of the electromagnetic spectrum inaccessible from the Earth's surface. Notable amongst them is the International Ultraviolet Explorer (IUE) telescope which has performed UV spectroscopy over the wavelength range 115 nm to 320 nm since 1978. This was joined in 1990 by the Hubble Space Telescope (HST) which is also used for UV observations. X-ray astronomy has seen a significant number of highly successful space telescopes,

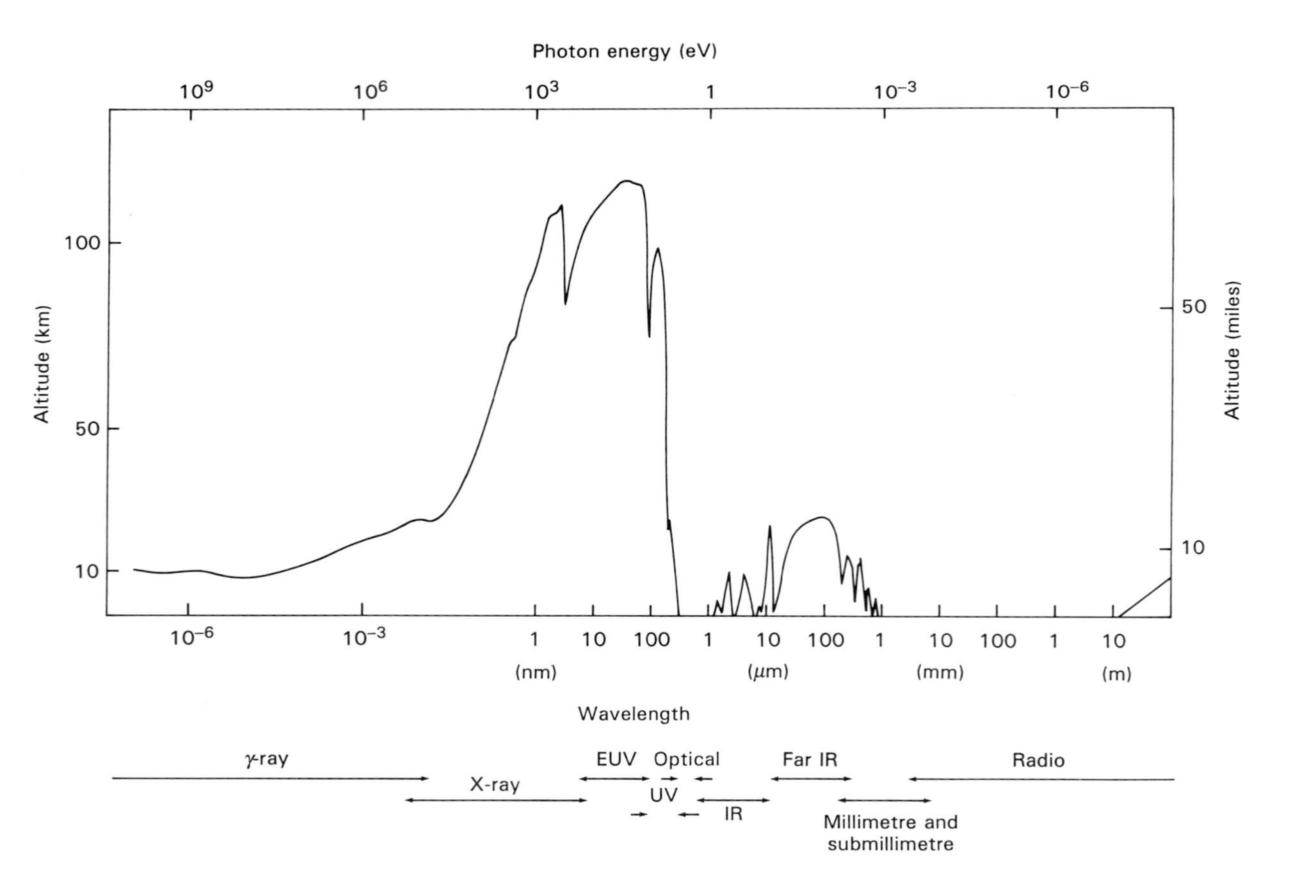

Fig. 2.4. Schematic diagram showing the transmission of the Earth's atmosphere to e-m radiation from space. The solid line shows the height at which 90% of the incoming energy has been absorbed.

notably the Einstein, EXOSAT, ROSAT, Ginga and ASCA satellites. Infrared astronomy came of age with the tremendously successful but short-lived IRAS infrared satellite, which was launched in 1983. Gamma-ray astronomy has now emerged as a fully fledged observational regime with the highly successful and multi-functional Compton Gamma Ray Observatory (CGRO) launched in 1991. An excellent discussion of astronomical space observatory platforms is *Satellite Astronomy* by Dr. John Davies.

In passing we note that space-borne astronomy is very expensive; each major observatory mission now costs many hundreds of millions of dollars, which is why most are collaborative ventures between many nations. To attempt to bridge the gap between wavelength regimes inaccessible from the ground but at a cost dramatically less than going into space, telescopes have been carried to altitudes of around 14 km by aircraft. The most famous and most successful of these is NASA's Kuiper Airborne Observatory (KAO), housing an 0.6 m diameter telescope in a Lockheed C141 aircraft and dedicated to far-infrared astronomy. The interplay between ground and space-based telescopes is interesting. Relevant factors include cost and the diameter of the telescope, and in the UV, optical and infrared, only very small diameter telescopes can be flown in space. The Hubble Space Telescope is probably the most famous space telescope, and although having only a relatively small collecting area, provides observations in the ultraviolet, something not possible using even the largest ground-based telescopes. But the HST has another major role, something with which it can out-perform even giant ground-based optical telescopes, and this is its very high angular resolution ability.

This nicely brings us to the second prime function of a telescope. Besides flux collection, telescopes improve the angular resolution and better determine spatial details. The angular resolution is the ability to discern fine spatial details and is directly proportional to the diameter of the collecting optic. The standard measure of the resolution, R, obtained by a circular collecting optic of diameter, D, is given in terms of the Rayleigh criterion for resolution of

$$R = \frac{1.22\,\lambda}{D} \quad \text{radians,} \tag{2.22}$$

where λ is the wavelength of observations (which must be expressed in the same units as the diameter when using eqn. 2.22). We can convert the units to arcseconds by multiplying by 206,265. Rayleigh's criterion says that two point sources are just resolved if the centre of the diffraction disk of one coincides with the first minimum of the diffraction disk of the second. In less technical language, Rayleigh's criterion is a useful format for being able to differentiate two point images as opposed to seeing a single blurred image. As an example, the dark-adapted human eyeball with a maximum pupil diameter $D \sim 8$ mm working at 550 nm has a theoretical resolution of 17.3 arcseconds but the retina limits this to about 1 arcminute in practice. This is about 1/30 times smaller than the diameter of the full Moon. It is also in theory sufficient to differentiate two car headlamps 1.5 m apart from the single light of a motorbike at a distance of ~ 5 km. A 3-inch refracting telescope operating at the same wavelength has a resolution of 1.82 arcseconds, whereas a 4 m diameter telescope has a theoretical resolution of about 0.035 arcseconds.

Unfortunately all is not straightforward. Turbulence in the Earth's atmosphere limits the resolution achievable by telescopes to only ~1 arcsecond at good sites and to 0.3 arcseconds at the very best sites (such as the mountain-top observatories in Hawaii, the Canary Islands and Chile). This turbulence limitation is called *atmospheric seeing* and it limits the spatial resolutions that can be achieved with large ground-based optical telescopes. This is one of the major reasons why the Hubble Space Telescope was built, to observe in space unhindered by the seeing of the atmosphere and hence achieve diffraction-limited resolution as given by eqn. 2.22. It is also the case that great strides are now being undertaken to enable large telescopes to 'beat' the atmospheric seeing. This is possible by using special observing techniques (speckle interferometry is an example but can only be used for bright sources) and technological innovations (such as adaptive optics). The drive for high spatial resolution in the optical/infrared regime is crucial for the study of active galaxies and is probably the next major development for ground-based astronomy.

2.4.3 Optical detectors

It is not possible to overstress the importance of detectors in an observational science. They are so crucial to much of the probing of extragalactic astronomy and the work on active galaxies that it is well worth dwelling on their performance characteristics in some detail. Any detector used in astronomy is merely a device to convert the incoming electromagnetic radiation from space (visible light, radio, X-rays, etc.) into a signal that can be recognized and measured with ancillary instruments. Since the time of Galileo the first generations of astronomers worked in the visible part of the spectrum making observations with the naked eye. The eyeball is a detector of electromagnetic radiation and not surprisingly it has served humanity very well. But for astronomy it has severe limitations.

It may be a surprise to learn that the spectral response of the eyeball is not uniform; even your own two eyeballs differ in sensitivity. Additionally, the eyeball is very difficult to calibrate. Even worse for science, the record of an observation is pertinent only to the individual who made it (memory) and is not available for inspection by the rest of the scientific community. It is well known that the eye can play tricks on even the most cautious of observers. However, the ultimate limitation of the human eyeball for astronomy is its intrinsic inability to integrate the signal and see faint sources.

What does this mean? Although integration sounds complex and mathematical, in our context it is very simple. The eye–brain combination takes a new picture about every one-fifteenth of a second. Expressed another way, it operates at a rate of 15 frames a second. The old image is wiped clear before a new image arrives and all that is left is memory. Therefore, once the eyeball is dark-adapted, no matter how long you stare at an image, no further detail will be discerned because the eye takes a new picture every 1/15th of a second. The eye is said to have an integration time of 1/15th of a second and there is nothing we can do to change this. So unlike a camera, which has the ability to change both the aperture (size of the collecting area) and the integration time (shutter speed), the eye has a fixed integration time. In daytime conditions the exposure is controlled by changes of the aperture (pupil stop). When we are in a bright light the iris closes (stops down) to a diameter of about 3 mm or even less to limit the light entering the retina. In very low light conditions, such as night-time, it opens up to its maximum diameter of ~8 mm to admit the maximum amount of energy into the eye. However a second effect, which has an even greater impact comes into play. This is a direct change in the detection sensitivity of the eye and is due to the rods in the retina taking over from the cones. The rods are much more sensitive to low light levels but are poor at wavelength selection, therefore we do not see colour at low flux levels. The cones, on the other hand, provide us with excellent colour recognition in medium to high flux levels. This ability to detect over a very high dynamic range of intensity is still unmatched by photographic film. However, the key point is that the integration time and the number of new pictures the brain perceives remains constant.

It is fun to imagine what it would be like if the integration time of the eye could be increased at will. On a dark night, as long as the eyeball was fixed in one direction, the picture could be built up until a clear image was seen; we could see in the dark. We might at first think that this would have been great for our distant forefathers hunting in the forest at night or dusk. But think again. If the hunted animal turned and charged, what good would a long integration time be? Useless, because the animal would have moved a considerable distance towards the hapless hunter before the picture was read out to the brain. The image would be a picture of a blurred charging animal. The result: brain to limbs; move; take cover; too late! Hence the integration time of the eye and the diameter of the pupil stop are an obvious compromise of our evolution.

Hand-in-hand with the introduction of reflecting telescopes came two of the most crucial advances in observational astronomy: the invention of photography and the spectrometer. The first good photographic plate detectors were used for astronomy around 1870 and provided the potential for opening up our view of the Universe. Photography enabled much fainter objects to be detected due to the photographic plate's ability to integrate the incoming signal for many hours. Furthermore, it provided a permanent record of images that were available for subsequent scientific scrutiny and archiving, a vitally important aspect when searching for variability in objects, as we shall see when we learn about the discovery of quasars. The spectrometer proved to be an invaluable tool in astronomy and its introduction brought about an entirely new technique of scientific study called spectroscopy. These studies were epitomized by Bunsen and Kirchhoff around 1870 (see section 2.6). Spectroscopy and the information gained from astronomical studies using this technique will form an important component of this book.

The intrinsic sensitivity of early photographic plates was still poor by modern standards. Nevertheless, the ability to expose the plate to incoming light for as long as the source could be tracked to the required accuracy by the telescope, or until the sky background started to fog the emulsion, resulted in great gains over what could be seen using the eye alone. Even if the first plates were a factor of ten less sensitive to light than the eyeball, instead of having an integration time of only 0.067 s, the plate could be exposed for an integration time of many hours. This resulted in much fainter objects being recorded than could ever be seen by the eyeball. Photographic plates became essential detection tools, capturing the images from the giant telescopes. Exposure times ranging up to many hours were required to detect the outlines of the distant galaxies.

Film emulsions eventually became as sensitive as the human eye and with further developments in a range of technology, entirely new detectors appeared on the astronomical scene. The introduction of the photomultiplier tube in the 1950s provided a significant increase in detection sensitivity. This possessed a detection quantum efficiency (DQE) of around 10% compared with that of the photographic plate of only about 1%. Like the photographic plate, the photomultiplier tube, or PMT, is a 'blue' device, which means it has maximum sensitivity in the blue part of the spectrum. Although sensitive, the PMT is a single pixel device, and measures the energy coming from a single spot on the sky. This is usually determined by a circular aperture placed in front of the detecting element, which is called the photocathode. To obtain a map of an object using a PMT, many individual measurements have to be made sampling the complete area of the object and the image subsequently reconstructed. This is clearly very time-consuming and the PMT's forte as a detector was in astronomical photometry of point sources, a field in which it reigned supreme from its introduction in the 1960s until the mid 1980s.

A range of two-dimensional devices next appeared. These were ideal for imaging extended objects and for spectroscopy. They mainly took the form of highly intensified television cameras and culminated in the Image Photon Counting System developed by Alec Boksenberg. This was sufficiently sensitive to detect individual photons and was to prove the dominant force in astronomical detectors, especially when fed by the output of a spectrometer. It featured highly in the study of emission and absorption lines in active galaxies throughout the 1980s.

The introduction of charge-coupled detectors (CCDs) in the 1980s produced a revolution for optical detection techniques. The CCD can be likened to a solid state version of the photographic emulsion. It has a quantum efficiency approaching 80% in the red part of the spectrum where they are most sensitive. Their sensitivity falls by over a factor of two in the blue but they are still much more sensitive than photographic plates or photomultiplier tubes. CCDs have not only revolutionized observational optical astronomy but they have brought a new consumer boom. These solid-state devices are the detecting elements in the ubiquitous compact video recording cameras (camcorders). However, the CCDs used for astronomical work are very different from their low-cost domestic camcorder counterparts. They are designed to be used only under very low light conditions and in addition are cooled to maximize sensitivity. They also require sophisticated readout electronics and associated computer control, all of which adds up to them being vastly more expensive.

The early CCD detectors had roughly 500 × 500 picture elements, or pixels. The pixel is the modern equivalent of the grain on a photographic emulsion except that the pixels are all of uniform size. Each pixel, usually about 25 μm square, gives an image of a small square piece of sky. The benefits of CCDs to astronomers are numerous. They are extremely sensitive (high DQE), have a wide dynamic range (brightest to faintest pixel), are linear devices (output is directly proportional to input), and are more suitable to computer interfacing for data analysis. The only drawback was in their original rather small size. Technology has come to the rescue. The latest CCDs have a very acceptable format of 2000 × 2000 pixels. The problem now is data storage.

In spite of their limiting field of view, CCDs have taken over virtually all the imaging work previously undertaken by photography. It was the CCD that opened up the investigation of the underlying galaxies in quasars; before, such studies were impossible.

This device now reigns supreme for single object work (photometry), imaging of extended objects and spectroscopy. However this miracle of modern technology has failed to displace the photographic emulsion from one important area, the domain of wide-field sky survey work using Schmidt telescopes. Here the photographic emulsion continues to dominate, as we shall see in section 2.5.

Throughout this book we will see the dramatic importance of spectroscopy. A recent development has been the use of optical fibres to transfer light from the images of a number of objects in the field of view of a large optical telescope to the input slit of a spectrometer. Each fibre carries the light from a single object. The resulting individual spectra from the objects in the field of view of the telescope are stacked in a vertical pattern and detected by a CCD. This gives a huge improvement in telescope efficiency. Instead of two hours being required to record the spectrum of a single object, nowadays over 200 objects can be studied simultaneously in a single observation. Of course great care must be taken to ensure that the fibres are very precisely positioned to match the positions of the astronomical sources in the focal plane. This technique, which has extended the use of large scale studies of the Universe by using quasars as probes, was pioneered by astronomers from the University of Durham in the UK.

Before leaving this section, we should consider the implications of the very high quantum efficiency of the CCD. Because this is now close to the theoretical maximum, only very small improvements in detection sensitivity will be achievable though increases in the efficiency of the detector. Therefore, to obtain major improvements in detection capability requires a drive to increase the flux collection capability. This is manifest by the latest generation of large telescopes with a number of 8 to 10 m diameter telescopes already built or under construction. The era of the 4 m reflector is over and we are now entering the realm of the true giant. The twin 10 m Keck telescopes, the Japanese 8 m Subaru and one of the two multi-national Gemini 8 m telescopes are occupants of the summit of Mauna Kea. The second of the Gemini telescopes will be constructed in Chile along with an ambitious European project of 4 × 8 m telescopes called the VLT. This has optical and infrared interferometry firmly in mind.

2.4.4 Radio astronomical techniques

Radio astronomy began in earnest in the late 1950s following the development of the technique of radar in World War II. Detectors for radio astronomy are called heterodyne systems because they work by *beating* the high frequency signal from space with a similar frequency signal from a local oscillator. These signals are 'mixed' in the detector and the resulting lower frequency (called an intermediate frequency) output signal can be amplified and measured. Technology has allowed heterodyne techniques to be employed over a very wide frequency range, even extending into the infrared. However, these very high-frequency devices are extremely specialized and require a battery of lasers to provide the local oscillator power.

Heterodyne detectors have a very small intrinsic bandwidth at any one time. A fixed frequency device would be a great drawback for radio astronomy and so the local oscillator is swept through a range of frequencies, allowing a range of observing frequencies (wavelengths) to be sampled. Radio astronomy operates from frequencies of a few MHz, that is, wavelengths of metres, up to 1,000 GHz, which is in the submillimetre regime with wavelengths of a few hundred micrometres (often shortened to microns). In this regime, the techniques overlap with those of far-infrared astronomy. These high frequency receivers are the current state of the art. They require cryogenic cooling for their superconducting-insulator-superconducting (SIS) detectors along with specialized telescopes at high-altitude locations to counter the absorbing properties of the atmosphere (fig. 2.4). The 15 m diameter James Clerk Maxwell Telescope and the smaller Caltech Submillimetre Observatory, neighbours on Mauna Kea, are the two major exponents of the highest-frequency radio observations.

Radio astronomy has a particular problem when considering spatial resolution. Because radio wavelengths are large, to achieve a high resolution the telescope must be enormous. For example, a telescope operating at a wavelength of 6 cm (5 GHz) would need to be 15 km in diameter to achieve a resolution of 1 arcsecond. Clearly this is not a practical proposition. Large single dish telescopes, such as the 76 m diameter reflector at Jodrell Bank in the UK, the 100 m diameter telescope of the Max Planck Institute at Bonn and the 64 m diameter Australian Parkes telescope represent the limit of what is achievable in terms of construction. Even so, resolutions of only an arcminute or so are obtained at their normal operating frequencies. It should also be remembered that with such massive structures, engineering constraints are a major consideration as witnessed by the catastrophic collapse of the Greenbank (West Virginia) dish in 1987. The largest radio telescope, the 1,000 ft diameter Arecibo reflecting dish in Puerto Rico, falls into a special category. This is because it is constructed within the rim of a natural hollow in the ground and as such it is a fixed dish. It can only see objects close to overhead and operates by using the rotation of the Earth to observe the sky passing over the fixed telescope. It can track objects to some extent by moving the prime focus feed.

To overcome this limitation on angular resolution, radio astronomers have been incredibly clever. They produced a scheme whereby the outputs from a number of adjacent telescopes are linked into a single control room and electronically combined. Because the relative spacings of the telescopes are precisely known, both the amplitude and phase of the incoming radiation from space can be derived from the signals, allowing linear interferometry to be undertaken from the resulting fringe patterns. Furthermore, the telescope spacing can be altered by moving the dishes, which is why most interferometers have their telescopes mounted on huge bogies running along railway tracks. Effectively this allows the interferometer to measure all the spatial frequencies required to simulate a single dish of linear dimension the length of the largest spacing between the ends of the linear array. This technique is called aperture synthesis. A series of measurements using differing spacings of the telescopes in the array can 'synthesize' a single large telescope, thereby producing a much higher spatial resolution than for any single telescope of the array. This development revolutionized radio astronomy and one of the pioneers, Sir Martin Ryle of Cambridge UK, was awarded the 1979 Nobel Prize in physics for his work. The famous 5 km synthesis array at Cambridge, now named the Ryle Telescope, made important measurements in the study of the structure of radio galaxies.

A further development was to extend this technique to a two-dimensional array. The Very Large Array (VLA) in New Mexico is the best example. This uses 27 telescopes on a series of railway tracks to give spatial resolutions of better than an arcsecond at the highest observing frequency of around 30 GHz (~1 cm). In the UK, this technique was further extended by using very widely spaced telescopes, linked not by cables, but by microwave links. This is called MERLIN and its seven telescopes are controlled from Jodrell Bank. This array produces images with resolutions better than 0.05 arcsecond at the highest observing frequency of ~ 22 GHz (1.4 cm).

Radio astronomers can also link large radio telescopes on differing continents to give the highest spatial resolutions achievable in astronomy. This is a technique known as very long baseline interferometry (VLBI). Here, telescopes on different continents work independently but make simultaneous measurements of the same astronomical source. Each telescope records data onto a magnetic tape synchronized with its own hydrogen maser atomic clock. In the early days, the data and clocks were all brought together at a common data analysis centre where the extremely complex synchronization, fringe pattern determination and resulting analysis were undertaken. Nowadays, global positioning satellites give sufficiently accurate overall timing for all the telescope locations that the atomic clocks no longer need to be brought to the common analysis centre, just the huge magnetic data tapes. This incredible technique now operates up to a very high frequency, corresponding to a wavelength of just over 1 mm, although only the brightest sources can yet be observed. World-wide VLBI studies have achieved spatial resolutions as small as 50 micro-acrseconds and observations from VLBI have provided key results for studies of radio jets from quasars (see chapter 8). The discovery of superluminal motion and the exotic nature of central engines in AGNs derive from such observational techniques.

The latest dedicated VLBI array, the VLBA, was commissioned in the early 1990s. Ten 25 m diameter telescopes are sited in a number of locations spanning North America, ranging from the US Virgin Islands in the east (where one just survived a 126 mph hurricane) to Mauna Kea in Hawaii in the west. These precision telescopes operate even at short millimetre wavelengths and are linked together to provide a radio telescope with a spatial resolution of better than 1 milli-arcseconds. Future possibilities enabling VLBI to provide even higher spatial resolution include locating a telescope in space and linking it with others on the ground. Such a project is planned for 1996 (called VSOP) and this will provide an expected resolution of 60 micro-arcseconds.

2.4.5 UV, X- and gamma-ray regimes

Optical photometry utilizes a U-band filter which, although referred to as an ultraviolet band, is only on the very long wavelength side of the ultraviolet domain. True ultraviolet astronomy, along with X-ray and gamma-ray observations, must be undertaken using space-borne telescopes. Ultraviolet telescopes are just like optical telescopes in that the reflecting surface is a mirror, bringing the light to a focus in the focal plane where a battery of UV detectors is arranged. The only difference is in the coating of the mirror surface, chosen to ensure a high degree of reflectivity for the UV radiation. UV telescopes have been flown successfully in space for many years; the International Ultraviolet Explorer (IUE) must be considered one of the most successful space observatories of all time. It has produced much important work that has a direct bearing on active galaxy research. The 2.4 m diameter Hubble Space Telescope is the successor to the IUE. It is designed to operate in the ultraviolet as well as the optical and near-infrared. Unfortunately, the well-publicized optical problems of the primary mirror and stabilization servo severely degraded its first few years of observing capability. Those projects which were least affected were then allocated priority to ensure useful results were still obtained while the solution to the problem was found. Following refurbishment with the COSTAR correcting system in late 1993, the HST is now producing fabulous diffraction-limited imaging, providing spectacular information for the study of active galaxies and other branches of astronomy. A sixteen-day mission in early 1995 by the Space Shuttle Endeavour, carrying three UV telescopes on-board demonstrated the flexibility and potential for Shuttle-based telescopes.

X-ray astronomy is even older than space-borne UV astronomy, but it is fair to say that it took a number of years before the discipline became well established and the telescopes had sufficient spatial resolution to obtain clear identifications of the X-ray sources. This lack of imaging capability has long plagued both X-ray and gamma-ray astronomy. It arises because of the inability to use normal incidence reflection techniques (like a mirror) to bring the short wavelength radiation to a focus. Instead, the technique of grazing incidence is used to construct an imaging X-ray telescope. This is an amazing feat that was first used on the short-lived Einstein Observatory launched in 1978, the same year as the IUE satellite. A relatively large number of X-ray telescopes have been flown in space and one of the latest and largest satellites, ROSAT, launched in 1990, has conducted an all-sky survey in the low-energy X-ray regime. The recent (1993) ASCA satellite is targeted towards spectroscopy and has made a number of crucial discoveries for AGN studies.

X-ray detectors often use particle detection techniques. The high-energy photons cause secondary events in the detector and it is these which are subsequently detected and analysed to provide information about the incoming photon (energy and direction). Proportional counters, scintillators, microchannel plate analysers and positional sensitive advances of these devices make up the detector payload of X-ray telescopes. The usefulness of X-ray observations cannot be overemphasized in active galaxy research as we shall see later. Any significant X-ray luminosity from a galaxy demonstrates the presence of a very hot or energetic energy source. Whether the X-ray emission comes from annihilation of electron–positron pairs, thermal emission from an accretion disk surrounding a supermassive black hole or alternative scenarios, will be discussed in the following chapters.

The X-ray region is usually broken up into the 'soft', 'medium' and 'hard' sub-regimes by workers in the field. However, to the outsider these names can often be confusing, particularly when X-ray astronomers also refer to the spectral slope of the emission by the terms 'soft' (meaning steep) and 'hard' (meaning flat). We shall try to avoid this confusion and use terms such as low-energy and high-energy. The former range from around 0.1 to a few kiloelecton volts (keV), which, in wavelength terms, is 12 nm to just under 1 nm. The high-energy X-ray region extends from around 10 keV to a few hundred keV where it merges into the gamma-ray regime.

If X-ray astronomy is thought to be difficult then gamma-ray astronomy looks nigh on impossible, yet huge progress has been made. To distinguish gamma-rays from the numerous high-energy cosmic ray particles requires techniques such as anti-coincidence timing being used. Furthermore, because gamma-ray photons are so energetic, the detector must be large and heavy to provide enough material in which the energy of the photon can be converted into the secondary products that are actually detected and measured. This type of arrangement is often called a calorimeter as it measures the total energy deposited by the photon as it loses energy by interactions in the bulk of the detector. Although the rate of photons captured by the detectors is small (observations taking many days to ensure an adequate signal-to-noise ratio), each photon carries over a million times more energy than an optical photon, hence the gamma-ray luminosity is far from negligible.

Gamma-ray telescopes are rare and they all have poor spatial resolution of tens of arcminutes to about a degree, hence source identification is extremely difficult. The US Compton Gamma Ray Observatory (CGRO) contains a number of individual detectors, and for our purposes the Energetic Gamma Ray Experiment Telescope (EGRET) observing in the 30 MeV to 30 GeV range provides most of the data. The CGRO has already made a number of very significant observations of AGNs. This is one of the new fields of discovery for active galaxy research and the results will be used in later chapters.

2.4.6 Infrared and far-infrared astronomy

Infrared astronomy began in the 1960s and over the past few years it has become an extended form of optical astronomy, frequently being carried out on standard optical telescopes. In the near infrared, the region between 1 and 3 microns, similar detectors to the optical are used, the most ubiquitous being the infrared version of the CCD. As the wavelength increases to the mid- and far-infrared, life becomes more difficult and both telescopes and the observing sites need to become more specialized. This is because the thermal emission from the Earth's atmosphere dramatically dominates the astronomical signal from space and so special techniques of removing this background signal are employed. This consists of wobbling the secondary mirror of the telescope backwards and forwards at a rate of around 10 Hz. The detector alternately sees two adjacent patches of sky, both of which contain equal amounts of radiating sky but one of which additionally contains the source. The difference signal between these is measured by the electronics and represents the signal from the astronomical source. The high background has been removed in the process.

Because of this requirement to minimize the background radiation, including that from the telescope itself, there are only a small number of dedicated infrared telescopes. The two most prominent of these are the United Kingdom Infrared Telescope (UKIRT) with a 3.8 m diameter mirror and the smaller NASA Infrared Telescope Facility (IRTF). Both reside on the summit of Mauna Kea. Although not easy to see on fig. 2.4, the atmosphere has a small number of semi-transparent 'windows' through which mid-infrared radiation can penetrate; these lie around 5, 10 and 20 μm. To undertake observations in the far-infrared (30–300 μm) airborne (the KAO) or space platforms are required. Beyond 300 μm the atmosphere again becomes semi-transparent in selected regions for ground-based study and we enter the submillimetre regime. This is the domain of very high-frequency radio astronomy. All mid- and far-infrared detectors must be cooled to liquid helium temperatures and those operating in the submillimetre use complex liquid helium-three refrigerators to obtain detector temperatures less than

100 mK for maximum sensitivity.

The launch of the Infrared Astronomical Satellite (IRAS) in 1983 marked a milestone in far-infrared astronomy, producing the first all-sky survey and many discoveries. For our purposes, the most important was the unexpectedly large number of strong far-infrared emitting galaxies (FIRGs), many of which are also in the starburst category. A follow-up telescope called ISO, just launched on November 17th 1995, is eagerly anticipated by astronomers as it will make much more detailed observations with higher sensitivity. It has a range of complex instruments ranging from spectrometers to imaging cameras and will produce the largest quantum gain in the field for a decade. Active galaxies are a prime target for these observations.

2.5 ALL-SKY SURVEYS

2.5.1 The first optical surveys

Optical surveys have always been an incredibly powerful tool for astronomers interested in a wide variety of problems. Some of the most challenging studies in the determination of the large-scale structure of the Universe, cosmology, dark matter and the formation and evolution of galaxies are currently being tackled by powerful sky surveys. From our previous discussions we can see that using sky surveys to search for active galaxies selected by specific criteria is an obviously powerful technique, but one in which there are probably inherent selectivity biases.

In the optical regime, sky surveys are the province of special telescopes called Schmidt telescopes. These were designed forty years ago with one aim in mind, to photograph a wide field of view with spatial resolution at least as good as the seeing at the site in question (~ 1″). The need for a special wide-field telescope is that a standard reflecting telescope, with a combination of parabolic or hyperbolic primary and secondary mirrors, has an extremely limited field of view. This is due to the optical aberrations introduced by the primary and secondary limiting the field of view to at best about 1 arcminute. This can be increased to about 1 degree by using highly complex optical corrector systems in front of the prime focus position. In comparison, a Schmidt telescope can cover a field of view of 6 degrees by 6 degrees.

The Schmidt telescope achieves this by using a spherical primary mirror; the resulting spherical aberration is corrected by a very thin and precisely figured glass plate located at the front of the entire tube. Special photographic plates are manufactured to record the image. These are very large, measuring about 0.35 m by 0.35 m and a single exposure covers almost 40 square degrees of sky. The special photographic plates used today have very fine grain, are blue-sensitive and will record images as good as the atmospheric seeing over a wavelength range extending from the short-wavelength atmospheric cut-off at 320 nm to about 520 nm. A Schmidt plate has a tremendous information content. Emulsions are now available for working in a range of colours in the optical (V) and far red (mistakenly called the infrared for these plates).

The first sky survey was undertaken by the 48-inch Schmidt at Mount Palomar, the size referring to the diameter of the glass corrector lens (the prime mirror being some 72 inches in diameter). The survey is called the Palomar Sky Survey and was taken in two ‘colours’ using emulsions selected to give blue and yellow responses. Because the

Mount Palomar Schmidt could not see all the southern sky, a second survey was undertaken by another 48-inch Schmidt at the Anglo Australian Observatory at Siding Spring, New South Wales. This survey, the SERC Southern Survey, was in four colours, extending the range into the red and the far red. Copies of these survey prints are vital equipment for all observatories and astrophysics research groups and provide a wonderful source of material for student projects and practical experiments. Teaching packs containing Schmidt survey prints and classroom exercises can be obtained from the Royal Observatory Edinburgh. The limiting magnitude of a Schmidt plate is not as deep as for a prime focus plate of a 4 m class telescope. Nevertheless, the latest plates will easily reach fainter than 22nd magnitude and even 23rd magnitude under the very best conditions.

Photographic plates are in general most sensitive in the blue part of the spectrum and it was in this waveband that some of the first sky surveys using Schmidt telescopes were carried out. Their results had dramatic consequences for active galaxy research. One of the most extensive surveys was that undertaken by Soviet astronomers beginning in the mid 1960s and still continuing today. Led by B.E. Markarian at the Byurakan Observatory in Armenia, the astronomers began a major systematic survey using a Schmidt telescope and a novel form of slitless low-resolution spectroscopic technique; the resulting images were recorded on blue-sensitive photographic emulsions. The 'spectrometer' was a thin wedge of glass placed in front of the glass corrector lens of the telescope and is an example of the long-established low-resolution spectroscopic technique practised with refracting telescopes known as objective prism spectroscopy. This produces low resolution spectra but has the enormous advantage that when used with a Schmidt telescope, a very large area of sky is covered in a single exposure. The objective prism spreads out the image of each object into a low resolution spectrum and the relative brightness of the ultraviolet part of the image compared to the blue or yellow can be seen. Prominent absorption or emission lines can also be readily identified. To obtain quantifiable data, however, is much more difficult and requires extensive calibration.

It turned out that many of the first active galaxies discovered, as well as all the first quasars, showed a blue-UV excess when compared to the output from normal galaxies. The use of objective prisms was then immediately employed to search for active galaxies by looking for blue-excess objects. However, this introduced a major selection effect. The emphasis on the blue-UV part of the spectrum discriminated against objects with very high redshifts, whose intrinsic blue excess was shifted into the visible or even red part of the spectrum. Furthermore, there was serious discrimination against those active galaxies that did not possess a blue excess, examples being BL Lac objects and those having very dusty nuclei.

To overcome the blue-continuum bias and to search specifically for signs of activity in galaxies a further selection criterion was introduced to Schmidt slitless surveys. Instead of the continuum emission, attention now focused on the presence of broad emission lines, which were a common factor in the spectra of the original lists of active galaxies. In the mid 1970s, Pat Osmer and Malcolm Smith began a survey using the 48-inch Schmidt at the Inter-American Observatory at Cerro Tololo in Chile. They elected to concentrate on the emission lines rather than a selection effect based on colour.

However, this in turn was a biased survey due to the redshift range over which the most prominent emission lines were best matched to the photographic emulsion. The astronomers were well aware of this bias and took it into account during the subsequent analysis. The limiting magnitude of a Schmidt plate is reduced to about 19 or perhaps 20 when an objective prism is used, the loss being due to the point-like image being spread out into a spectrum.

Another method of searching for active galaxies is to make a form of colour image by recording the same area of sky through different filters (such as U, B, V). These are inserted in front of the photographic plate to give two filtered images, with the telescope being moved slightly between exposures. Any blue-UV excess can be matched with typical colours of active galaxies to remove blue stars and white dwarfs. In the case of high redshift quasars, again there is the problem of the redshift moving the light into other regions of the spectrum in which the quasars are more difficult to distinguish from stars on the plates. Much effort is currently expended on techniques of identifying high-redshift objects.

Because of the huge sky coverage (nearly 40 square degrees per plate), searching for specific objects or features on a Schmidt plate is a huge task. Obvious features, such as a galaxy, will possess extended images and can be easily identified with as little as a hand-magnifier as an aid to scanning the plate with the naked eye. However, as in the military photo-reconnaissance world, a carefully trained eye is required in order not to miss candidates. Computer technology was brought into play to scan the plate with high-precision accuracy using a CCD camera and to store the subsequent data. Nowadays the scanning and searching of Schmidt plates has become the province of sophisticated computer-controlled measuring machines. It is a highly specialized technique practised by a very small number of groups in the world. These sophisticated machines can register positions to an accuracy of a few microns and can be programmed to look for specific features, such as extended shapes or strong emission lines. The machine then allocates the position on the plate, makes a probable identification based on its input parameters and in the case of a spectral plate, gives a first hint at a probable redshift. Measuring machines are vital when determining the brightness of all the 'stellar' objects on plates, calculating colour differences, and, finding potential quasar candidates when the UBV colour difference is used.

The scanning of Schmidt plates is the main technique by which the most distant quasars have been found. The search for the highest redshift quasar is a competition between astronomers and brings out much friendly rivalry. The current two protagonist telescopes are the Schmidts in Chile and Australia, both using red-sensitive emulsions to cover the redshifted emission lines for these high-redshift quasars. The 'world record' for the highest-redshift quasar stood for a long time at $z = 3.78$, and then within the space of a few years shot up to over 4.0 and then quickly to over 4.5. The end-point of this search is highly important for cosmology and the theory of galaxy formation as we shall see in chapter 7.

2.5.2 Radio surveys

The University of Cambridge in the UK began the first of their series of radio sky surveys in the late 1950s and early 1960s. However, the first of the surveys to stand the test of time in terms of quality and sensitivity was the 3C survey. This detected about 500 radio sources at a wavelength of 169 cm (178 MHz) to a flux level of 9 Jy. It was superseded in 1962 by the 3CR catalogue. This has more accurate flux values and covers all the northern sky. There are 328 sources in the 3CR catalogue and it is the best and most complete unbiased radio sky survey to date. Substantial work has been undertaken to identify all the sources; most of these (about 75%) are radio galaxies, while quasars and Seyfert galaxies make up the rest. Various other surveys (e.g., Ohio) have covered selected regions of the sky at a number of wavelengths.

As the wavelength decreases, the number of surveys becomes more restricted and there is only one high frequency survey, the 90 GHz survey which produced the Kuhr catalogue of radio sources brighter than 1 Jy. It should be noted that these surveys were all based on northern hemisphere telescopes; the equivalent 3CR survey has not been made of the southern sky. However, the survey from the Australian Parkes telescope produced a lengthy list of radio sources, most of which are active galaxies. We should not forget that the beamsizes used in these surveys were of order many arcminutes; hence, the major problem was source identification with optical images on the Palomar Sky Survey. This will become apparent when we investigate the story of the discovery of quasars in the next chapter.

Because the radio surveys were made with different telescopes, each produced its own cataloguing system (examples being 3C273, OJ287, PKS2000-330 and so on). Identifying a single object from its many different names is still one of the problems in active galaxy research (in common with other branches of astronomy). In the case of relatively nearby optical galaxies, the NGC (New General Catalogue) number is usually used, e.g., NGC5128. Very few active galaxies are contained in Messier's list, the most notable being M87 and M77 (the prototypical Seyfert 2 galaxy NGC1068). However, for distant radio sources another very sensible convention has emerged, called the IAU format. Here, the sources are labelled according to their position in the sky in terms of right ascension and declination, the celestial equivalent of longitude and latitude. In the case of the first quasar to be discovered, 3C273, its coordinate designation is 1226+023, meaning that its right ascension is 12 hours 26 minutes and its declination is 02.3 degrees north of the celestial equator in 1950.0 coordinates. This is very helpful for observational astronomers; they can figure out when a source might be observable without having to resort to a catalogue or an astronomical ephemeris.

2.5.3 The infrared sky

It might be thought that an infrared sky survey is not possible using a ground-based telescope because of the highly restricted fields of view. But in fact a survey was undertaken in the early 1960s by one of the founders of infrared astronomy, the young Gerry Neugebauer working with Bob Leighton. They used a small, 1.5 m diameter home-made telescope and a crude (by modern standards) detector and mapped most of the sky visible from their Los Angeles site. The survey took six years to complete and listed almost six thousand sources, most of which were stars. This was an amazing achievement. Subsequently, a series of rocket flights was carried out by the US Defense Department and extended the wavelengths surveyed to ~20 μm. These used very small telescopes and surveyed much of the sky, but even though they were relatively crude surveys, they were exploring a virtually unknown part of the e-m spectrum. Indeed for a number of years the results were classified, much to the frustration of astronomers, who knew the flights had been undertaken due to the inevitable 'leaks' but could not gain access to the new astronomical discoveries. The results were eventually released to the community and the catalogue (the Air Force Geophysical Laboratory catalogue) contained some 2000 bright mid-IR (10 μm) emitters, many of which did not appear in the 2.2 μm survey of Neugebauer and Leighton.

The latest, and by far the best in terms of sensitivity and sky coverage, was the survey undertaken in 1984 by the US–UK–Netherlands Infrared Astronomical Satellite (IRAS). During its 12-month lifetime, it produced a four-colour (12, 25, 60 and 100 μm) survey covering virtually all the sky. Furthermore, the mode of operation enabled regions of sky to be re-sampled at epochs of days, weeks and months. This allowed fainter sources to be detected by a process of co-addition and source variability to be tested. The IRAS sky survey is a major source of data for current astronomical research and as we have already noted, one of the highlights of the mission was the discovery and cataloguing of far-infrared bright galaxies, often referred to as IRAS galaxies. We shall discuss these later in the context of buried AGNs and starburst galaxies.

2.5.4 X-ray surveys

In the 1960s, the first rocket flights made very crude surveys, even more primitive than the infrared, but only the brightest of sources were detected. The first true survey was by the Uhuru satellite in 1971. The results contained a number of highly exciting discoveries and although most of the sources were Galactic in origin, some galaxies were found to be strong X-ray emitters. A plethora of X-ray satellites has subsequently been launched; those with a major survey component included the US-UK Ariel 5, and the US HEAO1. Many of the other telescopes surveyed only small patches of the sky, their primary mission being to improve detection sensitivity and to point and integrate at known or suspected sources. The earlier missions tended to operate at X-ray energies of around 10 keV and it is only recently (1990–91) that the first all-sky survey has been undertaken in the low-energy X-ray region of 0.1–2 keV. This was the ROSAT mission that has detected over 60,000 sources. This spectral regime is important for active galaxy research because it is the region in which very hot accretion disks might be expected to emit most of their luminosity.

2.6 THE POWER OF SPECTROSCOPY

2.6.1 Spectroscopy and Kirchhoff's radiation laws

Because line emission is one of the main criteria for classifying active galaxies, we will now present a summary of spectral lines and basic atomic physics before embarking on the details of classification in chapter 3. Students already familiar with Kirchhoff's laws of radiation, line widths and broadening, basic atomic physics of the Bohr atom, line series, and forbidden and permitted lines can safely continue directly to chapter 3.

The science of spectroscopy has been an intimate part of astronomy since its conception and indeed it was towards astronomical bodies that some of the first experiments in spectroscopy were aimed. Spectroscopy is the science of measuring the positions, widths and strengths of spectral lines. From these, crucial deductions can be made about the physical process involved in the production of the lines, either in the source itself, or perhaps an intervening medium. We have already seen that the first active galaxies to be discovered possessed very strange spectra, having prominent and broad emission lines from highly ionized species of atoms. This immediately differentiated them from normal galaxies. These emission lines hold a particular importance. This is not just that they are easy to observe and hence identify an active galaxy, but because they reveal detailed clues to the physical processes involved in the emission region. We cannot hope to progress much further in the study of active galaxies without a basic understanding of the physical processes causing emission and absorption lines.

The story of spectroscopy began in the latter half of the nineteenth century, around the year 1857. The puzzle of absorption and emission lines seen in laboratory and astronomical objects was then put on a firm scientific footing by the work of the great German chemist, Gustav Kirchhoff, working with Robert Bunsen. They discovered that hot gases gave rise to emission line spectra, heated bodies gave rise to a bright continuous spectrum, and a gas lying in front of a hotter solid body produced an absorption line spectrum superimposed on the otherwise bright continuum. As a result of many experiments they further deduced that the emission and absorption lines were due to the presence of particular *elements,* each possessing its own unique identifying fingerprint pattern of lines.

The implications were profound and it was instantly realized that the spectrum of an astronomical body gave information on its elemental composition. This was a staggering achievement of science considering that there would never be a possibility of visiting such distant and extremely hostile worlds. As well as being able to identify the chemical make-up of the object in question, the spectral lines also give information about the temperature of the body. This is another remarkable outcome that produced a much better understanding of the physical conditions inside stars and later on, their evolution. It is hard to overstate the importance of these discoveries. It could well be argued that the introduction of spectroscopy paved the way for the new atomic theory, introduced by Niels Bohr in 1913.

Kirchhoff produced three laws describing the nature of radiation emitted from heated bodies. They are illustrated in fig. 2.5 and are as follows:

(1) Solid bodies and high density gases when heated give rise to a continuous emission spectrum.
(2) A low density gas which is heated gives rise to a bright line emission spectrum, the presence and strength of the lines in the spectrum being determined by the chemical composition of the gas and the temperature to which it is heated.
(3) When a gas is placed between a hotter source of continuum radiation and the observer, the continuum is crossed by dark absorption lines produced by the gas, the presence and strength of the lines being dependent on the chemical composition and temperature of the intervening gas.

The discovery of well-known elements in stars and even hydrocarbons in a comet firmly established astronomical spectroscopy as a crucial tool for probing the composition and physical parameters of remote bodies. As an aside, in 1868 the previously unknown element helium was discovered in the spectrum of the Sun, from which its name is derived, coming from helios, the Greek name for the Sun. It was almost thirty years later that helium was finally isolated in the laboratory. Consequently, astrophysics very rapidly demonstrated its ability to make discoveries that had implications for us on Earth. The work of spectroscopy and the study of the periodic table of the elements eventually culminated in our concept of the composition and workings of the atom.

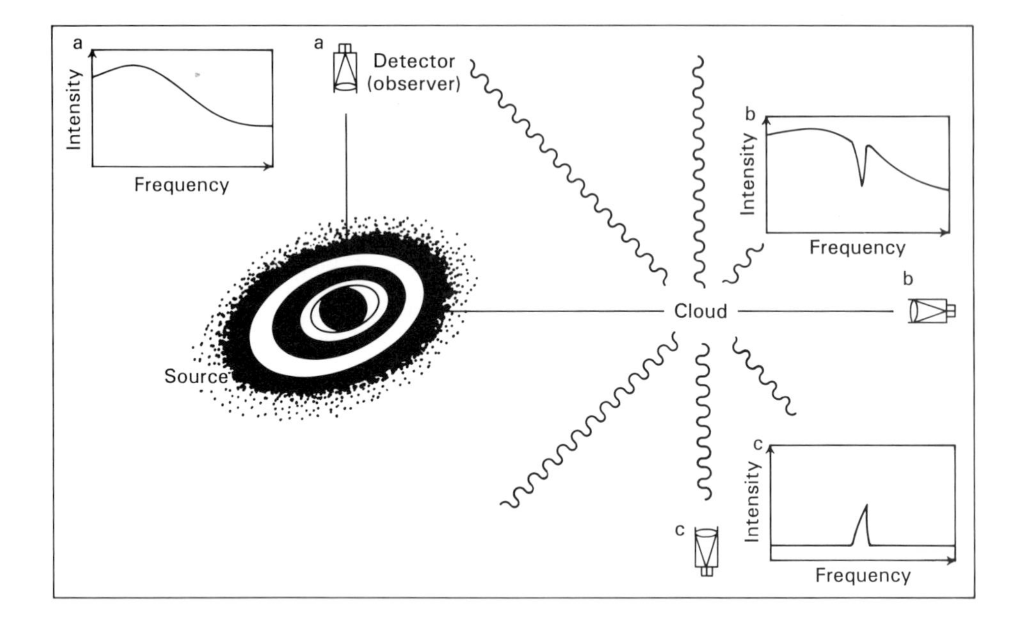

Fig. 2.5. Schematic illustrating Kirchhoff's radiation laws.

2.6.2 Atomic structure

A neutral atom has a nucleus of N neutrons and Z protons surrounded by a cloud of Z electrons. Z is termed the *atomic number* of the element. The electrons are assigned only specific energies, they cannot take up any arbitrary energy; the energy levels are said to be quantized. In the simple picture of the Bohr atom (which will suffice for our purposes), an energy level diagram can be drawn showing the states of all the available energy levels and the number of electrons in each (fig. 2.6). When energy is input to an atom, either by a photon from a radiation field or by collision due to another atom or a charged particle, an electron may be moved from its normal energy level, called the ground state, to a higher energy level. The atom is then said to be in an excited state.

If the energy of the photon or collision is sufficiently high, the electron may gain enough energy to be completely removed from the atom. The atom is then said to be singly ionized. When an electron that has been excited into a higher energy level returns to a lower energy level, a photon of energy equal to the energy difference between the two levels is emitted. This is called radiative emission, or radiative de-excitation, and results in a very narrow emission line being produced. The commonest example of radiative emission is the emission spectrum of hydrogen. This is composed of many series of lines; the four most common series are the Lyman, Balmer, Paschen and Brackett. The Lyman series occurs when an electron falls from an upper state to the first energy level; the Balmer series is when an electron falls to the second energy level; the Paschen series is for the third energy level and so on. These are shown in fig. 2.6.

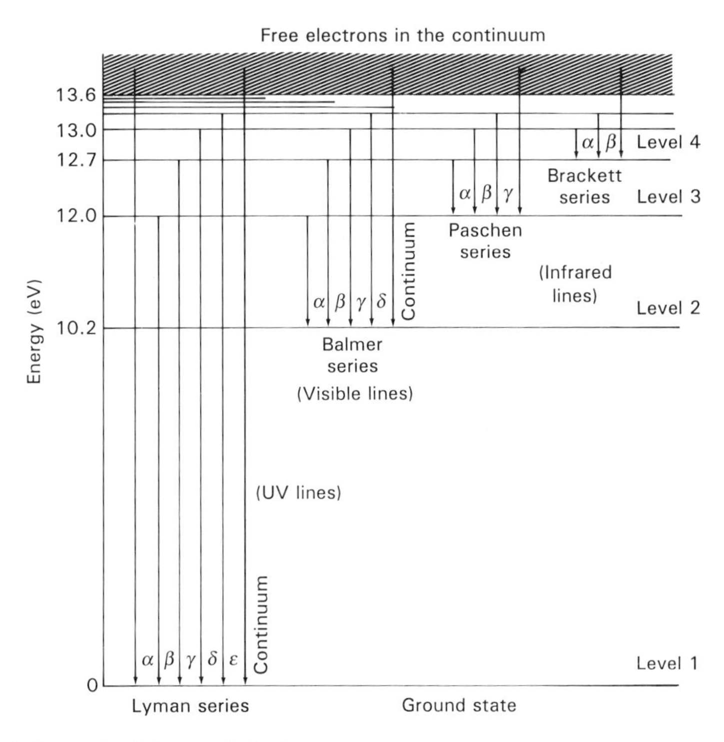

Fig. 2.6. Energy level diagram for hydrogen.

2.6.3 Emission lines from AGNs

Two types of emission line are observed from a number of astronomical objects. These are called *permitted* and *forbidden* lines. The radiative decays for hydrogen discussed above produce permitted lines and because these are by far the most common in laboratory situations, the word permitted is usually omitted, being taken for granted. In radiative decay, the atom is de-excited by the emission of photons as the electron returns to the ground state, either in a single step, or perhaps in a number of steps if there are intermediate levels.

In contrast, we can consider the case of emission caused by collisional de-excitation. In this process, an excited atom loses some, or all, of its excess energy above the ground state by means of collision with another atom or ion. The neighbouring atom carries off some or all of the energy in the form of increased kinetic energy, and the original atom is reduced to a state of lower energy. The rules for these processes are strictly governed by quantum mechanics and are extremely complicated and dealt with in specialized texts. We do not have to worry about them here, but there is an important result that refers to forbidden lines and is pertinent to our study of active galaxies.

Forbidden lines are produced by an atom when it de-excites to a lower state in a particular environment, such as the rarefied vacuum of space. Under normal laboratory pressures and densities this forbidden line is never observed because there are sufficient neighbouring atoms to ensure that the excited atom loses all its energy by collisional processes before the emission of the 'forbidden' line. So what does forbidden really mean? A forbidden line is only forbidden in the sense that it is never observed under the usual laboratory regimes of densities and pressures. If, on the other hand, the density and pressure are much reduced, then the possibility of radiative de-excitation, as opposed to collisional de-excitation, is increased. As the density is further reduced, the shortage of neighbouring atoms means that eventually all the de-excitations will occur by radiative rather than by collisional processes, producing a so-called forbidden line.

In fact the situation is actually more complex than the above description implies and we will present the following discussion in preparation for the detailed treatment of emission lines in chapter 4. The key factor is that per unit mass of gas, permitted lines have an emissivity that increases linearly with the density of the gas, whereas above a *critical density*, forbidden lines have a constant emissivity. As the density of the gas increases, although collisional de-excitations are increased, these are directly matched by a corresponding increase in collisional excitations. Therefore, the ratio of permitted to forbidden line flux increases with increasing density and it becomes more difficult to see the forbidden lines. However, an important point to bear in mind (and one which is often not appreciated) is that forbidden lines are indeed present in a high-density gas and are stronger than for a low-density gas, but because of the much bigger increase in the flux from the permitted lines, are much more difficult to detect. We will return to discuss this further when we consider the critical density in chapter 4.

To differentiate forbidden from permitted lines, identification of forbidden lines is given by the ionized species enclosed in square brackets followed by the wavelength in nanometres. An example is for the doubly ionized oxygen line given by [OIII] 500.7. In fact this line is a doublet, the 500.7 emission is the stronger of the two lines with another transition occurring at 495.9 nm. There is an interesting history to the [OIII] line emission in that early photographs of nebulae often revealed a dominant green colour. When spectra were taken of these nebulae they were found to be due to a strong emission line around 500 nm. Better spectrometers then showed this to be a doublet, but as it was believed to be a permitted line at the time, it was of unknown origin. The newly discovered element was named 'nebulium' for obvious reasons. Almost 75 years passed before the puzzle was finally solved in the early 1930s, when emission from forbidden lines was grasped and the [OIII] doublet was identified. The need for nebulium immediately vanished.

Forbidden line transitions are well observed both astrophysically and under specialized laboratory conditions and serve as useful tests for quantum theories. The presence of forbidden lines gives valuable information about the density and pressure of the medium in which the atom exists. The observation of permitted and forbidden lines forms one of the key methods by which temperatures and densities of gas in space are determined as we shall see when we probe into the broad and narrow line regions of AGNs in chapter 6. A further aspect of the forbidden lines seen in the optical part of the spectrum is that they are all produced by collisional processes. This collisional excitation must not be confused with the de-excitation process. The collisional excitation occurs because these lines arise from energy levels within a few electron volts of the ground state and so can be excited by thermal electrons in the environment.

Another crucial importance of emission lines is the relative strengths of certain species, usually ionized or highly excited. This gives clues to the mechanism that produced the ionization or excitation. In this book we shall be interested in only two mechanisms, radiative and shock excitations. Although these sound rather difficult and are often wrapped up in complex jargon, they basically mean that an ensemble of atoms are excited or ionized either by a photon field (radiative) or by collisions from other atoms (shocks). As far as active galaxies are concerned, the former could be the intense radiation field from a very hot accretion disk surrounding a supermassive black hole. Once photoionization is recognized as a source of line excitation, the temperature of the radiation field can be determined and hence the temperature of the radiating source.

Spectral emission lines are described by three parameters, the wavelength, linewidth, and line strength. The total energy emitted by the line is the integral of the energy versus wavelength, the area under the line shape. This is termed the line strength. However this by itself does not differentiate between a line that is very strong and narrow, or very broad and weak. Therefore we also have to consider the width of a spectral line. This is usually expressed in terms of the full width at half maximum (FWHM), which is the width of the line at a position of half its peak strength. Another term in common use is the full width at zero intensity (FWZI). These help to distinguish between the width of the line core and possible broad wings (fig. 2.7). When discussing Seyfert galaxies in the next chapter, we shall see that the width of the core and wings are powerful discriminators of emission processes and classifications. This gives rise to a two-component line, which is shown in figs. 2.7(b) and 9.5. It is usually the case that the broad wing component of the emission line is from a different physical region of the active galaxy than the core component.

When an ensemble of stationary atoms de-excite, the intrinsic width of the emitted line is extremely narrow, much too narrow to measure with any astronomical spectrometer. However, there are a number of ways by which this intrinsically narrow line is broadened in nature. The most common of these is due to the Doppler effect, leading to what is called Doppler broadening. This is caused by emission from an ensemble of gas atoms in *motion*, each atom having a slightly differing velocity or radial component of velocity. (Remember that the Doppler shift is produced by radial motions with respect to the observer, not by transverse motions.) The spread in velocity for an ensemble of gas atoms can come about due to a variety of means, but we shall consider only two: gravity and temperature induced. The former is the most important in our study of AGNs and the clouds of gas in the broad-line and narrow-line regions. Indeed, Doppler broadening is the main line-broadening mechanism we shall meet in this book.

In the case of temperature, the velocities and velocity spread of the atoms depend directly on the temperature of the gas; the greater the temperature, the greater the velocity and the range of velocity spread. Individual atoms emit a narrow specific wavelength, the value depending on their relative motion with respect to the mean of the ensemble of atoms. The outcome from the ensemble of atoms is the sum of all the 'narrow lines', which is a single, but now much broadened, emission line. The line is said to be Doppler broadened. The broadening can be easily calculated given the velocity distribution of the emitting gas. Alternatively, from observation of the width of the line, the velocity spread of the atoms can be determined.

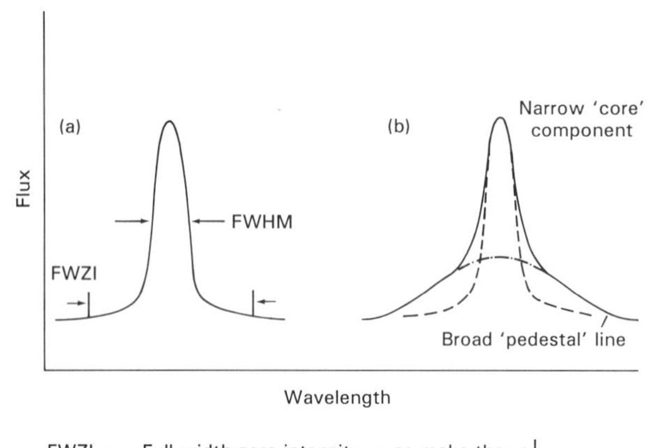

Fig. 2.7. (a) Illustration of linewidths. (b) The blending of a strong narrow line with a weak broad line and to give the 'pedestal and core' appearance.

Let us now consider Doppler line broadening due to the bulk motion of gas in a gravitational field. We can take the case of gas in a spiral galaxy viewed at an inclination angle of 45 degrees. The galaxy possesses copious amounts of neutral hydrogen gas in the spiral arms and this hydrogen has the property of emitting a sharp line at a radio wavelength of 21 cm. This line is due to the hydrogen atoms being in an excited state, but, in this case the excitation is not due to the electron being in a higher energy level than the ground state. The electrons are in the ground state, which has two slightly different levels due to a subtle interaction between the spins of the proton and the orbiting electron. (As well as having a charge, the proton and electron have a quantity referred to as spin, and the spins of the two can be aligned parallel or anti-parallel.) In the higher of the two energy states the spins are parallel, and the hydrogen atom can drop to the lower state, called a spin-flip transition, by the spins becoming anti-parallel. The very small energy difference between the two states (referred to as hyperfine splitting) is only 6×10^{-6} eV, and is emitted as a photon of wavelength 21.2 cm. This is a forbidden transition and is only seen in the rarefied conditions of space: it is the universal signature of atomic hydrogen. The atoms are mainly in the spin-excited state due to collisions.

Returning to our spiral galaxy, we note that it is rotating, not as a solid body like a compact disk but with differential motion, like the planets orbiting the Sun. The galaxy will also be receding from us at some velocity according to Hubble's Law. This will cause the 21 cm radiation from the central regions of the galaxy to be redshifted to a slightly longer wavelength. However, the rotational motion of the gas in the galaxy is not negligible. Gas at the side of the galaxy that is rotating towards the observer will give a slightly blueshifted line compared to that from the gas at the centre, while gas from the side moving away from us will be slightly redshifted. Therefore if we view the gas from the entire galaxy we will observe that the 21 cm line is broadened due to the spread of velocities along the line-of-sight. The width of the line is a direct measure of the maximum bulk rotational velocity of the gas in the galaxy. We chose to use the 21 cm line for convenience, (and also as a convenient time to introduce it) but the same principle of line broadening applies to optical, infrared or any other spectral line emitted by radiating material in the galaxy. It is apparent that for ensembles of gas atoms orbiting very close to the centre of an active galaxy, such as the gas in the broad-line clouds, the velocity spread due to Doppler broadening can become very large.

Before leaving this section it is necessary to note and to apologize for the fact that astronomers use a non-standard technique for labelling ionized species. Singly ionized helium would be labelled as HeI by a physicist and He^+ by a chemist, but unfortunately astronomers use the notation HeII. Using the same convention, fully ionized hydrogen (i.e., a single proton) is therefore labelled HII by astronomers, not HI as one might have thought, which represents the neutral state to the astronomer! Therefore, to obtain the ionization state of an element when reading astronomical notation, subtract 1 from the Roman numerals following the element to arrive at the number of electrons missing: e.g., NIV is triply ionized nitrogen, OIII is doubly ionized oxygen. I am personally not in favour of this peculiar notation but as it is universally used we shall employ it in this text, and we will usually note precisely what is meant for the non-astronomy specialist.

As an aside we should note that this convention is not the only peculiarity in astronomical notation. The standard international (SI) physical units are now widely used, but many (sadly most) astronomers (especially spectroscopists) continue to use the angstrom (1 Å = 0.1 nm). We must also not forget that Hipparchus had a lot to answer for when he defined the magnitude scale backwards, whereby the larger numbers refer to fainter objects!

Finally we need to draw attention to another emission process which will figure later in this book. This is X-ray emission from 'heavy' elements, and it is iron that will be important to our discussions. As X-ray facilities improved in their spectral resolution and sensitivity, especially with the Ginga and ASCA satellites, the identification of X-ray line emission took on much greater importance and respectability. This X-ray line emission comes from photoelectric absorption of energetic photons from a radiating source by the innermost electrons of the atom, resulting in them gaining energy and making the atom either excited or ionized. The resulting rearrangement of the electrons in the atom gives rise to photon emission. When the electrons are excited from the ground-state and subsequently return to their original energy states, this is an example of a process called fluorescence and the result for a heavy element is the emission of X-rays. For iron, there is a particular emission line centred at ~6.4 keV which has become very important and is referred to as the K-α line. We will not worry about the details of the nomenclature (the K refers to the inner, n = 1, energy level of the atom), but note that the line is believed to originate in material very close to the central engine. Indeed, the latest ASCA observation of this X-ray fluorescent iron line probably gives the strongest evidence for the existence of a supermassive black hole (see chapters 5 and 6).

2.7 SUMMARY

In this chapter we were introduced to the concepts of brightness, reflected in the terms *flux* and *luminosity*; the former is distance dependent whereas the latter is the power output of the body in question. Astronomers use measurement scales of *magnitudes* for apparent brightness and luminosity, and these are related by an important and useful equation termed the *distance* modulus (eqn. 2.15). In most cases, the accurate measurement of distances in astronomy is difficult. Another parameter that has special importance is the *intensity* of radiation emitted by a body, and like the luminosity this is an intrinsic property of the body and is distance-independent. When we deal with intensities we need to use *solid angles* and units such as the *steradian*. The measurement of the continuum radiation from an astronomical body is termed *photometry* and is an important tool in understanding the radiation mechanisms of these objects. It will figure prominently in the study of active galaxies.

In any measurement process there is always the possibility of the measurement system (telescope, detector, etc.) introducing a bias into the observations. We have seen some of the pitfalls that can occur for astronomy and issued a general health warning about such problems. All astronomical measurement processes require two fundamental tools: a collector to gather the energy from space (usually referred to as a telescope) and a detector to convert the energy into a recognizable signal for analysis. Generally, but not always, the telescope provides some sort of focusing of the energy so that an image can be formed. Whether an image can be directly recorded to show a two-dimensional picture depends on whether the detector is a panoramic device (such as a CCD) or a single pixel device (such as a radio receiver). Ancillary optical devices can be used to split the energy into smaller wavelength components: perhaps passband filters, or even better, a spectrometer. Spectroscopy is crucial in understanding the workings of the Universe. The information contained in spectral lines allows, us in principle, to identify the chemical composition of the radiating material, gives information about the state of ionization or excitation of the material (relating to its temperature), to obtain information about the velocity field of the atoms (temperature) and their motion along the line-of-sight. We presented a brief introduction to spectroscopy and a review of atomic physics and the structure of the atom and this will stand us in good stead for subsequent chapters. Observations across the electromagnetic spectrum require a range of techniques, telescopes, observing platforms and detectors. Dramatic advances in these facilities have led to the enormous progress in the study of active galaxies reported in this book.

The importance of performing all-sky surveys at a range of wavelengths was discussed and some of the most important surveys were highlighted. Without such surveys, the problems of observational bias are all too great. A number of amazing discoveries have come about from these surveys. We are now well placed to venture into the labyrinth of the secrets of AGNs.

2.8 FURTHER READING

General and reviews

Asimov,I., *Eyes on the Universe—a history of the telescope*, Andre Deutsch, 1975. (A wonderful book about the history of telescopes, highly recommended.)

Bowyer,S., 'Extreme ultraviolet astronomy', *Scientific American*, p32, August, 1994.

Bradt,H., Ohashi,T., & Pounds,K.A., 'X-ray astronomy missions', *Ann.Rev.Astron. Astrophys.*, **30**, 391 1992. (A comprehensive review.)

Cornwell,T.J., & Perley,R.A., 'Radio interferometry, theory, techniques and applications', *Ast.Soc.Pacific.Conf.Series*, **19**, 1991. (Very comprehensive.)

Davies,J.K., *Satellite astronomy,* Ellis Horwood, 1988. (The who's who of satellites, and missions; very comprehensive.)

Davis,R.J., & Booth,R.S., *Sub–arcsecond radio astronomy'*, Cambridge University Press, 1993. (Contains excellent reviews and up to date.)

Elvis,M., Ed., *Imaging X-ray astronomy'*, Cambridge University Press, 1990. (A good review of the imaging facilities for X-ray astronomy.)

Gehrels,N., *et al.*, 'The Compton Gamma-ray Observatory', *Scientific American*, p68, December, 1993. (An easy to read description of this major observatory.)

Giovanelli,R., & Haynes,M.P., 'Redshift surveys of galaxies', *Ann.Rev.Astron. Astrophys.*, **29**, 499, 1991.

Maran,S., 'Hubble illuminates the Universe', *Sky & Telescope*, June, 1992.

Mclean,I.S., *Electronic and computer aided astronomy,* Ellis Horwood 1988. (All you ever wanted to know about CCDs; a good student text.)

Ramana Murthy,P.V., & Wolfendale, A.W., *Gamma-ray astronomy'*, Cambridge University Press, 1986. (A review of all aspects of gamma-ray astronomy before the Compton Gamma Ray Observatory.)

Soifer,B.T., Houck,J.R., & Neugebauer,G. 'The IRAS view of the infrared sky', *Ann.Rev.Astron.Astrophys.*, **25**, 187, 1987.

Tucker,W., & Tucker,K., *The cosmic enquiries, modern telescopes and their makers*, Harvard University Press, 1986. (A selection of telescopes across the e-m spectrum including the VLA, Einstein, IRAS etc.)

Turver,T., 'The Universe through gamma-ray eyes', *New Scientist*, 27 March, 1993. (Gamma-ray astronomy after the CGRO.)

Verschuur,G.L., *The invisible Universe, the story of radio astronomy*, Springer–Verlag, 1987. (Very readable history of radio astronomy.)

Walker,G., *Astronomical observations,* Cambridge University Press, 1987. (An excellent review of optical instrumentation and techniques.)

Wall.J.V., & Boksenberg,A., Eds., *Modern technology and its influence on astronomy,* Cambridge University Press, 1990. (Advanced and up to date, an excellent read.)

Weedman,D.A., 'Seyfert Galaxies', *Ann.Rev.Astron.Astrophys.*, **15**, 69, 1977. (A classic review of the early years of Seyfert galaxy research, well worth reading.)

3

Observational properties of active galaxies

3.1 INTRODUCTION

In this chapter we present a compressed history of the discovery of the various categories of active galaxy along with a description of the major observational properties of the classes. As we will discover, virtually all the classes of active galaxy have a number of pathological cases that do not fit-in with at least one of the major criteria of classification. This gives us a clue to the value and limitations of such schemes. They point the way to further work, and for us, that is unification.

The classes of active galaxy introduced below are not presented in a strictly historical time-frame of discovery, but rather one which reflects observational similarities or technological breakthroughs. We will also hint at those areas whereby the original classification scheme might have resulted through an accident of the technique rather than revealing something truly fundamental or unique about the class thereafter named. Our ultimate goal is to reorganize the current classification scheme and replace it by a *picture of specific physical processes* that are believed to occur in various galaxies. This is the goal of unified models of AGNs and it must be able to explain why these processes produce the particular classes of active galaxy that we observe.

This chapter gives an overview of the global properties of AGNs. The physical principles which are presented in the next chapter will then put us in good stead for our detailed investigations of the processes taking place, leading up to the detailed understanding of the workings of AGNs in the remainder of the book.

3.2 SEYFERT GALAXIES

In 1943, Carl Seyfert completed a special study of six galaxies (NGC1068, NGC1275, NGC3516, NGC4051, NGC4151 and NGC7469) he had selected from the plate archives of the redshift surveys of the Mt. Wilson Observatory. Compared to normal galaxies whose spectra were generally devoid of emission lines, these galaxies revealed broad emission lines (fig. 1.8) and formed the basis for an entire class of active galaxy. They are now known as 'Seyfert galaxies' and as we shall see, the definition has changed over time. The majority of these six galaxies were clearly spiral in nature but had an unusually bright and starlike nucleus. However, one of them, NGC1275, showed a morphology that was irregular and peculiar in many respects. It was certainly not spiral, but nevertheless NGC1275 showed the same emission line property as its companions. This is a classic example of the dichotomy that pervaded the classification of active galaxies for a number of years. Do we look for a similar form in terms of the morphology (spiral galaxies) or another property (such as emission lines). We will tend to use whichever observation tells us about the physics of what is going on, and in this example, it is the emission lines. Indeed, Seyfert's adopted classification for his galaxies were those which had a bright, semi-stellar nucleus and broad emission lines covering a wide range of ionization states.

Fig. 3.1. The Seyfert 1.5 galaxy NGC1566. This is an example of an intermediate Sbc galaxy and also shows vestiges of a faint bar. (Courtesy of David Malin, Anglo-Australian Telescope.)

The surveys of Markarian and co-workers (see section 3.3) led to many more Seyfert galaxies being identified. As spectroscopic techniques improved in the early 1970s, the picture for Seyferts became more complicated and in 1974, Khachikian and Dan Weedman modified the basic classification scheme. The new classification was based solely on the spectroscopic properties and was introduced to take account of the fact that in one group of Seyfert galaxies both the permitted and forbidden emission lines had the same width, whereas in a second group there was a marked difference between the widths of the permitted and forbidden lines.

The 1974 classification subdivided Seyferts into two categories: Seyfert 1 and Seyfert 2. In Seyfert 1 galaxies, the permitted lines, originating mainly from hydrogen but also accompanied by lines from species such as HeI (neutral helium), HeII (singly ionized helium) and FeII (singly ionized iron), were very broad with full widths at half maxima [FWHM] corresponding to velocities in the range 1–10,000 km s^{-1}. The forbidden lines from species such as [OIII] (doubly ionized oxygen) lacked the very broad wings and had FWHM corresponding to velocities only up to ~1,000 km s^{-1}. In Seyfert 2s on the other hand, both permitted and forbidden lines were about equal, with velocity widths ranging up to about 1,000 km s^{-1}.

As we saw in section 2.6.3, line widths give clues to the origin of the lines, at least in terms of the velocity dispersion of the gas that is emitting the line. A simple interpretation is that when lines from the same element have the same width, they are probably formed in the same region. *So in the case of Seyfert 2s, we are left with the impression that both the forbidden and permitted lines are formed in the same region of the nucleus, whilst for Seyfert 1s, the permitted lines and forbidden lines originate from distinctly differing regions.*

The very broad lines, of linewidths of order thousands of kilometres per second, delineate what has become known as the broad-line region (BLR) of the emission zone. The current belief is that the broad-line clouds are caused by photoionization due to a very hot accretion disk surrounding a supermassive black hole in the core of the galaxy. They are located only ~1 pc from the central engine. The lines with linewidths only up to a thousand km s^{-1} come from the so-called narrow line region (NLR). This is believed to lie at a distance from ~10 pc to ~1 kpc from the central engine. (The NLR and BLR are discussed in detail in chapter 6.)

A shorthand notation has now arisen (for reasons dealing with unification models which will become clearer later) in that objects showing spectra of a BLR are called Type 1 objects (from Seyfert 1s as the prototype but ignoring anything to do with classifications based on luminosity). Type 2 objects on the other hand are those which show spectra representative of the NLR (and devoid of any BLR component).

Further improvement in spectroscopic capability showed that even some of the narrow permitted lines have two components, differing in line shapes and widths. The line is in fact formed from a strong narrow core superimposed on a very much broader but fainter base. The classification scheme was extended to take this into account and resulted in Seyfert 1.5, 1.8 and 1.9 galaxies. Seyfert 1.5s are those galaxies whose hydrogen emission lines are clearly made up of the two components and an example is shown in fig. 3.3. Seyfert 1.8 galaxies have an easily identifiable broad component visible in Hα, but only a very weak broad component in Hβ. Similarly, the 1.9 classification extends this further in the direction of the Seyfert 2 class; the broadening in the Hβ line is not seen. This gives a clear view of just how complex the classification can become, a mind-numbing array of parameters. We shall avoid becoming bogged down in the details of this sub-classification picture and rapidly progress. We must always remember that we are looking for the underlying physical processes and a simple picture that brings together the various types of active galaxy.

This classification picture for Seyferts based on the categories 1, 1.5, 1.8, 1.9 and 2, leads to the inevitable conclusion that Seyfert 2 galaxies do not possess a broad-line region. But in fact new techniques in observational ability have demonstrated that at least some of them do (see section 6.5). This was a breakthrough and represents one of the most crucial observational discoveries in AGN research, paving the way for the unified model of AGN activity. This leads us to believe that in Seyfert 2s the apparent absence of a BLR is, on the whole, the result of an observational selection effect. During the decade before the discovery of hidden BLRs, understanding the global properties of AGNs was severely hindered. What was this selection effect? Let us give some thought to this. What could prevent us from seeing a BLR in Seyfert 2s? Dust is an excellent

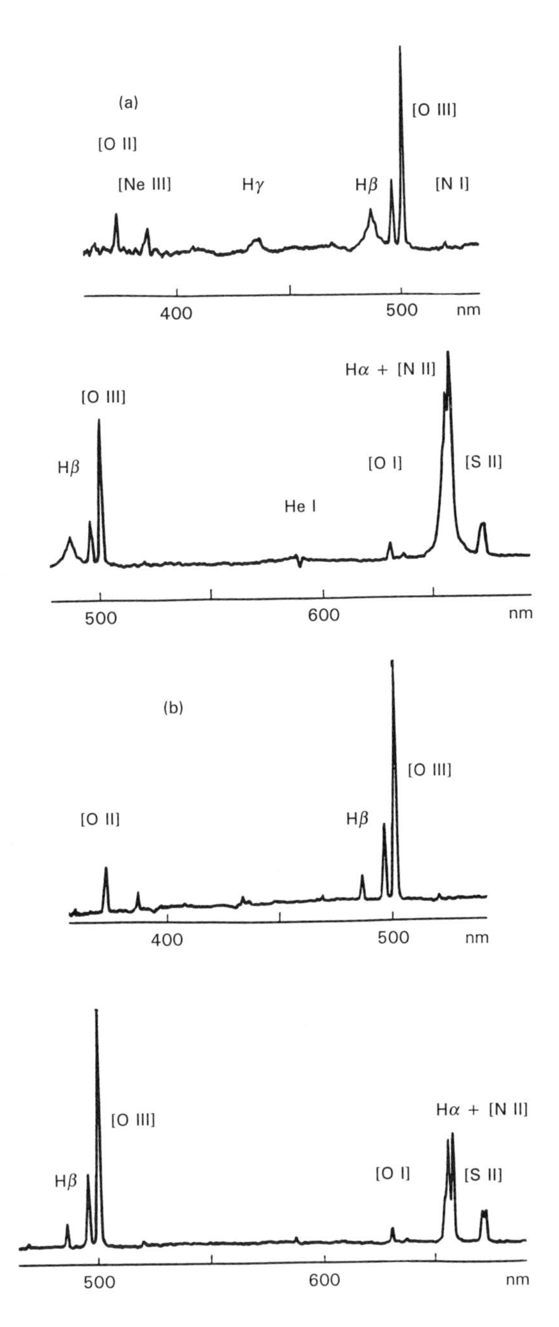

Fig. 3.2. Optical spectra of (a) a Seyfert 1 galaxy NGC3227 and (b) a Seyfert 2 galaxy Mkn1157. The flux is plotted on a linear scale against wavelength. Note how the permitted lines of Hα and Hβ differ in their FWHM. (Adapted from Osterbrock,D. *Astrophysics of Gaseous Nebulae and Active Galaxies*, University Science Books, 1989.)

candidate for hiding objects and so if Seyfert 2s had dust enshrouded cores, then this would do the trick. However, the cores cannot be totally buried by dust, otherwise we would not classify the galaxy as a Seyfert, because we would not see a NLR. But, if we make the dusty zone sufficiently small, such that it lies between the BLR and the NLR, then this might do the trick. The BLR and the central engine would be obscured, but the NLR would be visible.

Because at least one Seyfert 2 (NGC1068) had been shown to possess a BLR, albeit only in polarized light (and hence by reflection), it was only a small step to replace the spherically symmetric obscuring dust zone by a small but thick disk (or torus) surrounding the central engine (section 6.5). We are then left with the elegant and appealing possibility that the differences we see in AGN type might be due to, among other factors, the orientation of the nuclear zone and torus with respect to us. If they were lying more edge-on than face-on then this torus would hide the central zones and at some angle would hide the entire BLR. The much more extended narrow line region would remain visible, however. This is the foundation of the unified model that is a central and underlying theme to this book. Probably the key champion of this is 'Ski' Antonucci and we will return to discuss the unified model in detail in chapters 6 and 9. However, there may be some Seyfert 2 galaxies without a BLR and this is a hot topic of current research.

This detailed sub-classification for Seyferts also brings about one of the new challenges facing astrophysicists working in the field of active galaxies. Moving through the Seyfert 2 classification we find that as the overall luminosity of the galaxy decreases the linewidths also decrease. At what point is the galaxy still classed as a Seyfert, or merely an otherwise relatively normal galaxy but showing low intensity emission-lines? Such galaxies are known to exist (see section 3.7) and this blurring between degrees of activity, and even between what in the past were termed normal galaxies, is indeed suggestive that there is a smooth progression of activity and hence a unified scheme has obvious attractions. It is also clear that at some level the degree of activity is related to the luminosity of the nuclear regions of the galaxy. For low-luminosity objects, there is the added difficulty of separating the nuclear emission from that of the surrounding galaxy. Therefore we might immediately speculate that there are good grounds for suspecting this selection effect will discriminate against finding low-luminosity AGNs. This theme of contrast between nuclear emission and emission from the body of the galaxy will feature prominently and should not be forgotten.

In terms of Seyfert characteristics of active galaxies, the latest data suggest that Seyferts make up a few per cent of bright galaxies, or at least this number show some Seyfert characteristic. In the early days, most of the Seyfert galaxies for which a morphological type could be determined were found to be spirals, and for those that are sufficiently close to be well resolved, they seem to favour barred spirals. Additionally, Seyferts tend to populate the so-called early type spirals (Sa and Sb) and are rare in Sc and later categories which have much smaller and less pronounced nuclear zones. There appeared to be a deficiency of Seyferts in ellipticals, but this again comes down to agreements of classification. If we just take Type 1 objects, then we now know that ~10% occur in galaxies that are elliptical (and many of these are Markarian galaxies). Although originally there were believed to be around twice as many Seyfert 1s as 2s, this has turned out to be another example of observational bias and the latest data suggest at least a few times as many Seyfert 2s as 1s.

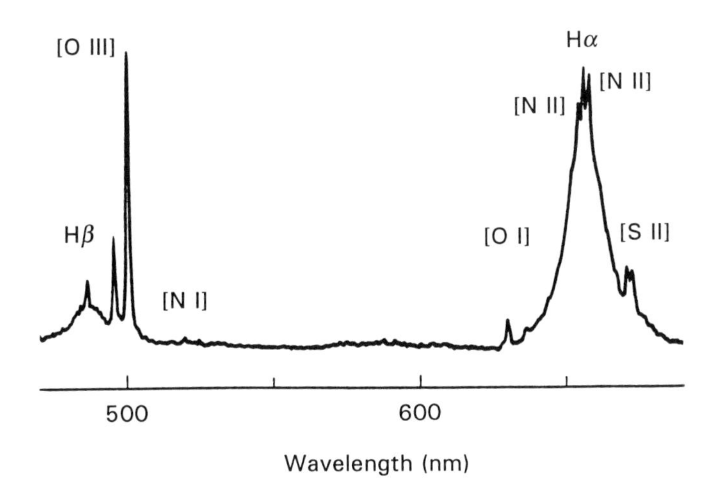

Fig. 3.3. Optical spectrum of the Seyfert 1.5 galaxy Mkn926 plotted on the same scale as fig. 3.2. (From Osterbrock,D. *Astrophysics of Gaseous Nebulae and Active Galaxies*.)

The continuum emission from Seyferts shows an increase from the shortest wavelengths (X-ray) through the optical with a peak in the output spectrum somewhere around 100 μm. The spectrum then falls rapidly to the radio. Subtle differences can be observed between Seyfert 1s and 2s which will be discussed in chapter 4; suffice to say at this point that the near-IR through far-IR emission from Seyferts is currently believed to be thermal emission from heated dust. Few Seyferts are strong in the radio region, and when they are, they are often peculiar in another context (e.g., NGC1275). The luminosities of Seyfert galaxies range from over 10^{11} up to $\sim 5 \times 10^{12}$ $L_{\odot}$ (but see later for clarification). The brightest (corresponding to $\sim 2 \times 10^{36}$ W) is less than ten times the luminosity of the brightest spiral galaxies. On the present evidence, there seems to be some correlation between the degree of Seyfert activity and absolute magnitude of the underlying galaxy; the more luminous the galaxy, the more powerful the active nucleus.

3.3 MARKARIAN GALAXIES

In 1963, B.E. Markarian of the Byurakan Astrophysical Observatory in Armenia published the results of his studies of 41 peculiar galaxies. He claimed that some of the nuclei showed evidence for non-thermal emission, by which he meant that the emission was not coming directly from stellar photospheres. All eight of Carl Seyfert's galaxies were included. As we saw in section 2.5.1, Markarian then began an objective prism slitless survey using the newly completed 1m Schmidt telescope of the Byurakan Observatory. Objects were selected primarily because of a blue-UV excess in their continuum and it was these that formed the basis of the famous lists of galaxies emanating from this observatory. The first survey took over ten years to complete and the 1500 or so galaxies in the published lists are referred to as Markarian galaxies. The survey is believed to be complete down to a limiting magnitude of about 16 and Markarian galaxies have one thing in common: unusually bright UV continua as required by the initial selection criterion.

An obvious question at the time was whether these galaxies were unique and formed a class all to themselves, or were merely a collection of objects selected according to the particular criterion of blue–UV excess. As we noted above, follow-up studies using spectrometers on large telescopes showed that about 10% of the Markarian galaxies were Seyferts, suggesting that they were a mix of objects, many at low redshift, rather than a unique class. It is now agreed that Markarian galaxies are not a unique category, but show one aspect of galaxian activity, the blue–UV excess. We will not discuss Markarian galaxies as a class of AGN. Using the data from the Markarian survey, extrapolations suggest that about 5% of all galaxies are Markarian-like in appearance in that they possess a strong blue–UV continuum.

3.4 QUASARS

3.4.1 The discovery of quasars

The mysterious quasars were finally identified in 1963 in the spirit of a true detective story. Although they had been discovered a number of years earlier by radio astronomers, it was optical astronomers who provided the clinching observations that would eventually reveal their true nature. Even so, although optical astronomers had obtained the first spectra in which the major clue to interpreting quasars was staring them in the face, the obvious conclusion from the data was so implausible that it was a while before the truth finally dawned. Only then did astronomers realize what exotic objects they had discovered. Quasars are the giants in the AGN league and as we noted in chapter 1, their discovery immediately brought into question one of the fundamental assumptions on which astronomers' study of the Universe had been based, namely the use of the Hubble Law to obtain the distances of galaxies.

The story of the discovery of quasars is fascinating and well worth recounting for the benefit of the student of active galaxies. Radio astronomers from the Mullard Radio Astronomy Laboratory at Cambridge in the UK made their first comprehensive all-sky radio maps in the late 1950s, charting the positions and extent of the newly discovered radio sources. Most of these radio sources were found to be extended in size and were identified with either supernova remnants in our Galaxy or, more excitingly, distant 'radio galaxies'. The early radio telescopes had large beamsizes (many arcminutes in diameter), and therefore any source subtending an angle less than this was an unresolved 'point source'. A small number of the unidentified radio sources fell into this category. It was tacitly assumed that because of their compact nature they were neither supernova remnants, star formation regions nor radio galaxies. On the other hand, they were strong radio sources and obvious high priority candidates for further investigation.

Radio stars were the explanation favoured by many astronomers, but a small renegade group believed they might be extragalactic in origin. This was supported to some extent by their distribution which appeared more isotropic than might be expected from a Galactic origin although the statistical significance was not overwhelming. To solve this puzzle it was essential to locate precisely the position of the radio source so that optical astronomers could use the new generation of giant reflecting telescope to obtain clues to their identifications by means of spectroscopy. However, the large beamsize of the radio telescopes meant that there was a significant patch of sky, called the error box, in which the compact radio object might be found. Optical photographs of these error boxes produced numerous possible candidates for the radio sources but none could be positively identified.

It turns out that this problem of source identification is nearly always faced by new branches of observational astronomy. As we saw in the previous chapter, in opening up a new discipline, an all-sky survey is undertaken as soon as possible in order to determine the numbers of sources (at least the brightest ones). But to produce a sky survey in a limited time, the telescopes are usually equipped with large beamsizes to enable them to map the entire sky. This was particularly the case for spaceborne telescopes and the first generation of sky surveys (see section 2.5). Precise source identification is always difficult. The logic is that once the brightest sources have been found, then second-generation telescopes with larger collecting areas, smaller fields of view and much more complex experimental apparatus will be able to make more detailed follow-up investigations.

Although this is entirely logical and in keeping with scientific methodology, like most scientists, astronomers are extremely impatient when faced with new discoveries. They want answers sooner rather than wait for a decade before a second-generation experiment is undertaken. Therefore as soon as a new field is opened up and the first sources discovered, albeit as in the radio surveys usually with very poor positional accuracy, astronomers immediately attempt to identify the unknown and exciting object with a corresponding optical object on a sky survey plate. This is standard practice and has been borne out by all the new wavelengths opened up.

However, a large error box of perhaps up to an arcminute square may contain up to a hundred potential candidates on an optical sky survey plate. Which, if any, of these is the object that sparked off the excitement by radiating strongly in the new wavelength regime? Not an easy task, but peculiar colours or unusual spectra suggest potential candidate objects. In some cases precise identification is both slow and very much hit-and-miss.

One of the compact radio sources attracted considerable interest because the Jodrell Bank astronomers had shown from radio interferometer observations using varying baselines that this source was extremely compact, less than 1 arcsecond in size. Because it was a very strong source, it must have a very high radio surface brightness and therefore it became the top candidate for further investigation. The source was called 3C48, the 48th source in the third Cambridge radio catalogue. Astronomers at the California Institute of Technology (Caltech) used a two-element radio interferometer to investigate 3C48 and in 1960 obtained a positional accuracy of about 5 arcseconds. Within a year this led to its optical identification on a photographic plate taken by Alan Sandage using the 5 m Hale reflector at Mount Palomar.

The candidate object appeared to be a 16th magnitude blue stellar object associated with a very faint nebulosity of low surface brightness measuring about 12 by 5 arcseconds in extent, with the point-like 3C48 lying not quite at the centre. The colours of 3C48 were obtained by using various filters (such as UBV) and were most unusual for a star. Spectra were subsequently taken and immediately the mystery deepened because many broad emission lines were clearly seen. As we are now aware, although broad emission lines are extremely rare in stellar spectra, they are not unknown. It was conceivable that a star that was sufficiently unusual to radiate strongly at radio wavelengths might also possess broad emission lines. However, identification of the emission lines with known species of chemical element and ionization state proved to be intractable. One line at 468.6 nm was suspected of being due to slightly redshifted HeII, and if this identification was correct, it corresponded to a radial velocity of the object with respect to the solar system of 100 km s^{-1}. Considering that the solar system orbits the centre of our Galaxy with a velocity of ~220 km s^{-1}, this figure was not unreasonable for a star in our Galaxy.

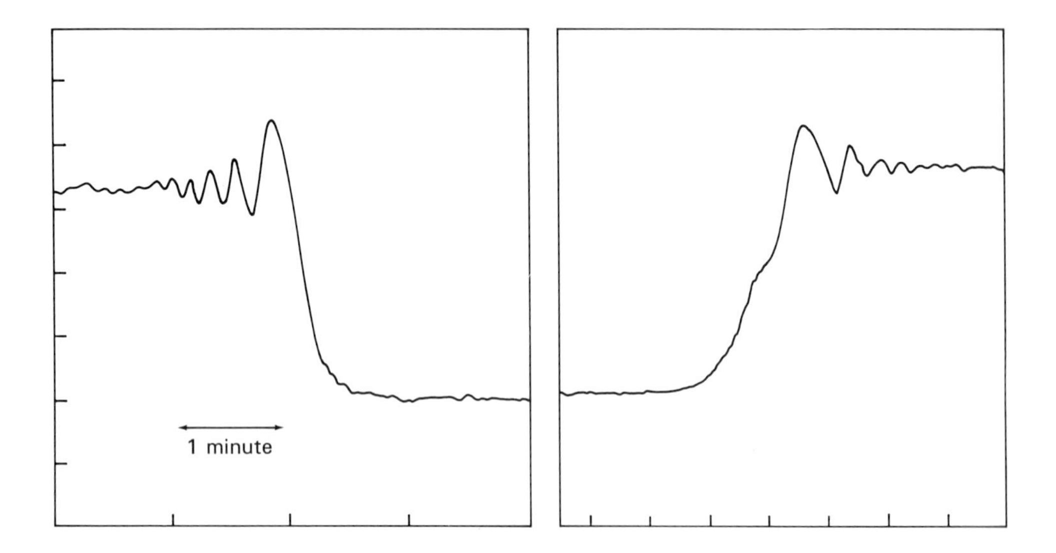

Fig. 3.4. The disappearance and re-appearance of 3C 273 behind the Moon on August 5th 1962. (Adapted from *Nature*, **197**, 1037, 1963.)

For the student of science, the study of 3C48 shows that although the correct solution to the mystery had not been obtained, nevertheless careful and well-thought-out scientific methodology had been applied. This is further demonstrated by the use of a later sky survey plate taken with the Palomar Schmidt, from which it was found that the proper motion of 3C48 had to be less than 0.05" y^{-1}. These measurements were consistent with no discernible proper motion at all; 3C48 was in exactly the same position on both plates even they were taken 11 years apart. With some assumptions, this suggested that 3C48 was probably, but by no means certainly, at a distance of at least a few hundred parsecs. The results of photometric data showed that 3C48 was variable in the optical on timescales of a year and surprisingly, also daily, although the amplitude of this latter variability was only a few hundredths of a magnitude. Special radio monitoring was undertaken and it was found that there were no corresponding radio variations. After all this extensive work, Thomas Matthews and Alan Sandage concluded that 3C48 was a peculiar star emitting radio emission; it was the first radio star ever discovered. This was in itself a novel discovery. Other suspected radio stars were soon identified and further studied, but the puzzle of their emission mechanism became no clearer.

The breakthrough and subsequent identification of quasars came with observations of another of these compact sources, 3C273. The astronomers at Jodrell Bank were again at the forefront of these new studies, conducting a friendly rivalry with their colleagues based in Cambridge. Along with their interferometric work, they had used the 250 ft diameter radio telescope (now renamed the Lovell telescope) to observe lunar occultations to determine better the position of at least one of these compact radio sources. Lunar occultation is a technique by which the Moon is likened to an obscuring disk, which blocks out objects that happen to lie on its track as it moves across the sky. The Moon's position is very precisely known and therefore by timing the exact moment of disappearance and re-appearance of radio signals as the Moon moved across the radio source, a very accurate position can be obtained. Additionally, the manner in which the radio signals faded can give some indication of the angular size of the radio emission (see fig. 3.4). The Jodrell Bank group happened to be strong supporters of the extragalactic location of these sources.

After returning to Jodrell Bank from a period in the armed forces, Cyril Hazard became interested in these strange radio sources. However, he soon moved to Sydney, Australia, to work on the Hanbury Brown–Twiss radio interferometer project, but he was determined to continue the work on lunar occultations. He had use of the 210 ft diameter radio telescope at Parkes in New South Wales and soon made an extremely important breakthrough. Hazard had calculated that one of these compact radio sources, 3C273, would be occulted by the moon on three separate occasions in 1962. The first measurement took place in April. Due to the particular way in which the telescope tracked, the observation had to be curtailed before 3C273 passed behind the Moon. Nevertheless, as 3C273 neared the limb of the Moon, the resulting diffraction fringe pattern showed that it must be smaller in extent than ~6 arcseconds.

Another occultation was coming up in August, but this was also at the extreme of the telescope tracking. Because of the importance of the measurement, the telescope was modified (by removing some panels) to ensure that the tracking limits could be extended to enable the full occultation to be observed. This observation was a tremendous success (fig. 3.4) and showed that the source was not only very compact but also double. A later occultation in October was observed at a higher frequency, confirming the double nature and providing a much better positional accuracy of about 1 arcsecond for each component of 3C 273.

One of those strange and extremely fortuitous coincidences then occurred. The famous Mount Palomar optical astronomer, Rudolph Minkowski, who was also working on compact radio sources, happened to be visiting Australia at the time. Hazard communicated his exciting results to Minkowski who showed him a photographic plate from the 200-inch telescope, on which he had tentatively identified 3C273 with a galaxy. Hazard could afford a chuckle because he knew this identification was incorrect; the galaxy was well distant from his new and very accurate position. So what was at the correct position? A 13th magnitude 'star' was coincident with one of the components and a faint elongated nebulosity with the other (Fig. 3.5). The excitement of the chase took hold and the new position was promptly forwarded to Maarten Schmidt at Mount Palomar.

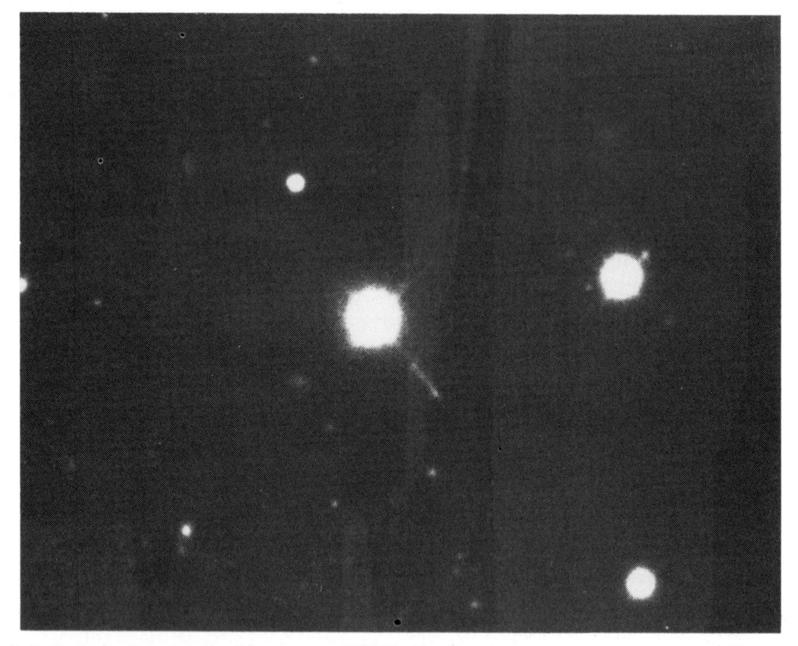

Fig. 3.5. Optical photograph of the quasar 3C273 showing the synchrotron jet. We now know that we only see the point-like active nucleus which dramatically outshines the stars in the underlying galaxy. (Courtesy of NOAO/Kitt Peak National Observatory.)

During an observing run in December 1962, Schmidt obtained a spectrum of 3C273; but to his dismay, the source was so bright compared to the very faint galaxies he had been regularly observing, that the spectrum was overexposed. During the next few nights he managed to obtain a perfectly exposed spectrum of the star-like component of 3C273. Even then, the story was not over. The spectrum showed broad and strong emission lines, just like 3C48, but these could not be identified with known elements. Deeply puzzled, Schmidt thought long and hard about these lines, and finally he recognized that they formed a pattern. He decided to experiment and see how they compared with the most famous pattern of stellar emission lines, the Balmer series of hydrogen. To his amazement, and presumably joy, the pattern was a perfect fit. He had discovered hydrogen in 3C273. Unfortunately this spectrum does not reproduce well but the interested reader is encouraged to refer to one of the discovery papers by Greenstein and Schmidt (1964).

Now at first sight this sounds rather absurd; what were these astronomers thinking about, being unable to identify hydrogen, the commonest element in the Universe with the best-known line spectrum. The problem was that the hydrogen lines were nowhere near where they were expected to have been in the spectrum; in fact they showed a redshift of nearly 0.16. Let us pause and repeat this redshift to understand its significance. At a redshift of $z = 0.16$, 3C273 is receding from us at the awesome velocity of 16% of the speed of light. Comparing this with stars in our Galaxy, which have maximum values of only a few hundred kilometres per second, 3C273 was presumably extragalactic in which case the Hubble Law (eqn. 1.3) could be used to determine its distance. The resulting distance of 948 Mpc (using a value for H_0 of 50 km s^{-1} Mpc^{-1}) was astounding because 3C273 was very bright, namely 13th magnitude (as had been very apparent to Schmidt when he was trying to photograph the spectrum). Therefore 3C273 must be extraordinarily luminous, by far the most luminous object ever observed.

Once the distance was known, the luminosity of 3C273 could be determined from the distance modulus equation (eqn. 2.15) which results in a value of $M_V = -26.9$. The Sun has an absolute $M_V = +4.8$, and so the visible luminosity of 3C273, L_q, in terms of the solar visible luminosity, $L_\odot$, is from eqn. 2.8

$$M_q - M_{V_\odot} = -2.5 \log\left(\frac{L_q}{L_\odot}\right) \tag{3.1}$$

giving $L_q = 4.8 \times 10^{12}\ L_\odot$. The most luminous stars have absolute V-band magnitudes of $M_V \sim -10$ and the most luminous galaxies $M_V \sim -23$, this latter figure corresponding to a luminosity of $\sim 10^{11}\ L_\odot$. If 3C273 was at its cosmologically deduced distance, then it had the power of almost fifty times that of the most luminous known galaxies. This marked the identification of quasars as the most powerful class of AGN discovered.

As soon as he had fitted the spectrum of hydrogen to 3C273, Schmidt showed the result to his colleague Jesse Greenstein, who had just submitted a paper on 3C48 arguing that it was a radio star. Greenstein no doubt felt that horrible sinking feeling because although he had identified a possible redshift in 3C48, he had discounted it. Sure enough, 3C48 was immediately seen to have a redshift of nearly 0.37. However, to be fair, it was much harder to spot the redshift in 3C48 as it was three magnitudes fainter than 3C273 and the much larger redshift moved more of the Balmer lines out of the range of the optical spectrometer (making the well-known series much harder to spot).

As an aside, it is interesting to note that in 1960 a source called Ton202 had been observed spectroscopically as a follow-up to the fact that it showed peculiar blue colours. It was interpreted as a peculiar white dwarf star. Six years later, Greenstein re-examined Ton202 and showed it was another quasar. Hence we learn the lesson that determining the type of object from a spectrum, particularly one that does not have a high signal-to-noise ratio, is by no means trivial. One should bear this in mind when looking at some of the spectroscopic classifications of unusual objects even today.

These ultraluminous objects represented an entirely new class and the search was on to find how many others inhabited the Universe. Because they appeared point-like on optical photographs, these sources were called quasi-stellar radio sources, abbreviated in

1965 to quasars by Chiu in a book called *Quasi-Stellar Sources and Gravitational Collapse*'. Through surveys of 'stars' with unusual colours more objects soon sprung up, and an even more bizarre property of quasars was discovered. Inspection of archival photographic plates revealed that many of the quasars were variable. The timescales of significant variability turned out to be as short as a year and on occasions even less. This variability timescale has the direct result that unless there was something very unusual in the geometry of the emission (such as departure from spherical symmetry or relativistic beaming – see later and chapter 7) then as we saw in section 1.2.1, the size of the emitting region is no larger than the time taken for the light to travel from one side of the region to the other. Hence the variability timescale gives an upper limit to the linear size of the varying region in terms of the light travel time. Therefore, for the variable quasars, the size of the emitting region was at most a light year.

As we noted in chapter 1, the size of the emission region deduced from the variability argument was the major cause of the negative reaction of some astronomers to the cosmological interpretation of the distances of quasars. Undeterred, the theoretical search for the power source immediately began and soon bore fruit in terms of super-massive black holes.

3.4.2 The properties of quasars

One of the most striking observational features of the new class of object was the presence of very broad optical emission lines. These lines had widths of up to 10,000 km s^{-1} and in particular showed strong emission from permitted lines of hydrogen along with other species. In many cases these line emission spectra resembled the most powerful of the Seyfert 1 spectra, but not much was made of this connection at the time and for over a decade and a half quasars were treated as totally separate entities from Seyferts.

The presence of strong, broad emission lines became a very useful parameter for discovering quasars with objective prism surveys (slitless spectroscopy) pioneered by the Byurakan astronomers. Soon, the radio surveys had reached a sufficient degree of sensitivity that all the radio-loud quasars had been found. Most quasars, comprising more than 90% of the total number, would later be discovered through optical astronomy (and even X-ray astronomy from the ROSAT survey). We now know that the overwhelming majority of quasars have weak radio emission compared to the optical and these are often referred to as radio-quiet quasars. Indeed, for the purposes of this book, quasar is the generic name applied to the entire category, although strictly speaking it was the name given to the radio-loud sources discovered by radio astronomy. For a while, the radio-quiet variety tended to be called quasi-stellar objects (QSOs), however, we shall use quasar for the entire class. When we need to refer to a quasar which has strong radio emission, we will call this a radio-loud quasar, otherwise the assumption should be that the object is radio-quiet.

As for Seyfert 1 galaxies the very broad emission lines in quasars indicates a broad-line region (BLR) of emitting gas deep within the core of the galaxy. An intriguing question arises in attempting to decide at what point a quasar ceases to become a quasar and becomes a Seyfert 1. It is by no means obvious from spectroscopy. A convention has arisen whereby if a broad-emission line object has an optical luminosity brighter than a V-band absolute magnitude of –23 (i.e., $M_V < -23$), then it is referred to as a quasar. However, there is nothing magical in this boundary and it highlights the lack of clarification in the differentiation between the two classes of AGN. Historically it seems that if the object was discovered from blue surveys of starlike objects and was found to be extremely luminous, with broad lines and a reasonable redshift, it became classified as a quasar. Obvious galaxies (therefore relatively local) in which a bright blue nucleus with broad emission lines was found, tended to fall into the Seyfert 1 classification.

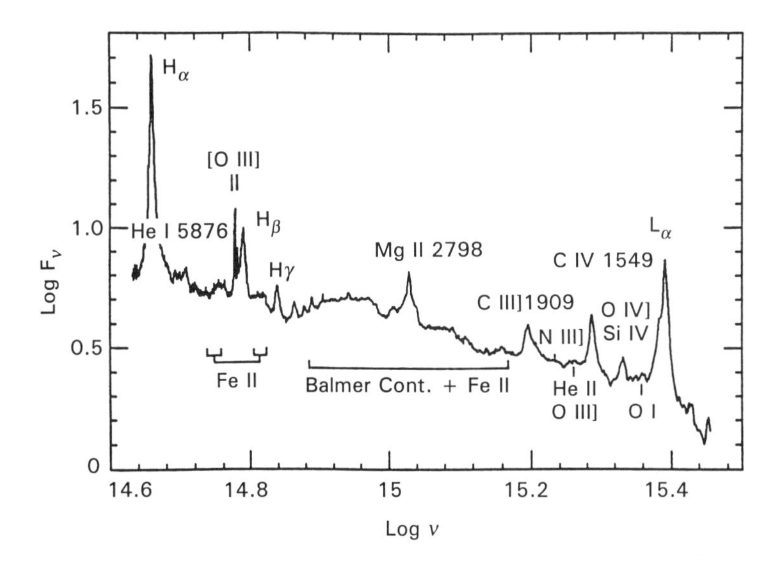

Fig. 3.6. A composite optical-UV spectrum of a typical quasar compiled by J. Baldwin and taken from Netzer in Saas-Fee Advanced Course 20, *Active Galactic Nuclei*. Note that in contrast to the earlier spectra, the flux is plotted on a logarithmic scale, as is the frequency. For comparison, Hα is at a wavelength of 656.3 nm and Lα is at 121.6 nm.

Astronomers have now accumulated a huge store of information on quasars. We know that they extend in redshift space from $z < 0.1$ to almost 5. (As noted above, redshift $z \leq 0.1$ quasars are often referred to as Seyfert 1s even if they have $M_V < -23$.) Quasars have luminosities extending up to 10^{13} $L_\odot$. It should be not surprising therefore to note that the lower level of luminosity for quasars overlaps that of Seyferts in that the most luminous Seyferts are more luminous than the least luminous quasars. This in itself reinforces the difficulty of differentiating a quasar from a Seyfert 1. Because there has never been any suggestion that Seyferts are anything other than galaxies with bright nuclei, there is no reason to suppose that the least luminous quasars are any different. Moreover, because it has not been possible to clearly separate the least luminous quasars from the most powerful quasars on anything but luminosity grounds, the argument extends to the entire quasar class. This is yet another strong reason why virtually all astronomers are comfortable with the concept of a quasar being a galaxy with an intense luminous nuclear region.

In terms of the underlying galaxy, in the mid- 1980s CCD images taken at sites with sub-arcsecond optical seeing produced images of the closest quasars in which the underlying fuzz was clearly revealed. This was a major breakthrough as optical photographs generally reveal only the point-like dominant core of the active nucleus. The images showed that the morphology of the host galaxies were frequently disturbed systems, showing multiple nuclei indicative of merged or interacting systems. In principle, determining whether quasars lie in particular types of galaxy can be done by either investigating the colours or by analysing the profiles of the underlying galaxy, but this has not been easy. We will discuss the host galaxy of radio-loud and radio-quiet quasars in section 5.2.3 and later. Spectroscopic studies of this fuzz remain extremely difficult however, and none have conclusively demonstrated stellar absorption lines typical of starlight from a galaxy. Nevertheless, it is confidently expected that observations with the new 8 m and 10 m telescopes will confirm once and for all that the nearby quasars are active nuclei embedded in a galaxy of stars.

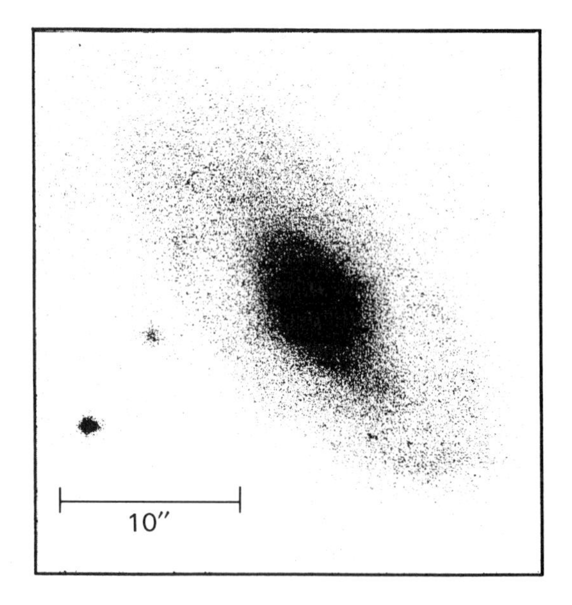

Fig. 3.7. I-band image of the radio-quiet quasar 2130+099 at a redshift of z = 0.061 showing the underlying spiral galaxy with the suggestion of a tidal tail to the lower right. The image was taken on the CFHT (Mauna Kea) with a special high resolution CCD camera and in 0.5 arcsecond seeing. From Hutchings & Neff, *Astron.J.*, **104**, 1, 1992, courtesy of John Hutchings.

Turning to the continuum emission from quasars we have already gathered that they can be split into two categories: those with strong radio emission comprising <10%, and the remainder that are radio weak compared to the optical. We shall see later in this chapter and in chapter 8 that the radio emission is associated with a jet of relativistic electrons. Both radio-quiet and radio-loud quasars have relatively steep infrared through optical continuum emission, the flux declining to shorter wavelengths, superimposed on which are the very broad emission lines. This IR-optical continuum follows a roughly power-law form (a straight line, usually of negative slope on a plot of log flux versus log frequency) and like the Seyferts, appears to peak somewhere in the region of 100 μm. By contrast, the radio-loud quasars do not show the prominent peak in the far-infrared/submillimetre region, although in some cases this is just because the radio emission dominates this thermal component which is probably still present. Overall, the radio-quiet quasars appear very similar to the radio-loud quasars minus the emission associated with the jet.

Some quasars, both radio-loud and radio-quiet, show an excess emission in the blue to UV parts of the spectrum. This has been termed the 'big blue bump' and has been interpreted as being the emission from a hot accretion disk of temperature around 30,000 K surrounding a super-massive black hole. On the other hand, it might be due to free-free emission. The quasar flux normally declines through the UV and X-ray regimes, but some quasars have significant X-ray excesses as we shall see later.

The variability properties of the quasar population are not well determined, although some, the radio-loud optically violently variable quasars (OVVs), vary on timescales of less than a day in the optical. In the X-ray region, variability timescales can be as short as hours. The radio-loud quasars are discussed in more detail below (in terms of the OVVs) and later in chapter 8. The polarization properties of quasars are not well measured apart from the radio-loud category, for which the polarization is strong and variable (see section 3.6.3). For the radio-quiet quasars, although the polarization is not well measured, it is tightly constrained in the optical, where it is found to be very low, as expected for thermal emission.

A further category of radio-quiet quasar is referred to as the Broad Absorption Line quasar, or BAL. As the name implies, these objects are characterised by showing broad absorption lines in their optical spectra. It is found that around 10% of radio-quiet quasars are BALs. The line widths indicate high Doppler velocity broadening, with velocities in the range 0.01 to 0.1 times that of light. This may be material which is being expelled from the central engine in what is presumably some form of radiation-driven outflow. We shall not discuss BAL quasars further except on those occasions when the information from the absorption lines gives us clues to the central engine (chapters 6 and 9).

3.5 LOW IONIZATION NUCLEAR EMISSION LINE REGION GALAXIES (LINERS)

The improvement in astronomical spectroscopic capability during the 1980s revealed that in addition to Seyferts and quasars many otherwise normal galaxies possessed emission lines. These lines were located in the nuclei of the galaxies but were not on the same scale of strength or ionization state as in Seyferts. The lines were generally much weaker and narrower (but note that in some starburst galaxies they can have higher equivalent widths than Seyferts and comparable luminosities). In morphological terms most of the galaxies were spirals and the emission lines appeared similar to the lines given off by the clouds of ionized hydrogen gas surrounding new regions of hot star formation. Such regions are well known to astronomers and are called ionized hydrogen regions, or HII regions for short, the most famous of which is M42, the Orion Nebula.

The galaxies in which such spectra are observed are called HII region galaxies and they are not 'active' in the sense that there is a central engine, rather that they have extensive and massive star formation occurring in them. As such they will not feature further in this book.

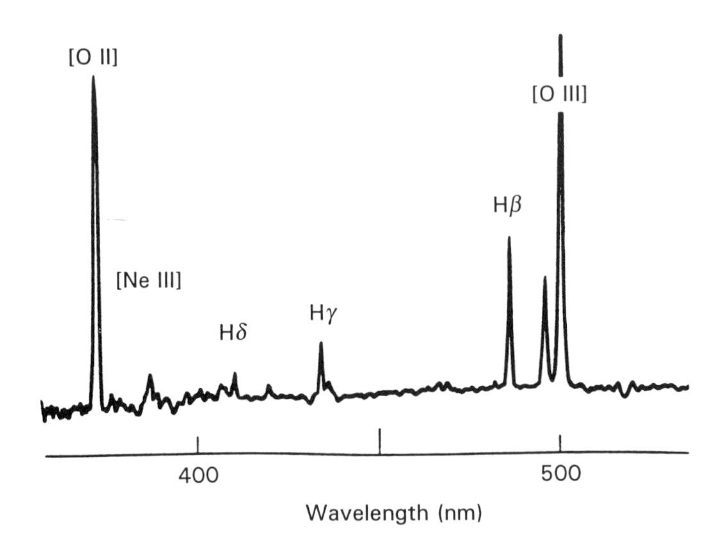

Fig. 3.8. The blue region of the optical spectrum of the LINER Mkn883 from Osterbrock,D. *Astrophysics of Gaseous Nebulae and Active Galaxies*. Although not shown here, this galaxy like a number of other LINERs shows a very weak but identifiable broad component to Hα.

However, lying somewhere between HII region galaxies and the weaker Seyferts are galaxies resembling normal galaxies but with Low Ionization Nuclear Emission line Regions, called LINERs. These are mainly found in the nuclei of Sa and Sb galaxies plus some in Sc and a subset of peculiar galaxies. Being a common phenomenon, it is therefore especially important to discover whether it is due to an active nucleus or some other mechanism. If LINERs are AGNs, then they might be thought of as mini-Seyferts. Alternatively, the 'non-active' hypothesis requires a different mechanism than a central engine to generate the energy to support the line emission. The most popular models in this category require extensive and massive star formation (the super HII region or Starburst concept) in the nuclear regions. So LINERs are intriguing objects and worthy of much greater discussion because they may form some missing link between the high-level activity seen in quasars and Seyferts to much lower levels of activity that are obviously common in many galaxies and which may in some cases be a transitory phase of galaxy evolution.

LINERs are classified spectroscopically and two main factors stand out; the presence of strong emission from neutral oxygen (OI) at 630 nm (also observed in Seyferts) and the weakness of lines from highly ionized species such as doubly ionized oxygen [OIII] and quadruply ionized neon (NeV), both of which are strong in the spectra of Seyfert nuclei. A spectrum of a typical LINER galaxy is shown in fig. 3.8. As far as their other properties are concerned, they are distinguishable from normal spiral galaxies by frequently having compact flat spectrum radio cores and some show point-like X-ray emission. In the optical however, the nuclear emission is weak compared to the surrounding starlight, precisely opposite to the situation for quasars in which the nuclear emission all but swamps out the starlight from the underlying galaxy. Again we see the critical effect of dilution and contrast; where the central nucleus is bright compared to the underlying galaxy, the object is clearly identifiable as an active galaxy (Seyfert or quasar), but where the strength of the nucleus is much weaker, the contrast between the active nucleus and the surrounding galaxy is much reduced and optical identification of activity is much harder. This is where radio and X-ray observations become an added discriminator. We shall return to LINERs when we look at the broad and narrow line regions and the unification picture.

3.6 BL LACERTAE OBJECTS, OVV QUASARS AND BLAZARS

3.6.1 Overview

We have grouped BL Lacertae objects, OVV quasars and blazars together because to a first approximation they show strong similarities of radio-loud flat spectra and variability. We are now confident that the flat radio spectrum is produced by a process called synchrotron emission (relativistic electrons in a magnetic field) from a powerful relativistic jet. Because of the spectacular nature of jets, chapter 8 is devoted to a detailed study of their properties.

In our discussion that follows we should remember an important factor. When we consider luminosity that is dominated by the emission from a jet, the luminosity of the source in question is always overestimated. Why is this? In determining the luminosity by measuring the flux from the object and using the calculations laid out in chapter 2.2, we have implicitly assumed that the object radiates isotropically, i.e., it radiates the same energy over all directions. This is the normal cause of events in the Universe; radiating bodies emit the same energy per second per unit surface area in all directions as long as the surface temperature is the same. For jet-like emission, this is no longer the case. For some objects we might be oriented preferentially and look directly down the jet. Our calculations would then assume the emission extended over the full surface of the object (i.e., it radiated uniformly into a solid angle of 4π steradians), whereas the cone angle might be very small, like the beam from a lighthouse. The solid angle of the cone could easily be less than 1% of the total solid angle of 4π and we would then have dramatically overestimated the luminosity of the source (by a factor of 100 in this example). Complicating this even further is that for many of the jets in question, relativistic effects come into play. These lead to a further overestimation of the luminosity. All these intriguing aspects of flat spectrum radio sources will be taken up in more detail in chapter 8.

3.6.2 BL Lac objects

BL Lac objects were named after the first member of the class discovered: an object previously suspected of being a variable star in our Galaxy and which had been catalogued under the name of BL Lacertae. This is the standard format for naming variable stars; they are given the name of the constellation in which they are found preceded by a series of letters and/or numbers to denote that they are variable stars. BL Lac came to prominence in 1968 because of the identification of a compact and highly variable radio source (VRO 42.22.01) coincident with the 'star' BL Lac.

Further investigations revealed that the so-called 'star' was suspiciously strange. The optical spectrum was featureless, showing neither absorption nor emission lines. Furthermore, the continuum emission was steeply rising to the red and infrared wavelengths and showed a power-law form. In addition, this emission was found to show strong linear polarization. These characteristics were unlike any previously found for a star but curiously they showed many, but not all, of the properties of the recently discovered radio-loud quasars! Eventually, further study and discovery of other similar objects led Peter Strittmatter and colleagues to propose that BL Lac and a handful of other sources were candidates for an entirely new class of extragalactic object. The class was named after the first member, hence BL Lacertae objects, or BL Lacs for short, and the list of active galaxy types was increased.

Classical BL Lacs are radio-loud objects with a featureless continuum spectrum which shows strong polarization and rapid variability (on timescales of days upwards). The continuum (synchrotron) emission rises steeply from UV through optical and infrared wavelengths until at a wavelength of around 1 mm it turns over and continues to longer wavelengths (fig. 3.9) with a strength which is almost independent of frequency ($S_\nu \approx$ constant). This type of radio emission is associated with a sub-group of extragalactic radio sources termed flat spectrum radio sources. The variability of BL Lacs is the most dramatic of all classes of active galaxy and can be measured on time-scales of hours (at the shortest wavelengths) upwards. When undergoing a major flare, the amplitude of variability can be as high as four magnitudes (which represents factors of forty in terms of flux or brightness). Although this variability makes them spectacular, classical BL Lacs are extremely rare; the total number known is only ~100. A strong clue to the origin of their extreme properties comes from the fact that all eleven BL Lacs so far observed at more than one epoch by very long baseline radio interferometry (VLBI) techniques show superluminal motion. As we shall discover in chapter 8, this is strong evidence of synchrotron emission beamed in a cone towards the observer (see also section 3.6.3).

With better detection techniques and more targeted observations, emission lines have been observed in the optical part of the spectra of a number of BL Lac objects, particularly when they are in a faint phase. These lines are redshifted and serve as proof that BL Lacs are indeed extragalactic. Once the distance to a BL Lac is known, the luminosity can be determined. With the probably naive assumption of isotropic emission, BL Lacs are found to have a range of luminosity, the upper end of which overlaps that of quasars. A number of BL Lacs whose distance has not been determined from the redshift of emission lines are known to lie at least a certain minimum distance from us because of the detection of an absorption line due to the light from the BL Lac object passing through intervening matter between it and ourselves. The intervening matter is probably the halo of a galaxy through which the line-of-sight passes. The redshift of these absorption lines then provides a minimum distance to the BL Lac and hence a minimum luminosity.

In general, BL Lacs have low redshifts ($z < 0.2$), but some are much more distant, and for these the link with Optically Violently Variable quasars and the unified hypothesis will be made in chapters 8 and 9. If BL Lacs are generally lower luminosity objects, then their apparent location at low redshifts only can easily be attributed to a detection sensitivity selection effect. We will see that for one or two sources (such as 0235+168) the passage of light through an intervening galaxy can give rise to gravitational lensing (see section 3.9), which has been suggested to be the case for all BL Lacs. However, although at first sight attractive, this speculative suggestion has been shown to be incorrect.

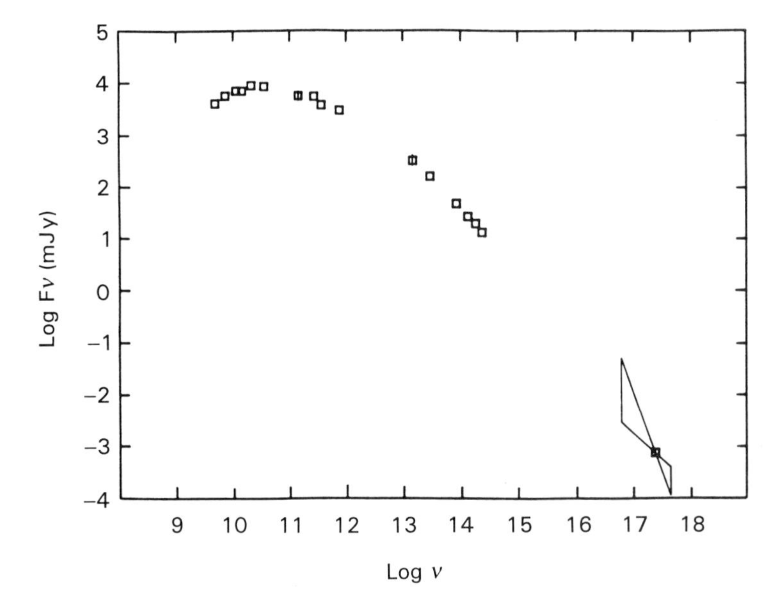

Fig. 3.9. The millimetre to X-ray continuum spectrum of the BL Lac object OJ287. The smooth single component synchrotron emission spectrum is clearly revealed.

X-ray observations showed that for nearly all detectable BL Lacs, the X-ray emission joins smoothly with an extrapolation of the optical emission. In some cases, the spectrum is steeper than an extrapolated power-law, indicative of radiative losses; the higher energy electrons lose energy through radiation emission more rapidly than it is replenished by new electrons (see section 4.3.1). With the arrival of high-sensitivity X-ray sky surveys, an entirely new population of objects has been discovered. These are called X-ray BL Lacs. They show strong X-ray emission along with an optical spectrum that is featureless in terms of strong emission lines. Interestingly, these X-ray selected BL Lacs have significantly steeper X-ray spectra than do the classical (radio-selected) BL Lacs.

Although the X-ray BL Lacs are radio-weak (but not radio-quiet) and significantly less radio-luminous than the radio-selected BL Lacs, nevertheless, their submillimetre–

far-infrared luminosity exceeds that of the optical. Overall, they are less luminous at all wavelengths than their classical radio-selected counterparts and their luminosity peaks in the UV to soft X-ray region. From redshift data of the stellar populations in the underlying galaxies, these objects are found to be on the whole much closer than the radio BL Lacs, and furthermore, much more numerous in terms of number of objects per unit volume of space (termed the number density). Some of them are variable at a level of a few per cent on timescales as short as minutes. To distinguish this new category of object, we shall introduce the term 'X-ray-selected' to differentiate them from the 'classical' radio-selected BL Lacs.

Returning to the 'classical' BL Lacs, CCD imaging has allowed significant progress to be made in underlying galaxy identification. In a number of examples, the faint surrounding nebulosity of a galaxy of stars has been detected and for the majority of these, the profile appears to be that of an elliptical galaxy. There are a small number of cases, such as 0537–441 and 2254+074, where the galaxy could be either elliptical or disk-like. For the object 1413+135, the profile is definitely disk-like, resembling a spiral galaxy. Therefore the tacit assumption found in most texts and review articles that BL Lacs are found only in elliptical galaxies is far from being a proven statement, although it would appear to apply to most of the well-studied objects so far. The link with elliptical galaxies is strengthened further by the similarity of the luminosity of the extended radio and optical emission with that produced by radio galaxies which are known to be elliptical galaxies.

Taking an overview, we shall assume that BL Lacs are usually associated with elliptical galaxies, have core radio emission produced by the synchrotron mechanism that is beamed towards the line of sight in a narrow cone-angle and effectively swamps the optical emission from the stars in the galaxy.

3.6.3 Optically violently variable quasars (OVVs)

In the mid- 1970s, it was noted that there was a category of quasar that, in some respects, resembled the BL Lacs. These were the so-called OVV quasars, a subset of the flat spectrum radio-loud quasars, but distinctive in that they were extremely variable in the optical and radio and showed strong polarized continuum spectra. The number of catalogued OVVs is less than a dozen. The variability has been found to be very erratic, often consisting of lengthy (timescales of years) periods of stability followed by rapid flaring of many magnitudes in timescales of weeks or even days followed by subsequent slower fading. The polarization properties, power-law optical through infrared continuum and flat radio spectra are very similar to the continuum spectra of the classical BL Lacs described above. On the other hand, what immediately differentiates the OVV quasars from the BL Lacs is the observational presence of a fundamental property of quasars, very broad and strong optical emission lines. This demonstrates that the OVVs have a prominent observable broad-line region (BLR). Another aspect in which they differ from BL Lacs is that the OVV quasars have medium redshifts ranging from $z \sim 0.1$ to >2 and are high luminosity objects (assuming isotropic emission).

At one time it was believed that there might be yet another class of quasar, the high polarization quasars (HPQs). However, it is now generally agreed that if sufficiently sensitive spectropolarimeters are used then all the flat spectrum radio-loud quasars reveal some degree of polarization. The degree of polarization generally decreases as the optical regime is approached from the infrared and this is easily explained by the dilution effects of starlight from the underlying galaxy. Extensive studies of the optical polarization of radio-loud quasars have now shown conclusively that there is excellent correlation between the strength of the optical polarization and the fraction of the 5 GHz radio emission observed in a VLBI core. As this fraction approaches unity, the degree of polarization increases dramatically (a trend also seen in radio galaxies). High frequency radio polarization also correlates extremely well with the optical polarization. Both of these studies support the notion that the polarization correlates with the compactness of the emitting component, a strong argument for a central engine and beaming. We shall not differentiate between high-polarization quasars and OVV quasars, and for the purposes of this book the names can be interchanged.

A further difference between OVVs and BL Lacs is that the former generally possess X-ray emission with a significantly flatter slope than that of the classical BL Lacs. Often, the X-ray emission in OVVs lies above an extrapolation of the infrared–optical spectrum. This is clearly indicative of another emission mechanism coming into play, perhaps from a very hot accretion disk or synchrotron-self-Compton emission (see sections 4.2 and 4.3).

Observations of the radio emission from OVV quasars using the first generation VLBI experiments (which began in 1971), produced the amazing discovery known as superluminal motion. Many astronomers were sceptical of this at the time and have only been convinced by the introduction of much better observational techniques (including self-calibration) in the latter part of the 1970s. Superluminal motion will be discussed in detail in chapter 8; suffice to say that the VLBI observations showed that the single-sided, parsec-scale jet structure was made up of an unresolved core and blobs of emission. Observations taken a few years later showed that the separation of the blobs from the core had increased. This apparent motion (the only example of extragalactic proper motion) was totally unexpected and produced amazing results. The distance of the source is obtained from the redshift and so the velocity of separation of the radio-emitting blobs is easily determined. However, to everyone's astonishment, this turned out to be many times the velocity of light, hence the term superluminal motion!

Violating a fundamental criterion of Einstein's special theory of relativity is not to be taken lightly and so astronomers came up with an alternative explanation of this apparent superluminal velocity: beaming. (Other suggestions, such as the 'lights on a Christmas tree' model were also made but soon rejected.) In chapter 8 we will see that if a narrow cone of radiation is beamed towards the observer but close to the line of sight, then an apparent superluminal motion can readily be explained provided that the velocity of the blob is close to the speed of light (such as $\approx 0.8c$ or higher). We noted in section 1.9 that although this beaming model has become widely accepted, the requirement of bulk motion of material travelling at least at $\sim 0.8c$ has always troubled some astrophysicists. Historically, it is another reason why non-cosmological redshifts were seen as some sort of salvation in the early study of quasars. We shall see later that additional evidence for superluminal motion is provided by distance-independent observations: the lack of X-ray and gamma-ray emission commensurate with an observed level of radio emission and a VLBI radio map (synchrotron-self-Compton radiation described in section 4.3.3). However, it was the VLBI investigation of the OVV quasars that gave relativistic beaming its firm foundation and this topic will figure strongly in the quest to unify the radio emission from classes of active galaxy.

3.6.4 Blazars

At a now-famous conference at Pittsburgh in 1978, devoted solely to the exchange of information and further understanding of the new category of BL Lac objects, it was suggested by Ed Spiegel at the conference dinner that the OVV quasars and BL Lacs had so many observational properties in common that they should be grouped together into a single category called 'Blazars'. The name presumably derives from a mix of 'BL' from BL Lac and 'azar' from quasar (with a substitution of 'z' for 's'); the resulting 'blazar' giving a vivid description of their dramatic luminosity changes in terms of flaring behaviour. The name seemed appropriate and stuck, becoming an accepted part of the active galaxy terminology. This again shows how unification themes have progressed through classifications, observation and re-classification.

However, within the past few years there has been a general agreement that blazar refers to a *phenomenon* rather than a category of object. This is a very helpful step forward and so although we will continue to use the term, we stress that OVVs and BL Lacs are very different objects, selected from different parent populations but possessing one thing in common, a jet produced in the core of a galaxy which is responsible for the non-thermal emission. Nowadays, a simple interpretation of a blazar is that it is an object that possesses beamed emission from a relativistic jet which is aligned roughly toward the line of sight to the observer. Beamed synchrotron emission from this jet dominates the radio through infrared spectrum.

Taking blazars as a category of phenomena, there are just over 200 known, comprising BL Lacs, OVVs and high-polarization quasars. We therefore see that blazars are very rare, and this must tell us that there is something very special about their properties. Catalogued Seyferts number around 1,000 while there are well over 5,000 quasars (most of which are radio-quiet) already listed.

Apart from supernovae, BL Lacs and OVVs are the most violently variable objects known in the Universe in terms of luminosity change. They are the focus of the efforts of teams of astronomers attempting to determine the precise nature and timescales of this variability. One finds however, that no matter on what timescale one observes blazars, some degree of variability is found. Major outbursts occur on timescales of years for OVV quasars and perhaps less for BL Lacs. There is hope that these flaring phenomena will provide insights into the heart of the source of the activity and reveal more about the exotic processes occurring in relativistic jets.

It is usually assumed, and with good cause, that the shortest timescales and most violent events probably occur closest to the central power-house of the activity. Detailed studies reveal that although the synchrotron emission is produced by the relativistic jet, this is not precisely spatially coincident with the central black hole and accretion disk. Although intimately linked, the radio emission from the jet may originate as much as a parsec from the black hole. On the other hand, the rapid X-ray flickering is another story as we shall see later.

At this point we can say that the relativistic jet is most probably produced by the interaction of magnetic fields and charged particles in the vicinity of an accretion disk of a supermassive black hole. Furthermore, as we shall see in the next chapter, these conditions give rise to inverse Compton radiation, which is manifest as strong X-ray or gamma-ray emission. Although this was not shown in fig. 3.9, fig. 5.3 provides an excellent example for the blazar 3C279 (but note the ordinate scales are not the same – see section 5.1).

3.7 POWERFUL RADIO SOURCES

Although radio galaxies were amongst the first type of active galaxy to be discovered, it is by no means clear that they either have a unique position or that they fit any clear pattern. The original discovery was somewhat haphazard and based solely on the fact that these objects were powerful emitters of radio radiation. However, one thing was clear from the outset: the radio emission was produced by the synchrotron process, whereby relativistic electrons interact with a magnetic field and in so doing lose energy by radiation. We now know that the extended lobes of synchrotron emission are fed by a jet of relativistic electrons, originating at the central engine. The precise means by which the electrons are transported to the outer zones, and whether re-acceleration takes place to compensate for the enormous synchrotron losses expected in travelling from the core to the lobes are still a matter of debate and will be discussed further in chapter 8.

We can distinguish two basic types of radio galaxy emission: (a) steep-spectrum, extended radio structures, and (b) compact, flat-spectrum, VLBI core-dominated sources. In fact classification of radio emission from galaxies is amazingly complex, and begins to compete with the Seyfert 1.5 to 1.9 sub-classification in detail. In keeping with a general theme of not cluttering up the big picture with unnecessary complexity, we shall treat only the most global of the parameters and leave the keen reader to follow-up the references given in section 3.11. We shall avoid reference to 'fat doubles' 'wide-tailed', 'narrow-tailed', and so on.

The advent of powerful radio arrays such as the VLA and MERLIN enabled a much greater dynamical range to be obtained in the radio images, providing much needed enhanced contrast between the brightest and faintest fluxes. The two categories have now been modified to (a) lobe-dominated and (b) core-dominated. For simplicity, in terms of (a) we shall look at the extreme of the lobe-dominated sources, the extended double radio sources, and use some of the more famous of these sources as hints to what we shall learn in chapter 8.

3.7.1 Extended, double radio sources

These objects are dominated by the radio emission from two extended lobes, symmetrically distributed about a central galaxy. Examples of this spectacular emission are shown below. The radio emission extends over many arcminutes on the plane of the sky. When the distances of the galaxies are taken into account, we find that the lobes extend many tens of kpc, and in the case of 3C236, the lobes extend almost 3 Mpc from the central galaxy (fig 3.10). These are the largest single structures in the Universe and for 3C236 for example, this is easily greater than the separations between galaxies in groups (such as our Local Group of galaxies).

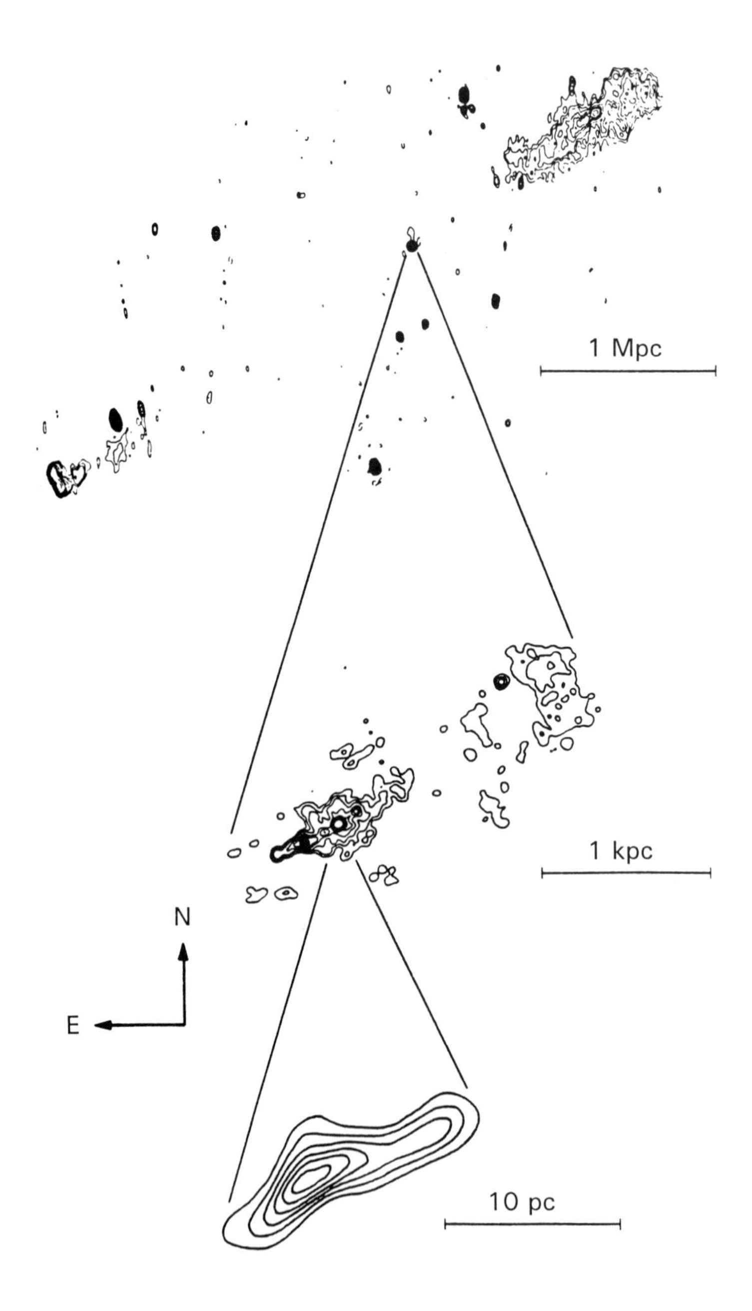

Fig. 3.10. The emission from the radio galaxy 3C236 seen on various scales. Although this lies on the borderline of the FRI/FRII divide in terms of luminosity, morphologically it is a clear FRII. Note the NW–SE extensions seen on megaparsec to parsec scales. (From Bartel *et al.*, *Astron. Astrophys.* **148**, 243, 1985.)

Although we shall discuss all extended radio doubles in a single section, it is useful to consider two sub-classes that have been found to be helpful in answering questions about parent populations. It has been shown that at a frequency of 178 MHz there is a distinct change in the properties of lobe dominated radio sources at a flux density level of 5×10^{25} W Hz^{-1}. This was first observed by the Cambridge astronomers Bernard Fanaroff and Julia Riley, and the sources are referred to as Fanaroff-Riley Class I and II, or more usually, just FRI and FRII for short.

The FRIs are the lower luminosity lobe-dominated radio sources. As a whole, they exhibit very extended twin lobe structures, the ends of which show the steepest radio spectra and appear faint on radio maps (called edge darkening by radio astronomers). The lobes are connected to the central optical galaxy by smooth and continuous double-sided jets. We shall reserve discussion of the details of the jet-lobe structures and parent galaxies until chapter 8.

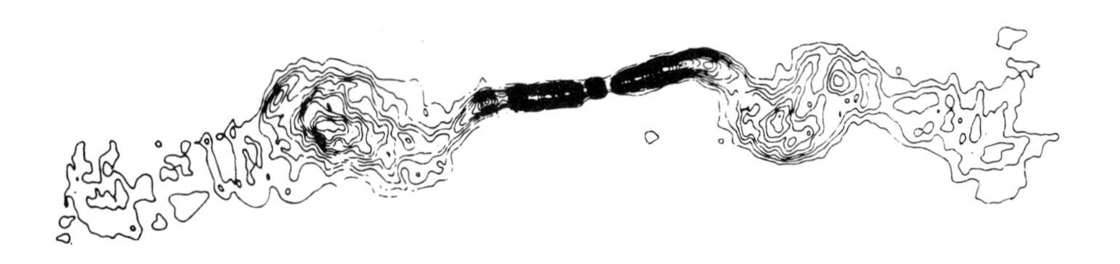

Fig. 3.11. A 20 cm VLA map of the FRI radio source 3C449. (From Perley, Willis & Scott, *Nature*, **281**, 437, 1979.)

The FRII sources are the more powerful radio lobe-dominated sources radiating greater than 10^{35} W of power at centimetre wavelengths, and this emission occurs on scales of kiloparsec size. Their lobes differ significantly from those of the FRIs in that the steepest spectrum radio emission is found at the inner regions. The ends of the lobes are frequently edge-brightened and show bright knots of emission or 'hot-spots' at their outer extremities. Unlike the FRIs, the jets are usually single-sided, or, when double, one side is many times brighter than the other. Viewed on radio 'pictures', the FRII jets seem at first sight to be less prominent than their FRI counterparts. In actual fact they are usually more luminous and merely appear less striking because of the higher luminosity of the FRII lobes and reduced contrast between the jet and lobes. The jets have a smaller opening angle than those in FRIs and are frequently dominated by bright knots rather than having a smooth appearance. An example of a FRII radio galaxy is shown in fig. 3.12.

The classification of FRI and FRII may appear to be somewhat arbitrary, and an important question is whether the two classes are different in some fundamental way. (Remember the Seyfert 1 and 2 discussion in section 3.1.) We now believe that they probably are, and the differences may be seen in two parts: the central galaxy and the physics of the jets. In the latter context, the separation may represent a transition in the fluid dynamics of the flow of material in the jets that feed the giant lobes, perhaps changing from a turbulent flow of the prominent jets of the FRIs to that of a smooth supersonic flow seen in FRIIs. The importance of jets and their link with lobes is clearly demonstrated in fig. 8.1 which shows synchrotron emission extending from less than 1 milli-arcsecond for the innermost part of the jet to nearly a degree for the double-lobed extended emission of the galaxy NGC6251.

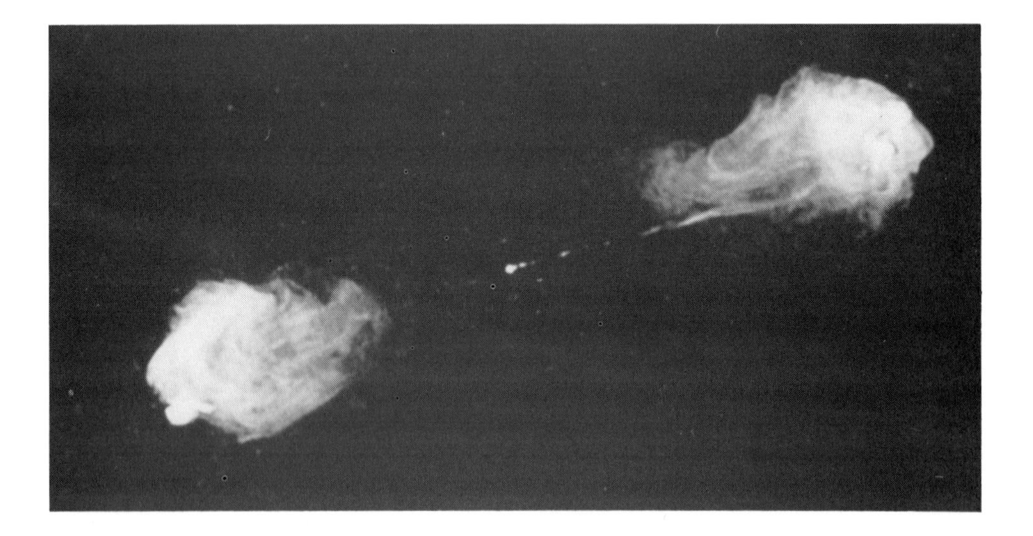

Fig. 3.12. VLA 6-cm image of the FRII radio galaxy Cygnus A. (Courtesy of NRAO/AUI).

But what about the central galaxies, do they differ in terms of luminosity, morphology etc.? Much work has been undertaken on this topic and the results seem to be consistent with the host galaxies of radio galaxies being elliptical. FRIIs are housed in normal, giant elliptical galaxies, but are generally not found in rich clusters of galaxies. On the other hand, the hosts of FRIs are, on the whole, significantly larger and more luminous galaxies, often of class D or cD, (the giant galaxies found in the central regions of rich clusters).

An interesting study showed that the total radio luminosity (core plus lobes) as well as the strength of the optical emission lines in FRIIs appear to be correlated with the total radio power emitted by the central galaxy. This brings up another important aspect of jets which is worth pointing out: the jet luminosity in the above correlation refers to the total power output of the jet, most of which is dominated by the mechanical energy rather than the radiated energy. Even apparently weak radio jets can have significant luminosity, although this is not seen in their e-m output. This also highlights the importance of the discovery of extended radio emission from galaxies and quasars and the contrast with the different information gleaned from the line studies. Although the jet radio luminosity is very much less than the line luminosity, we can make an analogy with the 'smoking gun' idea. The 'smoke' of the extended radio emission gives clear evidence about the presence of an unknown 'gun', an exotic central engine. The unseen jet is then the invisible path of the bullets from the gun. In this way the radio gives the big picture, while the emission lines provide the details.

Many of the lobe-dominated radio-loud quasars show one-sided jets which, although broadly similar to the FRII class of object, are significantly more luminous than the jets emanating from the giant elliptical galaxies of the FRIIs. Whether the double-lobed structure is due to the symmetric fuelling by continuous double-sided jets, or whether the jets are single-sided but flip by 180 degrees from time to time, is a discussion we will postpone until chapter 8.

Before leaving this section we will mention a nearby FRI galaxy, the well-known southern galaxy NGC5128, called Centaurus A by radio astronomers and shown in plate 2. Being only 4 Mpc distant, it can be studied in great detail. The twin-lobe radio emission is present on scale sizes from 0.02 pc to 400 kpc and follows approximately the same orientation over this length. The continuous nature of the emission shows that the fuelling has remained approximately the same level for at least a million years. Furthermore, the lack of change of orientation shows that the emission direction has been more or less fixed in space for a similar length of time. The most likely interpretation is that there is an object at the centre of NGC5128 that is stable with respect to a fixed orientation over these long timescales. It is extremely hard to think of any other mechanism apart from rotation that could produce such a stable configuration. A supermassive rotating black hole clearly fits this picture.

3.7.2 Core dominated radio sources

As the name suggests, these sources are dominated by strong radio emission from a compact core that is unresolved on the scale of an arcsecond. Usually, these sources show almost flat spectra extending to the highest radio frequencies of 300 GHz (1 mm) and in some cases into the submillimetre region. Additionally, they often show a prominent single-sided (kpc-scale) jet, extending over scales of arcseconds. Because these sources are bright, VLBI studies have been undertaken of many of them, with the result that a significant number have been shown to have milli-arcsecond jets that feed into the arcsecond scale structures. Also, many have been shown to demonstrate superluminal motion and are often OVV quasars and BL Lacs, typically the blazar family of section 3.5.4. We will return to discuss the physics of the jets and superluminal motion in chapter 8.

3.7.3 Pathological examples

Objects that fall into this category are either unusual and rare (M87) or are the result of the effects of an external medium on their emission structures. We shall tackle the latter first. There are a number of spectacular radio pictures of twin-lobed structures that are

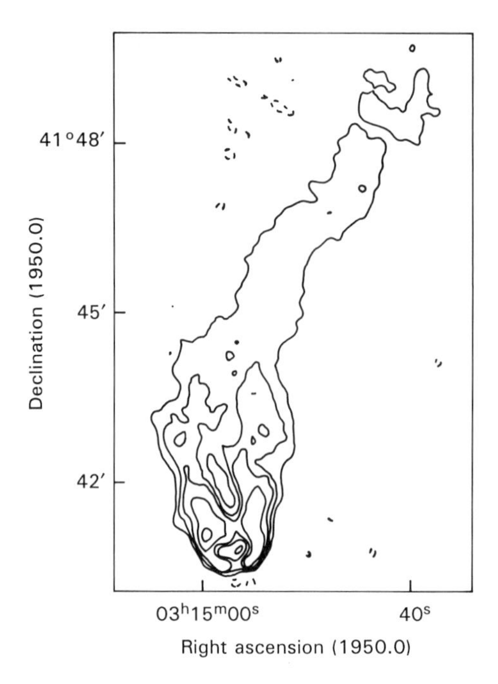

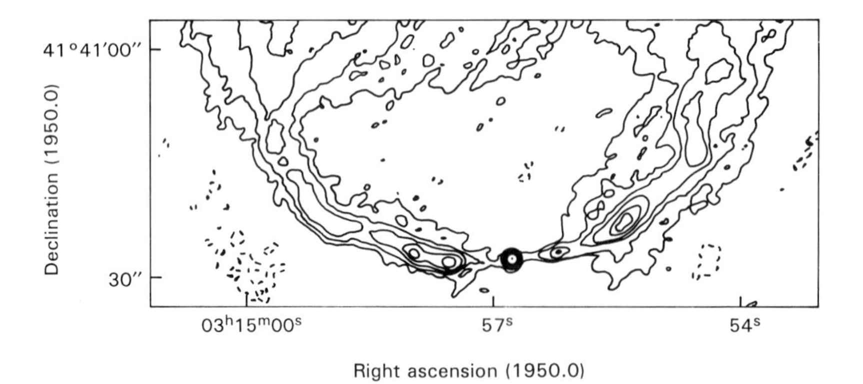

Fig. 3.13. The swept-back lobes of 21 cm radio emission from the galaxy NGC1265 showing (top) the large-scale extent and merging of the 'tails' and the region of the jets emanating from the central galaxy (bottom). (Adapted from O'Dea & Owen, *Astrophys.J.*, **301**, 841, 1986.)

not linear, but which show large degrees of bending. The best example is the elliptical galaxy NGC1265 in the Perseus cluster. The dramatic swept back lobes, shown in fig. 3.13, are the result of the parent galaxy NGC1265 moving through the hot gas of the intergalactic medium of the Perseus cluster, which causes a drag on the radio-emitting material.

We shall assume that all the bent lobe structures are a result of interaction with a cluster medium. The second of our examples focuses on the nearby giant elliptical galaxy M87, also known as Virgo A by radio astronomers. This galaxy lies about 15 Mpc distant and is one of the most luminous elliptical galaxies known. It is also the central galaxy of the Virgo cluster of galaxies. The radio emission from M87 is complex (figs. 8.12 and 8.13), comprising an inner double component of scale 5 kpc (about 1 arcminute on the sky) surrounded by a low surface brightness radio halo extending some 80 by 60 kpc. However, the most spectacular feature of M87 is a single-sided jet that joins the core of the galaxy with one of the lobes. This jet is 2 kpc in extent and is not only visible in the radio, but at infrared, optical, UV and X-ray wavelengths. The emission from the jet is synchrotron radiation, and the derived lifetimes immediately highlight one of the major areas of work in jet physics: how to get the electrons to the ends of the jet before they lose their energy by interactions with magnetic fields. We shall pursue this further in chapter 8, where the details of the jet of M87 will figure prominently.

3.8 OTHER CATEGORIES

In this section we list some of those categories that have been added during the development of the subject. Most of these were classified according to a single criterion, but as time progressed and with better observations they have now been recognized as examples or subgroups of one of the classes discussed above. Markarian galaxies are probably better placed in this category but because (and only because) they presented a prime example of the use of Schmidt telescopes to search for galaxian activity, they were discussed as a separate entity above.

X-ray galaxies and IRAS galaxies are examples of a wavelength selection categorization, but by the time these surveys were undertaken, a more mature approach to classification schemes had occurred and X-ray Seyferts became a more recognised descriptor. IRAS galaxies possess very strong far-infrared (60 to 100 μm) emission, which is usually associated with dust re-radiation from extensive star formation. This brings us to the class of object called starburst galaxies. These were recognized before the IRAS survey, but due to the difficulty of observations in this wavelength regime, were small in number. The two main examples were the nearby galaxies M82 and NGC253. The starburst phenomenon is precisely what is sounds, very extensive star formation underway in a galaxy of stars, usually in the central 1 or 2 kpc.

The IRAS far-infrared survey discovered many more far-infrared galaxies (FIRGs) and after further study, many of those which showed 'warmer temperature dust emission' were found to be Seyfert 2s. Some of the FIRGs were labelled 'ultraluminous infrared galaxies' or 'super-starburst' galaxies (but see below) and have luminosities exceeding $10^{12}\ L_{\odot}$, easily comparable to quasars. Again, there is only a small number of these ultraluminous infrared galaxies and two have been found to lie at redshifts exceeding 2. *If* their luminosity is driven by star formation, then the rate at which stars are forming is incredible, and would be expected to last for only a very short time in the lifetime of the galaxy. Hence high-redshift objects of this type are very important for understanding the rate at which stars and dust evolved in a galaxy in the early Universe. However, there are two cautionary aspects to this story. The prime candidate for this class of object has now been found to be gravitationally lensed (see below and chapter 9) and its apparent luminosity magnified in the process. Secondly, it is not clear that the extreme far-infrared luminosity in all of these objects derives from star formation. Near infrared observations have revealed that at least one or two of these ultraluminous FIRGs harbour an optically hidden broad-line region. Hence there is the strong belief that at least some, if not all of these are buried quasars. We shall pursue this topic further in chapter 9.

3.9 GRAVITATIONAL LENSING

This section deals with a property which can seriously confuse the investigation of the continuum emission from AGNs. Although a rare phenomenon, it has contributed to what was believed to be the most luminous object in the Universe (the ultraluminous galaxy IRAS FSC 10214+4724 being severely downgraded in luminosity). Although sometimes giving spectacular results, at other times its presence is difficult to spot as we shall see in chapter 9.

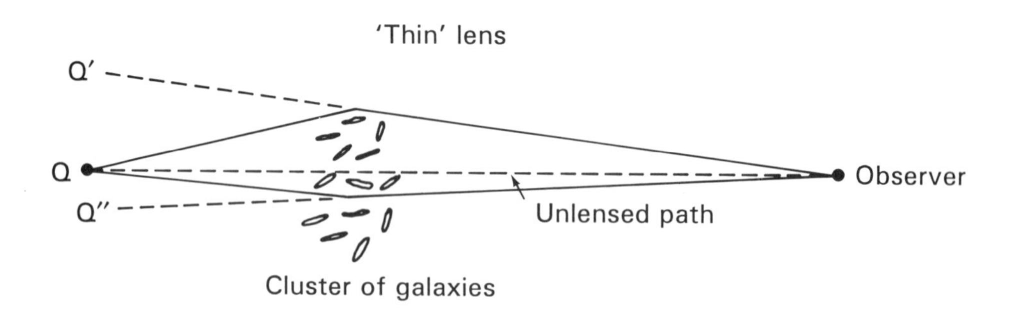

Fig. 3.14. Schematic of gravitational lensing. The light from a quasar Q, is deflected by a galaxy, or galaxies in a cluster to produce multiple images Q′ and Q″ etc., as seen by the observer. The angles shown are very greatly enlarged and depending on the geometry (relative distances) and the property of the 'lens', arcs or even a circle can be produced. Flux magnification is also a property of the lens.

The story of the discovery of gravitational lensing began in 1979, when Denis Walsh, Bob Carswell and Ray Weymann were investigating a radio-loud quasar. The quasar was known as 0957+561 and when it was studied in the optical it was found that there were in fact two quasars separated by ~6 arcseconds. This is an extremely rare occurrence and the story took on another twist when the astronomers discovered that although the quasars had differing brightness, they had identical spectra. We now know that there is in fact only one quasar, and the light from this very distant object ($z = 1.41$) has been deflected from its rectilinear propagation by the mass distribution of an intervening object, a galaxy at $z = 0.36$. The result is a multiple image, in this case a double image lying on opposite sides of the intervening galaxy. This phenomenon is a consequence of Einstein's theory of general relativity and is similar to the deflection of the light from distant stars during a solar eclipse, as described in section 7.1. The result of gravitational lensing can be multiple images, arcs, or even a ring. We will not discuss gravitational lensing in detail but refer the reader to suitable references. The concept of the geometry of the gravitational lens is shown in fig. 3.14, which is similar to that used to show the deflection of light by the Sun in Eddington's 1919 observations verifying Einstein's predictions.

Fig. 3.15. A spectacular example of gravitational lensing showing arcs of distant galaxies produced by the intervening cluster of galaxies Abell 2218. Some of the very small arcs on this HST picture are from galaxies which are fifty times fainter than can be detected with ground-based telescopes. (Courtesy STScI-PR95-14.)

How does gravitational lensing work and what does it imply for the background object? Basically, the intervening galaxy acts like a lens (but in this case a gravitational lens) which bends the light passing through it. The light is then brought to a 'focus' or a series of foci (fig. 3.14). The predictions suggest that not only a double image should be produced, but much more complex multiple images, and in certain circumstances, even circular arcs (fig. 3.15). These are sub-units of what are referred to as Einstein rings, which are predicted by consideration of special geometries and distances. Gravitational lenses have an important role in the study of distant material in the Universe.

In principle the gravitational lens effect can tell us about the distribution of matter in the lens, which is usually a cluster of galaxies between the even more distant quasars and ourselves.

A second aspect is that unobservable distant matter, which is otherwise too faint to detect can be revealed by another property of the gravitational lens, that of flux magnification. This boosts the apparent brightness of the background object, hence making it visible. If the geometry can be figured out and the properties of the lens are known, then valuable information on very distant galaxies can be gleaned, information which cannot be determined by any other means. However, we inject a note of caution in that the same process can cause an overestimate of the true luminosity of the background object if lensing is not appreciated. This has an impact for AGN studies and a number of cases of highly luminous objects have now been discovered to be artificially boosted by gravitational lensing We will meet this problem when we discuss evolution of objects and unification scenarios in sections 7 and 9.

3.10 SUMMARY

In this section we have met all the major classes of active galaxies and now have an overview of their global properties. We have seen how the various classes of active galaxy were discovered, and noted the somewhat chaotic nature of the early classification schemes. These assisted in grouping together certain observational parameters, but on the other hand tended to obscure the clarity of the 'big picture' because of the wealth of details contained within the classes. The dangers inherent in observational selection effects and biases were pointed out. A prime example is in terms of luminosity, where the least powerful quasars are less powerful than the most powerful Seyferts. Take a powerful Seyfert 1 to a distance where the light of the underlying galaxy becomes undetectable; would we still call it a Seyfert 1? In fact it would be called a quasar. This idea is highlighted by asking whether an optical spectroscopist, using spectroscopic data alone, would be able to separate a quasar from a Seyfert 1.

Likewise, objects that are classified according to some wavelength selection (Markarian galaxies, radio galaxies) are only manifestations of a particular physical phenomenon rather than a special type of galaxy. It is more pertinent to ask why some spiral galaxies become Seyferts, or why some elliptical galaxies become FRII radio sources while others become quasars. We recognize that blazars tell us about a phenomenon, not a class of galaxy. Herein lies the transition from the old-fashioned 'classification' view to the current 'physical process' concept. What physical mechanisms are under way, in what type of galaxy, and why? Suffice to say that quasars are the most powerful, followed by Seyferts and then LINERs.

As observational techniques improve still further, even lower luminosity AGNs might be found residing in many 'normal' galaxies and this shows the need for care in how we define activity. Many galaxies are now known to show emission lines in their nuclear regions, but these lines are much narrower and less ionized than in their active galaxy counterparts such as Seyferts. These 'HII galaxies' tend to be in later Hubble-type galaxies and hence the smaller nuclear gravitational potential explains the narrower linewidths. LINER galaxies may lie somewhere in between, and deserve much greater study to determine whether they are active or some form of starburst-powered galaxy. We will opt for the former but it highlights the need to be able to separate galaxies that have central engines from those that may have extensive star formation underway.

The starburst galaxies share a common property with the FIR galaxies: high far-infrared luminosities due to re-radiation from heated dust. Sometimes, the origin of this in a starburst is clear because of the spatial extent of the emission. For more distant sources, the distinction between a super-starburst and buried quasar is far more difficult to determine. The infrared search for a (dust shrouded) broad-line region is a key tool in this study. The presence of obscuring dust also hints at possibilities for an orientation-based unification scenario. A dusty molecular torus, lying in the plane of the sky, could easily obscure the broad-line region, thereby causing us to refer to an object as a Seyfert 2 because of the presence of a narrow-line region. Seen from a different orientation, the same object might easily be called a Seyfert 1. The presence of ultraluminous infrared galaxies having extremely high luminosities which are only found in quasars at redshifts greater than 2 pose severe problems for star formation models. However, it is now suspected that these objects may be gravitationally lensed by an intervening galaxy and their intrinsic luminosities are more like the FIRGs.

Radio emission, either from core-dominated jet sources or extended double-lobed structures, is a powerful tool in showing that non-thermal emission is taking place. Where the jet is clearly revealed we see dramatic evidence for particle acceleration mechanisms at work, a clue to the presence of a central engine. Even where the jets are invisible, or very weak, the extended radio emission provides a clue as in a 'smoking gun'. The radio emission tells us that it is jet produced and fed and this in turn points to a central engine of activity. Indeed, all the elements of non-thermal emission from AGNs can be related to the jet emission. For radio-quiet quasars, on the other hand, the emission is dominated at all wavelengths by thermal processes.

The giant lobe-dominated radio galaxies are some of the most spectacular structures in the Universe. These stretch out into space for distances larger than the sizes of some groups of galaxies, including our Local Group of galaxies. We have avoided being drawn into the intricate classification of their structures but noted that the lobes are likely to be affected by an external environment if they lie in a cluster of galaxies. The lobes are fed by highly collimated kiloparsec-scale jets, which start their life as parsec scale jets emanating from the core of the galaxy as demonstrated by VLBI observations. The physics of such jets, the explanation of superluminal motion and the solution of the problem of how to get the electrons from the core to distances of tens of kiloparsec without losing their energy, will all be left until section 8.

We have made strong hints about the presence of a central engine, these are strengthened by observations of symmetrical jets of synchrotron emission emanating from the nucleus of galaxies and radio-loud quasars. The lifetime and relative straightness of these structures argue for a long-lived and stable (in terms of spatial orientation) phenomenon to be present to explain such radical departures from spherical asymmetry. The idea of a rotating supermassive black hole fits comfortably into such a picture.

3.11 FURTHER READING

General and reviews

Angel,J.R.P., & Stockman,H.S., 'Optical and infrared polarizations of extragalactic sources', *Ann.Rev.Astron.Astrophys.*, **18**, 321, 1980. (Introduction of blazars.)

Blandford,R.D., Netzer,H., & Woltjer,L., *Active Galactic Nuclei' Saas-Fee Advanced Course 20*, Springer–Verlag, 1990. (A wonderful series of papers by three world experts giving detailed and specialized discussions of most aspects of AGNs including unification models—advanced but very readable and excellent.)

Blandford,R.D., & Narayan,R., 'Cosmological applications of gravitational lensing' *Ann.Rev.Astron.Astrophys.*, **30**, 311, 1992. (Non-trivial discussion of lensing.)

Beckman,J.E., Ed. *The nearest active galaxies*, Kleuwer, 1993. (Up to date review.)

Bridle,A.H., & Perley,R.A., 'Extragalactic radio jets', *Ann.Rev.Astron.Astrophys.*, **22**, 319, 1984. (Brilliant review—excellent source material for students.)

Burbidge,M.E., 'Quasi-stellar objects', *Ann.Rev.Astron.Astrophys.*, **5**, 399, 1967. (A most interesting review of the early years of quasar research after their discovery.)

Duschl,W.J., & Wagner,S.J., Eds., *Physics of active galactic nuclei*, Springer–Verlag, 1992. (A very good coverage of the subject.)

Dyson,J., Ed., *Active galactic Nuclei*, Manchester University Press, 1985. (Contains an excellent review by Hazard on the discovery of quasars.)

Finkbeiner,A., 'Active galactic nuclei, sorting out the mess', *Sky & Telescope*, August, 1992. (An excellent and easy to read review.)

Keel,W.C., 'Crashing galaxies, cosmic fireworks', *Sky & Telescope*, p18, January, 1989. (An easy to read account of galaxy interactions.)

Maraschi,L., Maccacaro,T., & Ulrich,M-H., *BL Lac objects: Lecture notes in Physics—334*, Springer–Verlag, 1989. (All you need to know about BL Lacs.)

Stein,W.A., O'Dell,S.L., & Stritmatter,P.A., 'The BL Lacertae objects', *Ann.Rev. Astron.Astrophys.*, **14**, 173, 1976. (The early days of BL Lacs.)

Stritmatter,P.A., & Williams,R.E., 'The line spectra of quasi-stellar objects', *Ann. Rev.Astron.Astrophys.*, **14**, 307, 1976. (Well laid-out descriptions.)

Weedman,D.W., *Quasi-stellar objects*, Cambridge University Press, 1986. (Good historical notes and an excellent overview.)

Weedman,D.W., *Quasar Astronomy*, Cambridge University Press, 1988. (A concise and excellent review on quasars, highly recommended.)

Weedman,D.W., 'Seyfert galaxies', *Ann.Rev.Astron.Astrophys.*, **15**, 69, 1977. (The introduction of the Seyfert sub-classifications.)

Weymann, R.J., Carswell,R.D., & Smith,M.J., 'Absorption lines in the spectra of quasistellar objects', *Ann.Rev.Astron.Astrophys.*, **19**, 41, 1981. (A good introduction to absorption lines and their implications - now somewhat out of date.)

Wolfe,A.M., Ed., *Pittsburgh Conference on BL Lac Objects*. Pittsburgh University Press, 1979. (An historical book, an excellent read and a must for students of blazars.)

Special Issue of Mercury commemorating the 25th Anniversary of the Discovery of the redshift and Bizarre properties of Quasars. *Mercury*, 27, 2-30 (Jan-Feb), 1988. (Contains an excellent, historical review which is well researched.)

Specialized

Boroson,T.Y., & Oke,B., 'Detection of the underlying galaxy in the QSO 3C48', *Nature*, **296**, 397, 1982. (A breakthrough observation.)

Hazard,C., Mackey,M.B., & Shimmins,A.J., 'Investigation of the radio occultations of 3C273 by the method of lunar occultations', *Nature*, **197**, 1037, 1963. (The discovery paper for quasars, plus the three following papers—a must for those interested in history of the subject.)

Heckman,T., 'An optical and radio survey of the nuclei of bright galaxies' *Astron. Astrophys.*, **87**, 152, 1980. (The introduction of LINERs.)

Filippenko,A., & Sargent,W., 'A search for "dwarf" Seyfert 1 nuclei. I. The initial data and results', *Astrophys.J.Suppl*, **57**, 503-522, 1985.

Greenstein,J., & Schmidt,M., 'The quasi-stellar radio sources 3C48 and 3C273' *Astrophys.J.*, **140**, 1, 1964. (A historical paper which was published simultaneous with the discovery paper above but contains much more information.)

Koratkar,A., *et al.*, 'Low luminosity active galaxies: are they similar to Seyfert galaxies?', *Astrophys.J.*, **440**, 132, 1995. (A good reference source for low luminosity galaxies.)

Matthews,T.A.K., & Sandage,A.R., 'Optical identifications of 3C48, 3C196 and 3C286 with stellar objects', *Astrophys.J.*, **138**, 30, 1963. (Well worth reading—beat the discovery paper, but incorrect and read the footnote in section VII.)

Maoz,D., *et al.*, 'Detection of compact UV nuclear emission in Liner galaxies' *Astrophys.J.*, **440**, 91, 1995. (A good source of references for LINERs.)

Seyfert,C.K., 'Nuclear emission in spiral nebulae', *Astrophys.J.*, **97**, 28, 1943. (The discovery paper for Seyfert galaxies—another must for those interested in history but it requires a good library for such an old journal.)

Véron-Cetty,M.-P., & Véron,P., *A catalogue of quasars and active nuclei—5th edition*, ESO Scientific Report no. 10, 1991. (The catalogue of AGNs.)

Weedman,D.W., 'High velocity gas motions in galactic nuclei', *Astrophys.J.*, **159**, 405, 1970. (First thoughts on Seyfert sub-classification.)

4

Emission processes in active galaxies

4.1 INTRODUCTION

This chapter discusses the physical processes and mechanisms that give rise to the phenomena we observe in active galaxies. By necessity it requires the use of fundamental physical ideas alongside some mathematical manipulation of equations. As such, much of the 'meat' of the subject is contained within this chapter. For the reader who is interested in the *phenomenon* of active galaxies and who is content to forgo the intricacies of the detailed physics, an appreciation of the flavour of the topics is given in summary sections liberally spread throughout this chapter. For the general reader, the message is 'don't panic'. For the student of astrophysics, there are a large number of texts detailing the specific treatments of the topics contained in this and other chapters. One of the very best and currently available is that by Frank Shu noted in section 4.6.

From the preceding chapters we have seen that there is strong persuasive evidence to conclude that *all* the aspects of active galaxy phenomena originate in the very central regions of a galaxy. We have hinted at possible reasons why this activity may be associated with a central engine, which we have tacitly assumed to be a massive black hole surrounded by an accretion disk. We have also noted that the manifestation of the different types of active galaxy may be dependent on such effects as the mass of the black hole, the gas supply and possible interaction of the host galaxy with neighbouring galaxies. We will now begin to put some flesh onto the bones of the outline sketches we have used up to now. We will first concentrate on those mechanisms that produce the continuum emission from radio through gamma-rays and then we shall tackle the topic of spectral lines. This will focus on the UV through IR wavelengths, sometimes referred to as UVOIR region.

The most frequently found processes producing continuum emission from astrophysical bodies are thermal, synchrotron, Compton and inverse Compton. It cannot be overstated that active galaxies radiate by electromagnetic radiation and to understand fully the intricacies of the detailed physics taking place within their nuclei, it is important that students feel comfortable with these emission concepts. Although not strictly essential for the casual or interested lay-reader, the student of AGNs *must* be able to appreciate the physical principles behind all the emission mechanisms. In the first part of this chapter we will review the basic processes involved in gaining this understanding.

The descriptions will be brief but adequate for our purposes and will not involve detailed mathematics. It should also be stressed that radiation is produced whenever a charge is accelerated. This is a fundamental concept and another way of looking at radiation processes is to investigate the different ways in which charged particles are accelerated.

Before we launch into these physical processes, a reminder is in order. It is undoubtedly the case that for virtually all the active galaxies we observe, we see a *blend of differing emission mechanisms*. We must be very careful to separate these in order to discover the underlying physical processes at work. As we shall see, this task is far from simple, and its solution occupies a significant amount of satellite and ground-based telescope time. The limitations imposed by lack of spatial resolution (apart from VLBI) remain a major drawback.

4.2 THERMAL EMISSION

Thermal emission is the most common form of continuum emission we meet both in everyday life and in astrophysics. It is a property of a radiating body. The light we see from the Sun, Moon, planets, stars, incandescent tungsten and quartz halogen light bulbs are examples of thermal emission. However, even here we must be very careful, because we have made an underlying assumption that we are talking about visible wavelengths and direct emission. For AGNs we need to consider the entire electromagnetic spectrum, and we should note that in the case of the Moon and planets, the visible light we see is merely reflected sunlight. This point deserves immediate clarification because although the Moon and planets shine of their own accord, they do so at wavelengths to which our eyes are insensitive, the infrared. Our eyes only register the reflected sunlight, and indeed, for most objects that we see around us, we are looking at reflected sunlight. The colours we see are the result of how the surface (or skin) of the object modifies the incident sunlight, a red surface absorbing the other colours and leaving the red to be reflected. We should also note that the familiar radiation we see from the neon, mercury and sodium lamps which light up our cities and towns are not examples of thermal continuum radiation, but of spectral line emission.

4.2.1 Blackbody emission

We begin our discussion of thermal emission by defining the concept of a blackbody radiator. A blackbody is a body in thermal equilibrium with its surroundings and is both a perfect absorber and perfect emitter of radiation. Blackbody radiation is isotropic, un-polarized and has a smooth continuous emission spectrum. A perfect blackbody is extremely difficult to build in the laboratory, but stars turn out to be reasonable approximations. A blackbody has a unique defining temperature that is expressed in units of the absolute temperature scale, the Kelvin (K). No radiation falling onto a blackbody is reflected and therefore the radiation we see from a blackbody is a property of its emission alone. This follows a uniquely defined form given by Planck's radiation equation

$$B_\nu(T) = \frac{2h\nu^3}{c^2\,(e^{h\nu/kT} - 1)} \tag{4.1}$$

in units of watts per square metre per unit solid angle per unit frequency interval (W m^{-2} sr^{-1} Hz^{-1}) and in this case the square metres refers to a unit of surface area and not distance. B_ν is the emitted intensity of radiation per unit frequency interval at a frequency ν, h is Planck's constant, c is the velocity of light in *vacuo*, k is Boltzmann's constant and T is the absolute temperature. This equation can also be written in terms of the wavelength and per unit wavelength interval by

$$B_\lambda(T) = \frac{2hc^2}{\lambda^5\,(e^{hc/\lambda kT} - 1)} \qquad (4.2)$$

which has units of W m^{-2} sr^{-1} m^{-1}. Eqn. 4.2 is shown graphically in fig. 4.1 and the resulting curve is known by a variety of names: the Planck spectrum, blackbody curve or blackbody spectrum. Figure 4.1 is an extremely important and useful diagram in physics and depicts the emission over all wavelengths (or frequencies) from a blackbody of absolute temperature T.

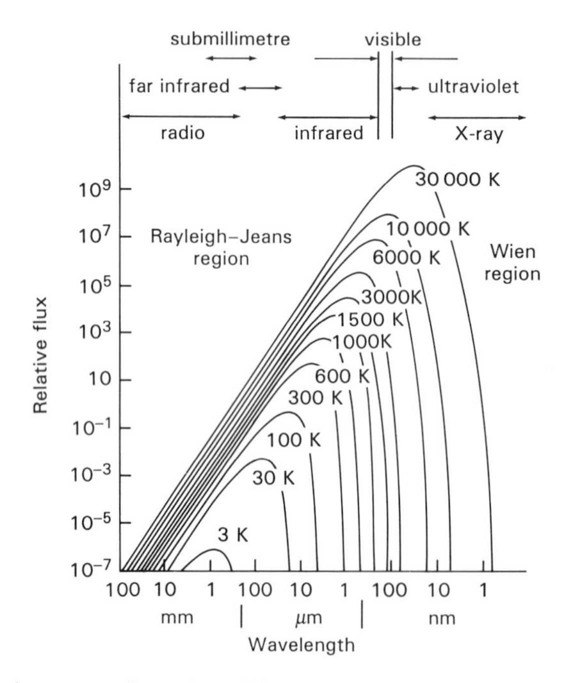

Fig 4.1. Emission spectra of a number of blackbody radiators of different temperatures. Note that a hotter temperature blackbody curve always encloses that of a cooler blackbody. The long-wavelength (Rayleigh–Jeans) and short-wavelength (Wien) regions are indicated.

Equation 4.2 can be simplified mathematically for the two regions shown in fig. 4.1, referred to as the Rayleigh–Jeans and Wien regions. The Rayleigh–Jeans regime lies at wavelengths longward of the peak of the emission and is characterized by $hc/\lambda kT \ll 1$. The exponential term can be expanded to give $e^{hc/\lambda kT} = 1 + hc/\lambda kT$ + higher powers in $hc/\lambda kT$ which we will ignore. Eqn. 4.2 then reduces to

$$B_\lambda(T) = \frac{2ckT}{\lambda^4} \tag{4.3}$$

or, in terms of frequency, ν,

$$B_\nu(T) = \frac{2\nu^2 kT}{c^2} \,. \tag{4.4}$$

When this is plotted on a logarithmic scale of log B_ν versus log ν, it is a straight line of slope 2. This is the Rayleigh–Jeans region and the usefulness of the approximation is obvious; life is much easier when working in this regime as eqns. 4.3 and 4.4 are significantly simpler to work with than eqns. 4.1 and 4.2. At the short wavelength side of the peak emission we find that $hc/\lambda kT >> 1$, and so we can make the approximation $[e^{hc/\lambda kT} - 1] \sim e^{hc/\lambda kT}$, giving

$$B_\lambda(T) = \frac{2hc^2}{\lambda^5 e^{hc/\lambda kt}} \tag{4.5}$$

and, in terms of frequency, ν,

$$B_\nu(T) = \frac{2h\nu^3}{c^2 e^{h\nu/kT}} \tag{4.6}$$

which explains the exponential shape of the curvature of the short wavelength part of the blackbody spectrum.

The luminosity, L, radiated by a spherical blackbody is given by a simple relation:

$$L = 4\pi r^2 \sigma T^4 \,, \quad \text{(W)} \tag{4.7}$$

where r is the radius of the body, σ is the Stefan–Boltzmann constant and T is the absolute temperature. Equation 4.7 can be thought of as being in two parts: the rate at which a blackbody radiates per unit surface area given by σT^4 (and is referred to as the Stefan–Boltzmann Law), and the total surface area of the body given by $4\pi r^2$. Note that the luminosity depends on the surface area and on the fourth power of the temperature. Therefore very luminous bodies should either be very hot, very large, or both.

We can now consider what happens to the emission as the temperature changes. It is well known from everyday experience, such as a heated fire poker, that hotter bodies radiate at shorter wavelengths and this is precisely explained by the Planck equation (eqn. 4.1). For blackbodies there is an exact expression, obtained by differentiating eqn. 4.2, relating the wavelength of maximum emission (λ_{max}) to the absolute temperature (T) of the body. This is called the Wien Displacement Law and is given by

$$\lambda_{max}\, T = 3 \quad \text{(mm deg)} \tag{4.8}$$

and is shown in fig. 4.2. Note the units of wavelength used in eqn. 4.8. We see that as the temperature falls the peak emission moves to longer wavelengths. Also, because the luminosity of the blackbody decreases with decreasing temperature, at any wavelength the emission for a cooler blackbody per unit surface area is always contained within the envelope of that of a hotter blackbody. Examples of the application of eqn. 4.8 are given in Table 4.1

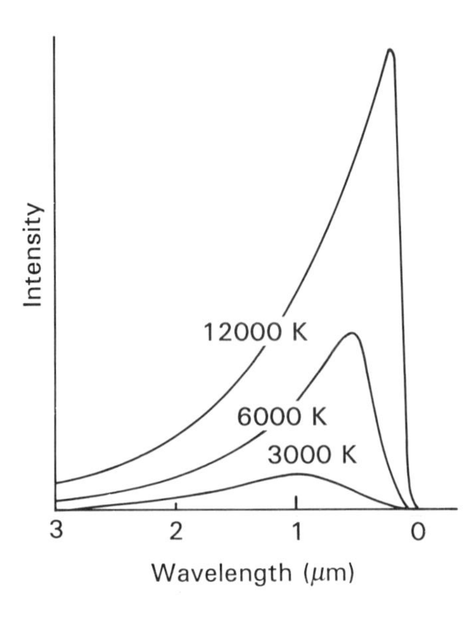

Fig. 4.2. The locus of the peak emission of a series of blackbodies of differing temperatures is shown by the solid line, called the Wien Displacement Law and given by $\lambda_{max}T = 3$ mm K. Note that, unlike fig. 4.1, this is plotted on a linear scale.

Example	Temperature T(K)	λ_{max}	Wavelength regime of peak emission
Coronal gas	1,000,000	3 nm	soft X-ray
Hot plasma	300,000	10 nm	EUV
O star	30,000	100 nm	UV
G star (Sun)	6,000	500 nm	visible
M star	3,000	1 μm	near infrared
Hot dust	1,500	2 μm	near infrared
The Earth	300	10 μm	mid-infrared
Cold dust	100	30 μm	submillimetre
2.7 K cosmic background radiation	2.7	~1 mm	millimetre

Table 4.1 Illustration of the Wien Displacement Law

In principle we can determine the approximate temperature of a radiating body by assuming that it emits as a blackbody and determining the wavelength of the peak output flux. Stars are not perfect blackbody radiators because of absorption of radiation by cooler gas in the outer layers beyond the photosphere. Nevertheless, this method gives reasonable estimates of stellar temperatures.

Let us now consider the production of a blackbody spectrum in a little more detail. This aspect is often glossed over and many students fail to appreciate the details of the emission process. We can pose the question of what causes a very cold, say 10 K, blackbody to radiate a Planck spectrum. We select this cold temperature for a good reason as it suitably clarifies the picture. Being a blackbody the atoms of the gas must be in thermal equilibrium with their surroundings, a necessary condition for a blackbody. Let us assume for the moment that the gas is immersed in a photon field. The atoms will have a speed distribution in keeping with their temperature and this is the Maxwellian distribution given by

$$N(v)\,dv \propto v^2 \exp\left(\frac{-mv^2}{2kT}\right) dv \tag{4.9}$$

where N(v) are the number of atoms with speed in the range v to $v + dv$, T is the absolute temperature and k is Boltzmann's constant $= 1.38 \times 10^{-23}$ J K^{-1}. By differentiating eqn. 4.9 and equating to zero we can find that there is a most probable speed (v_p) for the atoms of the gas given by $[(2kT/m)^{0.5}]$ and this corresponds to the absolute temperature defined for that ensemble of radiating atoms. It can also be shown that the mean square speed $\overline{v}^2$ is equal to 3kT/m and hence from the mean kinetic energy of the atoms, we can obtain the useful relation that $\frac{1}{2}m\overline{v}^2 = \frac{3}{2}kT$. This is a special case of a general theorem known as equipartition of energy. Finally, in passing we should note that eqn. 4.9 can alternatively be expressed in terms of the number of atoms with energy lying between E and E + dE.

But what causes the emission from the ensemble of atoms? The atoms are in relative motion and pass sufficiently close to each other that they interact. What precisely is this interaction? It is the electronic charge interaction between the electron cloud of one atom which 'feels' the Coulomb repulsion of the electron cloud of the other atom. In the interaction, the trajectory of each atoms is changed, the energy is altered, and the resultant change of energy is emitted as a photon. The result is a continuous spectrum because it is the atom that suffers a change in energy rather than an individual electron within an atom.

For the case of an isolated and very cold gas in equilibrium with its surroundings (which implies that there must be a large number of collisions within the gas to maintain the equilibrium situation), the results of the 'atomic collisions' and subsequent emission of photons results in the emission of a blackbody spectrum. The atoms of the gas must also be in thermal equilibrium with the surrounding photon field (otherwise energy exchange will cause warming of one and cooling of the other) and for our example of a 10 K gas cloud, the photon field has a radiation temperature of 10 K. Such temperatures are possible in regions of the Galaxy distant from hot stars and where the radiation field is due to the general diffuse starlight.

4.2.2 Interactions, opacity and radiation transport

It follows from our above descriptions that a blackbody is opaque, but what happens for bodies that are semi-transparent? This is a common event in astrophysical situations. To make progress in this topic we now need to discuss a fundamental aspect of all physical processes, interactions. The usual situations we shall consider are photon–particle or particle–particle interactions.

In regions where matter and radiation co-exist, there is a finite probability that a photon will interact with particles of the medium. (The particles can also interact with each other.) The probability of an interaction is described by a parameter called the interaction cross-section, σ_c. This can be related to a parameter that is somewhat easier to visualize, the mean free path, λ_m. The mean free path is the average distance that a photon or particle travels before suffering an interaction in the medium in question. The mean free path is a unit of length and is given by

$$\lambda_m = \frac{1}{N\,\sigma_c}\,, \tag{4.10}$$

where N is the number density of particles in the medium.

Interactions between photons and the atoms/ions of a gaseous medium occur in two main ways: absorption and scattering. Reflection can be thought of as a special form of scattering which occurs at the interface of two media, and results in the photon being deflected away from the medium instead of passing into it. Although this is somewhat of a gross over-simplification, it will suffice for our purposes as reflection is usually not an important phenomenon compared with absorption and scattering in most astrophysical situations. Therefore we will not pursue it further.

Absorption of radiation by matter is an everyday phenomenon. Our bodies absorb solar energy and are warmed. Solar radiation panels absorb sunlight, heating water and providing an alternative energy source to fossil fuels. Although the term absorption is used in everyday parlance, in physics it has a very specific meaning. Absorption occurs when an incident photon of radiation is completely absorbed by an atom or ion of the absorbing material. The photon is lost to the radiation field and by conservation of energy its energy is given to the absorbing medium. The absorbing material therefore reduces the intensity of radiation passing through it and, in the process, is heated. The absorption of energy by the medium is initially taken up by the atoms or ions.

The absorption of a photon by a single atom will raise the energy of that atom to a higher excitation state or even an ionized state. The atom will subsequently decay to a lower energy state according to the rules of quantum mechanics. This will be by the two mechanisms that we have already met in section 2.7, radiative and collisional de-excitation. In the former process, secondary photons of lower energy (longer wavelength) than the initial photon are emitted as de-excitation occurs when an electron drops through intermediate energy levels of the atom. If all these lower energy photons can escape directly from the medium then no overall heating is produced; the incident photon energy has been transferred to a number of lower energy photons that escape. However, because this emission from the medium is isotropic, the intensity (a directional quantity – see section 2.2.3) of the incident beam is reduced as fewer photons remain in the beam in the original direction.

Radiative de-excitation to the original energy level of the atom via intermediate energy levels is the usual mechanism in astrophysical situations and poses interesting (and complicated) problems that have to be solved in order to understand the heating and cooling processes of photon irradiated gaseous nebulae. Note that when photons escape from the medium, they remove energy and this therefore represents a cooling process. If, on the other hand, the lower energy photons are trapped within the medium due to further absorption by atoms; then the energy of the incident photon has successfully been transferred to internal heating of the gas, because more atoms are now in a higher energy state or have higher kinetic energies. In the case of collisional de-excitation, excess energy is removed from the atom by collisions with neighbouring atoms, which thereby gain energy in the process and the absorbing medium is heated. In both cases it is rare for all the secondary emission to escape, therefore absorption usually produces some degree of heating of the medium.

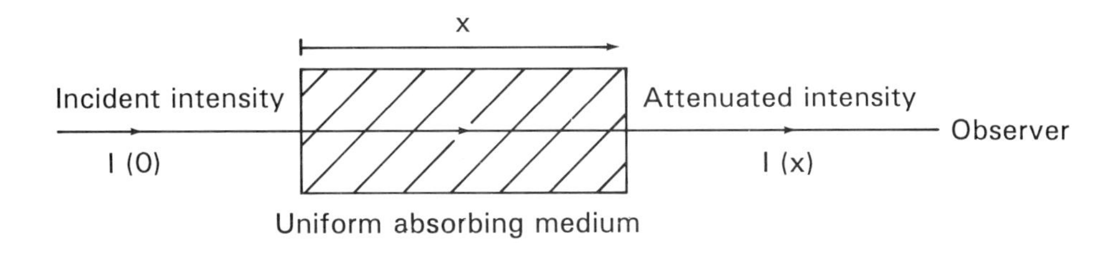

Fig. 4.3. Radiation transport schematic. If the observer is located within the absorbing medium, then the variable (x) is the distance from the front face of the absorbing slab.

The complex process by which radiation travelling in a specific direction interacts with a medium is known as *radiation transport*. Consider radiation passing through a slab of intervening material in the line-of-sight to an observer. We can assume for simplicity (even though it may be unlikely) that this material has a uniform density, i.e., it is homogeneous. If the slab is at a sufficient distance from the source then the radiation can be considered as a parallel beam of photons. The intensity measured by the observer, I(x), is related to the intensity entering the slab, I(0), by

$$I(x) = I(0)\, e^{-a x} \quad , \tag{4.11}$$

where x is the thickness of the material and a is the absorption coefficient per unit volume of the material. The attenuation in the medium is given by $e^{-a x}$. I(x) is the intensity that is transmitted by the medium, which (assuming no reflection) has absorbed the fraction [I(0)–I(x)]/I(0).

However, the case is still more complex because we have seen that the medium will become heated and will thermally re-emit, either by line emission or by increased continuum emission. The continuum emission will most probably have a different wavelength from the incident photon and is thereby distinguishable from the incident photon energy distribution. A common example is the emission of infrared blackbody radiation following the absorption of ultraviolet light by dust grains. In general, however, the process of measuring the output emission from a beam of radiation traversing a semi-transparent medium is far from simple.

Where the medium is nearly transparent, the problem is much simpler. We have a situation referred to as an optically thin medium, where the photon beam loses little energy in traversing the medium and we can ignore any heating and subsequent re-emission of the medium. (The term 'optical' does not refer only to the visible part of the spectrum; it is merely a hang-over from the times when all calculations were performed for visible photons.) The underlying concept is that of opacity, or optical depth, usually expressed as τ and inversely related to the transparency of the medium. A medium of high optical depth ($\tau >>1$) has high opacity, is opaque and the transmission is very small. On the other hand, an optically thin medium ($\tau <<1$) has a very low opacity, a low optical depth and a relatively high transmission. All these quantities are wavelength-dependent, often strongly.

Although the detailed treatment of radiation transport in space is highly complex, solutions are usually readily at hand for those occasions where astrophysicists restrict themselves to considering sources which are either optically thick ($\tau >>1$) or optically thin ($\tau <<1$). As we saw above, when considering the transmission through a medium, both absorption and scattering must be taken into account. In many cases, one of these dominates, making the calculations much simpler.

Let us look at a commonplace example of attenuation. Consider the radiation from the Sun when viewed through a V-band filter centred at 550 nm. The solar flux at the top of the Earth's atmosphere is 1.64 W m^{-2}. The atmospheric attenuation at V is about 0.2 magnitudes per airmass for a clear dust-free sky. For a location where the Sun is directly overhead, such as at noon on the equator at one of the equinoxes, the transmission is then almost 83%. (Converting the magnitude attenuation into a percentage transmission can be done by using eqn. 2.5.) Most of this loss at V-band is due to scattering. We can use this information to calculate the attenuation of the atmosphere through the filter by using the radiation transport equation, 4.11. As $I(x)/I(0) = 0.83$, we can write

$$0.83 \quad = \quad \exp^{-(a\,x)} \quad , \qquad (4.12)$$

and to a first approximation the atmosphere can be reduced to a slab consisting of nitrogen and oxygen of thickness $x \sim 8$ km. This then gives a mean absorption coefficient, $a \sim 0.023$ km^{-1}. As the observer moves away from the equator, the effective path-length through the atmosphere is increased and the attenuation is greater.

Absorption is usually a strong function of wavelength and in the case of the Earth's atmosphere this fact has now become common knowledge because of the greenhouse effect and the ozone hole. Solar radiation at wavelengths shorter than about 320 nm is absorbed by ozone in the atmosphere, the absorption coefficient being orders of magnitude greater than that for the V-band. Indeed, the U-band filter (which, on its short wavelength side is defined by the strong ozone atmospheric absorption) has an extinction coefficient of about 0.65 magnitudes per airmass. For shorter wavelengths the atmosphere used to be opaque but the ozone hole is gradually changing this.

The effects of scattering explain other everyday phenomena concerning the Sun and the sky. When the Sun is well away from the horizon, it is yellow in colour. This is due to its temperature (see Table 4.1) and the atmosphere has very low absorption at visible wavelengths. What do we see when the Sun approaches the horizon? Its colour changes from yellow to orange, becoming redder the closer the Sun moves to the horizon and we note that the colour of the sky also changes. What has happened is that the Sun is now seen through a much greater atmospheric path-length and the scattering is increased greatly. When it is five degrees above the horizon, the path-length has increased over eleven-fold from overhead. There are now many more atoms in the line-of-sight and because the scattering coefficient is highly wavelength-dependent (varying with wavelength as λ^{-4}), the blue light from the Sun is scattered with much greater efficiency than the red light. The Sun therefore appears orange or red in colour (having lost the blue wavelengths) and much dimmer due to more of all wavelengths being scattered. The deep blue colour of the daytime sky is also a direct result of scattering of sunlight by the molecules of the atmosphere. At directions far away from the Sun we see the scattered short wavelength photons, hence the sky looks blue when we observe it. If it were not for this scattering, the daytime sky would be black and we could see the stars.

The transport of energy in stellar interiors is an excellent example of the effects of scattering and absorption. Nuclear fusion provides the energy source at the cores of stars. The stars are stable because of hydrostatic equilibrium, in which the pressure of the overlying layers of gas is balanced by the gas pressure of the core (because the gas is very hot) or, for the most massive stars, the gas pressure is replaced by radiation pressure. How does the energy travel from the core, with a temperature of 15 million degrees in the case of the Sun, to the photosphere of temperature about 6,000 K?

Examination of this question provides an excellent illustration of widely differing interaction coefficients. In the nuclear fusion reactions, gamma-ray photons and neutrinos are emitted. Neutrinos are elementary particles that have either zero or a very small mass and travel at speeds very close to that of light. The interaction length (mean free path) for the high-energy (gamma-ray) photons is about 0.01 m, whereas for neutrinos it is many millions of times greater the diameter of the Sun. The photons in the core of the Sun therefore travel in a chaotic path, undergoing an interaction and being scattered on average every 10 mm, (leading to what is called the random, or drunkard's walk), whereas the neutrinos propagate outward totally unimpeded, reaching the solar photosphere some 2.3 seconds later and arriving at the Earth after a further 8.3 minutes. Because of their extremely small interaction cross-section and amazingly long mean free path (a few light years for the density of the Earth) they pass straight through the entire planet unimpeded.

As an aside we can note that Sir Arthur Eddington, amongst others, produced a set of equations that describe the complex processes of radiative transport. These must be solved in order to calculate the emergent flux seen by an observer. The equations depend on the density, pressure, temperature and elemental composition of the medium, many of which are often unknown. However, as with much of astrophysics, sensible approximations can be made which allow the equations to be solved relatively easily for most stars. Moving now from high density regions of stellar atmospheres to the very low regions of space, we will find that radiation transport calculations have an important bearing.

Where a source is optically thick, we cannot see very far into it at the wavelength in question (due to scattering or absorption). This also means that the same wavelength we see in emission comes from a region close to the outer surface of the medium (because of scattering or emission). On the other hand, an optically thin medium means that we can see well into it, perhaps even through it, and so the emissions we see come from deep within the outer boundary. These are just differing ways of considering the same basic phenomenon, but often students fail to appreciate this fact and remain deeply confused by opacity, absorption and transparency concepts. They are really simple when reduced to the fundamentals.

4.2.3 Greybody emission

Greybody emission refers to the case of thermal continuum emission from a non-perfect blackbody, the usual case in astrophysics. Let us consider this in a little more detail. We have already seen that the presence of an obscuring torus may explain the differences between the type 1 and type 2 AGN spectra. We will now consider a spherically symmetric envelope of dust grains heated by a central source of radiation. We can take this heating source to be either a hot star or, more appropriately for our studies, a very hot accretion disk surrounding a supermassive black hole. Let us consider only photon–particle collisions in the low energy (non-relativistic) domain. We can also make the approximation (not essential and a better approximation for a hot star) that the central source radiates like a blackbody and that the dust cloud is opaque to the central radiation field (i.e., none can escape directly).

In the innermost regions surrounding the source, the radiation field is sufficiently intense that the equilibrium temperature for a dust grain exceeds its evaporation temperature; this zone is therefore dust-free. At a greater radial distance, the dust grains can just survive and this point forms the inner boundary at the evaporation temperature of the dust. This is on average ~1,500 K but depends on the precise composition of the dust grains. Radiation transport calculations treat the surrounding cloud as a series of shells, each with a characteristic temperature and mass. The dust in the inner shell is heated directly by the central object and also by the large solid angle of the opposing mass of dust on the far side of the central object, a factor termed backwarming. The inner shell absorbs some fraction of the incident radiation which depends on the photon–particle interaction coefficient. The remainder of the incident energy penetrates deeper into the cloud and is absorbed farther from the central object. At shells more distant from the central source of energy, the radiation intensity is greatly reduced due to the absorption by the inner shells. The dust in the outer shells will therefore be heated less and will have a lower equilibrium temperature.

The dust radiates isotropically, and this contributes to further heating of the dust in the cloud. The process of UV absorption and infrared re-radiation continues until the long-wavelength photons emitted by a certain shell can escape freely into space. The source is now optically thin at these wavelengths. If the region which emitted these photons lies deep within the cloud, then this means we can probe to a significant depth into the cloud. On the other hand, if this region is close to the outer edge, then the cloud is optically thick. Note how similar this is to the treatment of the emission of energy from a star at various wavelengths; the physical processes are identical. Because the dust will be immersed in a cloud of gas atoms and molecules, we will also need to consider the gas interactions and line emission, which will be given in section 4.4.

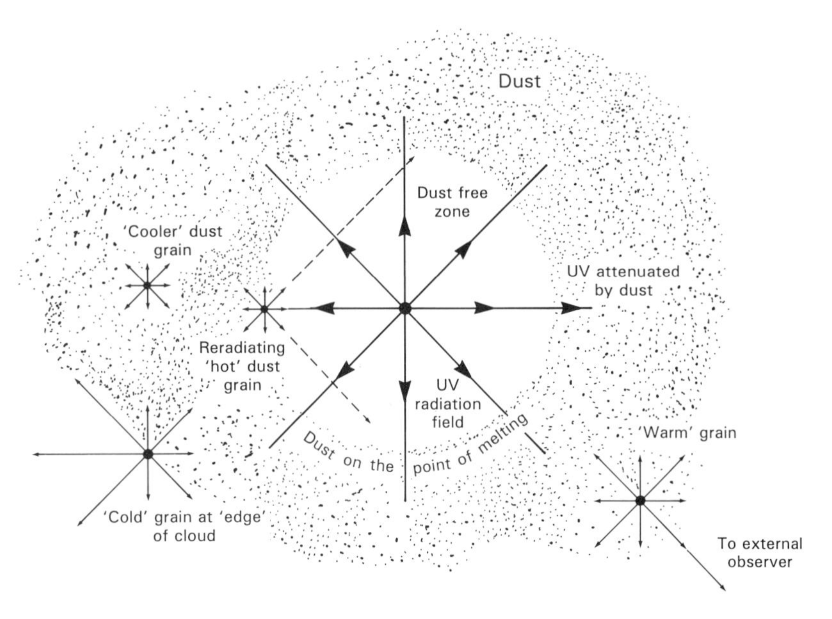

Fig. 4.4. A schematic showing the complex process of absorption and re-radiation by dust surrounding a central hot source. Individual dust grains radiate isotropically but because of the temperature gradient from the inner zone of the dust cloud to the 'edge' the net flow of radiation is outwards.

The opacity is a function of wavelength. This is well determined for emission from heated dust sources, from which astronomers nearly always observe an optically thin spectrum in the submillimetre. This means that the Planck spectrum is modified by an emissivity factor (which is a function of wavelength) for the dust material. A number of studies have shown that the emission from heated dust grains in the submillimetre is given in frequency terms by

$$I_\nu = \nu^{(\beta)} B_\nu (T) \tag{4.13}$$

where the emissivity term is contained in the expression $\nu^{(\beta)}$, and β is called the emissivity index. The far-infrared continuum emission is then given by $\nu^{(2+\beta)}$ (due to the use of the Rayleigh–Jeans approximation for this part of the spectrum). Observations of dust in HII regions and nearby FIRGs give values for β ranging from 1 to 2. For a value of $\beta = 2$, the resulting Rayleigh–Jeans spectrum is of the form ν^4 rather than the ν^2 expected from a blackbody (eqn. 4.4). The observed emergent flux, S_ν, from dust emission from a centrally heated source can be written as

$$S_\nu = \frac{2h\nu^3 \, \Omega \, [1 - e^{\tau(\nu/\nu_0)}]^\beta}{c^2 \, (e^{h\nu/kT_d} - 1)} \tag{4.14}$$

where ν is the observing frequency, Ω is the solid angle of the source, T_d is the temperature of the dust and τ is the optical depth. This is a very useful but complex equation to solve, given that there are usually more unknowns than can be determined by observation. In terms of our study of active galaxies we want to know the size (Ω) and mass of this dust-emitting zone to determine whether it comes from a highly condensed central regime of the galaxy or an extended disk. However, β is generally unknown, along with the dust temperature and the optical depth. Furthermore, these are all inter-related and eqn. 4.14 is usually solved by a complex iterative process. If sufficient spectral data points can be obtained, the resultant continuum emission spectrum can be fitted by varying the unknowns and obtaining the best fit from statistical tests. We will use this information in section 5.2.2.

We find that for many galactic and extragalactic sources the emission at wavelengths shorter than the peak bears little resemblance to the Wien distribution of a single temperature blackbody (fig. 5.6). For thermal sources this provides conclusive evidence that the dust is not radiating at a single temperature. The longer wavelength Rayleigh–Jeans part of the spectrum samples the emission from the coldest (the outermost component of the dust), but as might be expected, there are regions of hotter dust closer to the central heating source and as these peak at shorter wavelengths (eqn. 4.8) the overall emission spectrum reflects the range of temperatures for the radiating dust.

Before closing this section we should remember that an obvious source of thermal continuum emission from galaxies is the sum of the photospheric emission from the constituent stars. The output spectrum will obviously depend on the relative numbers of the various populations of stars present, and whether absorbing dust is playing a significant effect. This latter factor has an important role in that the highest luminosity stars are the young, massive, blue OB stars which occur preferentially within large clouds of gas and dust and so their contribution is dimmed. The galaxy will therefore appear redder and the stellar populations might be incorrectly apportioned. Likewise, we again point out that orientation effects can play a very big part. There is an industry within the astrophysics community that attempts to model the emergent spectrum of galaxies in terms of populations of stellar types, with and without dust absorption. In the case of active galaxies this area is not crucial because for the most powerful objects, the luminosity of the active nucleus greatly dominates the starlight from the surrounding galaxy. However it becomes important for lower luminosity objects such as LINERs and many Seyfert 2 galaxies.

4.2.4 Summary of thermal emission, opacities and radiation transport

A blackbody is a perfect emitter and absorber of radiation at all wavelengths/frequencies and is therefore of very high optical depth. It is in thermodynamic equilibrium with its surroundings and the atoms of the emitting body follow a Maxwellian velocity distribution. The emission from a blackbody follows a well-defined spectrum (eqns. 4.1 and 4.2), the peak of which is directly related to the absolute temperature of the body (eqn. 4.8).

Both absorption and scattering attenuate the energy in a beam of photons passing through a medium, the degree being given by the interaction cross-section and the path-length in the medium. In both cases, the original radiation intensity is reduced and this radiant energy is converted into heating the medium. The extent of the heating depends on whether the secondary photons can immediately escape or are trapped. Photons that can escape contribute directly to cooling the medium and both line and continuum emission are observed. Absorption and scattering are both wavelength-dependent processes.

In considering the full treatment of radiation passing through a medium, a complex set of equations must be used to determine the emergent spectrum as a function of wavelength. This is the realm of radiation transport that includes continuum and line processes. A number of simplifications are possible, usually reducing to the optically thick and thin situations. At a given wavelength, a transparent medium is referred to as optically thin with a low optical depth, whilst an opaque medium is optically thick with a high optical depth. A high opacity indicates multiple scattering and/or absorption whilst a low opacity indicates negligible scattering/absorption. For semi-opaque sources, such as dust envelopes around hot central objects (a star or central engine accretion disk), the emergent spectrum in the submillimetre and mid-infrared shows departures from a blackbody. The former, the Rayleigh–Jeans region, is due to the dust grains not radiating as blackbodies (they have a reflectivity greater than zero), while the latter is due to higher temperature grains from closer to the central heating source.

4.2.5 Free–free emission (bremsstrahlung)

As we noted earlier, the secret of understanding emission processes is to be able to reduce any situation to the specific photon–particle interactions taking place. In astrophysics, the electron is the commonest interacting particle and therefore electronic charge reactions are seen most often. Low-energy electron interactions include the case where a photon interacts with an electron bound to an atom leading to ionization (called a bound–free interaction), the inverse (free–bound = recombination) or, where free electrons of an ionized gas interact either with each other or the positive ions (free–free interactions). In all of these cases the emission spectrum is continuous. This is because a free electron can possess a large range of continuum energies; it is only when it is bound to an atom that it must occupy very specific (quantized) energy levels. *Quantized situations always produce line emission, non-quantized situations produce continuum emission.*

We now consider the case of free–free emission, a common occurrence in astrophysics that typifies the presence of a hot ionized gas. We will, however, limit our discussion of this radiation mechanism and refer to specialist texts for a detailed treatment. Furthermore, we will only consider the case of free–free emission from ionized hydrogen regions surrounding a central hot object. Another name for free–free emission is thermal bremsstrahlung, the latter word deriving from the German for braking radiation. This is a good description and comes from the de-acceleration of an electron in the field of another charged particle, radiating in the process. Observation of free–free emission is typical of an optically thin gas, although as we shall see, at some (long) wavelength the emission eventually becomes optically thick and self-absorbed.

The commonest examples a free–free emission come from the regions of ionized hydrogen surrounding the very hot O,B type stars. These regions are referred to as HII regions and radiate by the interactions of the free electron with the positively ionized hydrogen ions, i.e., protons. A typical free–free spectrum is shown in fig. 4.5 and extends from the optical to the radio depending on the density of the gas. We shall not dwell on the shape of the spectrum except to point out that the long-wavelength turnover corresponds to the wavelength at which the medium becomes optically thick to the radiation. The radiation is then self-absorbed at longer wavelengths. Self-absorption is a key feature of emission mechanisms; it occurs because no mechanism can produce more energy from particles than blackbody radiation (without amplification) and it produces the long-wavelength blackbody shape, the long-wavelength cut-off for free–free emission and also a distinctive cut-off for synchrotron emission which we shall investigate in some detail in section 4.3.1. The optically thick part of the free–free spectrum is given by

$$I_\nu = \frac{2kT}{c^2} \nu^2 , \tag{4.15}$$

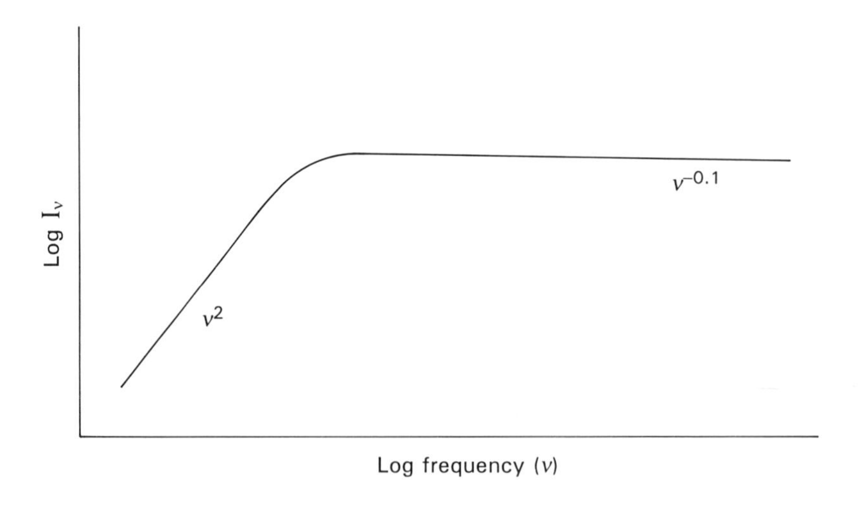

Fig. 4.5. Schematic of a free–free emission spectrum.

where I_ν is the emitted intensity as a function of frequency, ν. In this region the power-law spectral index, α, ($I_\nu \propto \nu^\alpha$) is equal to 2.0. In the optically thin part of the spectrum the intensity is only weakly dependent on frequency, with a spectral index of –0.1.

The HII region is another excellent demonstration of radiation transport. The star provides the source of ionizing photons, the hydrogen gas absorbs those of wavelengths shortward of 91.2 nm to produce the ionized nebula. The ionized hydrogen (HII) emits by free–free emission which escapes the region, thereby providing cooling. Recombination photons to higher atomic levels of hydrogen (which eventually lead to bound–bound transition such as Hα) can also escape the nebula, providing further cooling. Elements such as oxygen and nitrogen provide cooling as their line emission can completely escape the region. As we saw in section 2.6.3, line emission from [OIII] can be large and in fact gives the greenish hue to the colours of many HII regions when viewed through a telescope. (The eye is insensitive to the Hα emission which gives the dominating red colour when nebulae are photographed.)

As long as the hydrogen remains ionized, the interaction cross-section for photons is dominated by the classical electron cross-section and as this is very small, the mean free path is very large. Therefore the ionizing photons can maintain a large volume of ionized gas. However, at a particular distance from the star, depending on the stellar temperature and luminosity and the density of the gas, the intensity of ionizing photons is no longer sufficient to maintain the gas in a fully ionized state. The presence of neutral gas changes the photon interaction cross-section, which increases dramatically (to about the size of the hydrogen atom) and the mean free path is correspondingly reduced. This results in a rapid depletion of available ionizing photons and more of the gas becomes neutral. The region suddenly changes from being ionized (HII) to neutral (HI). The sharp transition delineating the fully ionized to neutral zones is characteristic of such regions and the thickness of this transition zone is approximately that of the reduced mean free path. The characteristic spherical shell of an HII region is referred to as the Strömgren sphere.

In principle, there is every reason to expect that free–free radiation might be important from ionized regions surrounding a hot central engine of an active galaxy. Indeed, there is a body of opinion which suggests that the big blue bump seen in some quasar continuum emission might be better explained by free–free emission than thermal radiation from an accretion disk. We will discuss this in more detail in subsequent chapters.

4.2.6 Summary of free–free emission

Free–free emission is indicative of hot gas and is produced wherever there is an adequate density of free electrons. There is a wide range of examples including the atmospheres of stars, hot dense plasmas (that may occur in the nucleus of an active galaxy), the ionized gas torus surrounding Jupiter's satellite Io, and the most widely known example—HII regions. An HII region surrounding a very hot star is formed by photoionization of the hydrogen by the UV photons from the star. Continuum emission from this region comes from free–free emission produced by the electrons of the gas. The free–free spectrum can extend from the optical to the radio region and is fairly flat in the optically thin regime.

4.3 HIGH-ENERGY PROCESSES

We now move to the realm of what is often called high-energy astrophysics. This considers high-energy processes involving either particles or photons. They all have one aspect in common as far as photon output is concerned: the processes are all non-thermal (in the sense that the particle distribution is non-Maxwellian). High-energy, X-ray and gamma-ray photons along with relativistic electrons are the order of the day.

4.3.1 Synchrotron radiation

Alven and Herlofson proposed the basic idea of synchrotron radiation in 1948. Seven years later the Soviet astrophysicist Iosef Shklovsky correctly proposed that the so-called non-thermal radio and optical emission originating from the Crab Nebula was of the same origin as that produced in man-made electron accelerators called synchrotrons. This radiation came to be called synchrotron radiation and it originates from the interaction between a magnetic field and a relativistic electron. Because the velocity of the electron is close to the speed of light, effects of special relativity must be taken into account in order to explain the process, and to predict such properties as the frequency of the emitted radiation, a key parameter for astronomers.

Let us begin by refreshing our memories of what happens to a low-energy electron (of velocity much less than the velocity of light) orbiting in a magnetic field. The classical electrical force (in newtons) on the electron is

$$F = B e v \ , \tag{4.16}$$

where B is the magnetic field strength in tesla, e is the charge of the electron and v is the velocity of the electron. The force, magnetic field and velocity are vector quantities (showing direction) but for our purposes we shall ignore this. The orbital (centrifugal) force is

$$F = \frac{mv^2}{r} \ , \tag{4.17}$$

where m is the mass of the electron in kilograms and r is the orbital radius in metres. Equating these and rearranging gives the gyro-radius

$$r_g = \frac{mv}{Be} \ . \tag{4.18}$$

The frequency of the emitted radiation and of the orbiting electron is given by

$$\nu_g = \frac{Be}{2\pi m} \ . \tag{4.19}$$

This is the frequency of the electromagnetic radiation emitted by the electron while spiralling in the magnetic field and it depends linearly on the field strength, B. For typical magnetic fields, the emitted radiation lies in the radio part of the electromagnetic spectrum. The radiation emitted by an orbiting electron has a dipole pattern, the radiation pattern being shown in fig. 4.6(a).

In the case of a relativistic electron the above argument is modified. Although too complex to discuss in detail here, it can be shown that the electron has a frequency of radiation that is much higher than that in the classical case. It is found that the emitted frequency of electromagnetic radiation, ν_s, for a relativistic electron is given by

$$\nu_s = \gamma^2 \nu_g \tag{4.20}$$

where the relativistic gamma factor, γ, is given by

$$\gamma = \frac{1}{\sqrt{1 - \frac{v^2}{c^2}}} \ . \tag{4.21}$$

Alternatively we can write

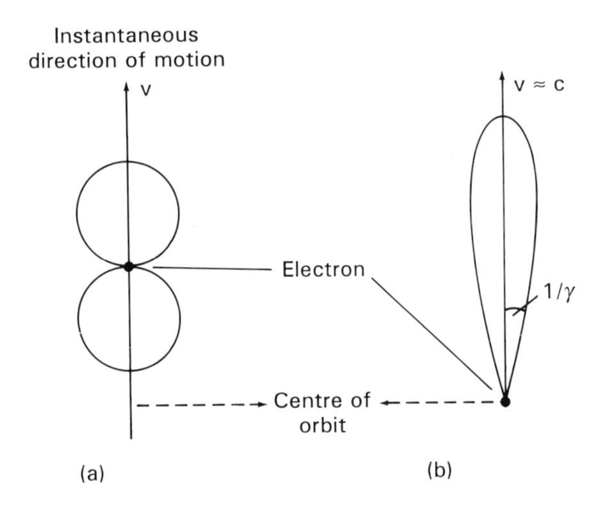

Fig. 4.6. (a) Dipole radiation pattern from a low-energy accelerating electron, (b) synchrotron radiation pattern from a relativistic electron. The radiation pattern is a dipole when seen *in the frame of reference of the electron.*

$$\gamma = \frac{1}{\sqrt{1 - \beta^2}} \tag{4.22}$$

with $$\beta = \frac{v}{c} \; . \tag{4.23}$$

The radiation is now no longer emitted as a dipole in the observer's reference frame, but is beamed in a cone of semi-opening angle approximately $1/\gamma$ in the instantaneous direction of travel as shown in fig. 4.6(b). The spiralling electron therefore sweeps this cone through a complete revolution, but an external observer sees only a brief flash as the cone is swept past. This is one reason why the frequency appears much higher; it is explained by the relativistic transformation of the electric field vector for a relativistic particle as seen from the frame of reference of an observer who is effectively stationary. Because ν_g is proportional to the magnetic field strength, B, and the electron energy is given by $E = \gamma\, mc^2$, eqn. 4.20 becomes

$$\nu_s \propto B\,E^2 \qquad \text{(Hz)} \; . \tag{4.24}$$

It is convenient at this point to re-introduce the unit for energy, the electron volt (eV), that we first met in section 2.2.1. Remember that this is the usual unit for describing the energy of a high-energy photon or particle, rather than the joule. We can make use of this in another method of defining the relativistic factor, γ. This comes from the ratio of the energy, E, of the particle and E_0, its 'rest mass' energy, i.e., the energy of the particle at rest (effectively low velocities). We therefore have

$$\gamma = \frac{E}{E_0} \; . \tag{4.25}$$

The rest mass energy of an electron is 0.511 MeV (from $E_0 = m_0c^2 \times 1.6 \times 10^{-19}$). In our above example of a synchrotron radiating electron with a γ of 10, the electron would have an energy of 5.11 MeV. Expressing the electron energy, E, in electron volts, eqn. 4.24 can be written numerically as

$$\nu_s = 0.06\,B\,E^2 \qquad \text{(Hz)} \ . \tag{4.26}$$

Because the electron is radiating, it must be losing energy, and synchrotron radiation is another means of 'cooling'. Remember, when we use the word cooling we specifically refer to an energy loss mechanism. The average rate at which energy is lost (averaging over all the pitch angles of the spiralling electrons) is given by

$$\frac{dE}{dt} = \frac{4}{3}\sigma_T\, c\, \gamma^2\, U_{mag} \ , \tag{4.27}$$

where σ_T is the Thompson scattering cross-section and U is the energy density in the magnetic field ($U_{mag} = B^2/(8\pi \times 10^{-7})$). Here we illustrate another important way of describing radiation mechanisms. The electron is essentially 'scattering' the magnetic field and thereby creating photons. As the electron is travelling near the speed of light, it sweeps out a volume of $\sigma_T\, c$ in one second and hence scatters $U_{mag}\,\sigma_T\, c$ of magnetic energy, boosting it by a factor $\sim \gamma^2$ in the process. We shall meet this again when we discuss inverse Compton emission.

In terms of variable quantities, eqn. 4.27 can be written as

$$\frac{dE}{dt} \propto B^2\, E^2 \ , \tag{4.28}$$

where B is the magnetic field strength and E is the electron energy ($E = \gamma m_0 c^2$). Numerically this becomes (for an average pitch angle)

$$\frac{dE}{dt} = -1.0\text{x}10^{-14}\, B^2\, \gamma^2 \qquad \text{W} \ , \tag{4.29}$$

with B measured in tesla. We see that the rate at which energy is lost increases in proportion to the squares of the magnetic field strength and the electron energy. This has important consequences as we shall see later in this section.

Although a single electron radiates at a particular frequency, ν_s, astrophysically we are more interested in what happens when an ensemble of electrons radiates. We can take an excellent example in cosmic rays. These are high-energy electrons and protons which permeate most of our Galaxy and to some extent intergalactic space (but that is another story). The protons have energies ranging up to at least 10^{20} eV and are the most energetic particles observed. [For comparison, particle accelerators on Earth can produce energies of $\sim 10^{12}$ eV.] At the lower cosmic ray energies, where observations are much more accurate, electrons are found to constitute ~1% the number of protons at any single energy. The energy spectrum of both protons and electrons follows a roughly power-law distribution shown in fig. 4.7. Protons are almost always unimportant in radiation mechanisms compared to electrons because being much more massive, they are that much harder to accelerate. The energy spectrum is expressed by the number of particles per energy range (N(E)dE) and is given by

$$N(E)dE \propto E^{-s}\,dE\,, \tag{4.30}$$

where s is the index of the electron power-law distribution. Technically, eqn. 4.30 refers to the differential energy spectrum as opposed to the integral energy spectrum, which is the total number of particles of energy exceeding energy E, i.e., N(>E). We will use the differential energy spectrum throughout this text.

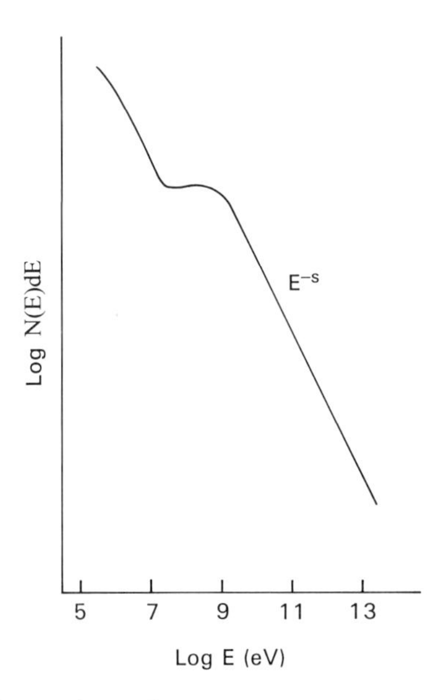

Fig. 4.7. Schematic of the cosmic ray electron energy spectrum.

To determine the electromagnetic radiation emission spectrum, consider an ensemble of relativistic electrons of differential energy spectrum given by eqn. 4.30. The emission coefficient η is the number of electrons radiating an energy dE/dt in energy range E to E + dE per unit time per unit volume of space. Assuming that the source of emission is optically thin, then the energy lost by the electrons per unit time must be manifest in photon emission of intensity $I(\nu)d\nu$ and we can write

$$I(\nu)d\nu \;=\; \eta(E)dE \tag{4.31}$$

But we know that $\eta(E)dE$ is given by $N(E)dE \times dE/dt$. Hence eqn. 4.31 becomes

$$I(\nu)d\nu \;=\; N(E)dE\,\frac{dE}{dt}\,, \tag{4.32}$$

and for the power-law energy distribution this becomes

$$I(\nu)d\nu \;=\; E^{-s}\,dE\,\frac{dE}{dt}\,. \tag{4.33}$$

We can now substitute for dE/dt from eqn. 4.28 to obtain

$$I(\nu)d\nu \;\propto\; E^{-s}\,E^{2}\,dE\,. \tag{4.34}$$

We also know from eqn. 4.26 that $\nu \propto E^2$, so 4.34 now becomes

$$I(\nu)d\nu \propto \nu^{-s/2}\,\nu\frac{\nu^{-0.5}}{2}\,d\nu \qquad (4.35)$$

or

$$I(\nu)d\nu \propto \nu^{(1-s)/2}\,d\nu\,, \qquad (4.36)$$

which we see is also a power-law in terms of the photon frequency. We can also express $I(\nu)d\nu$ in terms of the emitted electromagnetic radiation spectral index, written in the form $I(\nu)d\nu \propto \nu^{\alpha}$, where, as we saw earlier, α is the spectral index of the radiation spectrum. Hence we can finally write

$$\nu^{\alpha} = \nu^{(1-s)/2} \qquad (4.37)$$

or

$$\alpha = \frac{1-s}{2}\,. \qquad (4.38)$$

A very simple relation therefore exists between the index, s, of the electron energy distribution and the spectral index, α, of the emitted synchrotron radiation energy spectrum. This is very useful because a simple measurement of the spectral index from a synchrotron emitting source allows us to determine the value of the power-law index of the electron energy distribution.

So much for the theory, but does this simple relation appear to be borne out in astrophysical situations? In fact it does; the index, s, of fig. 4.7 is measured to be 2.55. These cosmic ray electrons permeate our Galaxy and interact with the weak magnetic field, producing synchrotron radiation that can be observed easily at radio wavelengths. The radiation is found to be a power-law with a spectral index, α, of –0.7, in excellent agreement with prediction from eqn. 4.38. Observations of synchrotron radiation from other astrophysical sources such as supernova remnants and the extended components of radio galaxies also give a synchrotron radiation spectral index close to –0.7. This strongly suggests that there is a fundamental acceleration mechanism that produces electron power-law energy spectra of index around +2.5.

Current theories favour a mechanism based on the Fermi process, whereby charged particles interact with (bounce off) sheaths of magnetic fields in space. Statistically speaking, a particle will undertake most of these interactions in the approaching direction rather than in the overtaking direction (just as being in a moving vehicle on a freeway counting the number of other vehicles passing per unit time in each direction). The approaching interactions produce particle accelerations and hence the cumulative effect is an energy increase for the electrons.

Let us now investigate further the power-law photon spectrum emitted by the synchrotron radiating electrons. The above treatment considered only the optically thin part of the spectrum and from our earlier discussions on radiation, we should expect that at some (lower) frequency, the photons emitted will become self-absorbed by the ensemble of electrons. This produces the characteristic self-absorption turnover in the spectrum as shown in fig. 4.8. For a homogeneous self-absorbed synchrotron source, it is quite easy to show that the spectral index, α, of the self-absorbed component is +2.5, although we shall not pursue this treatment here.

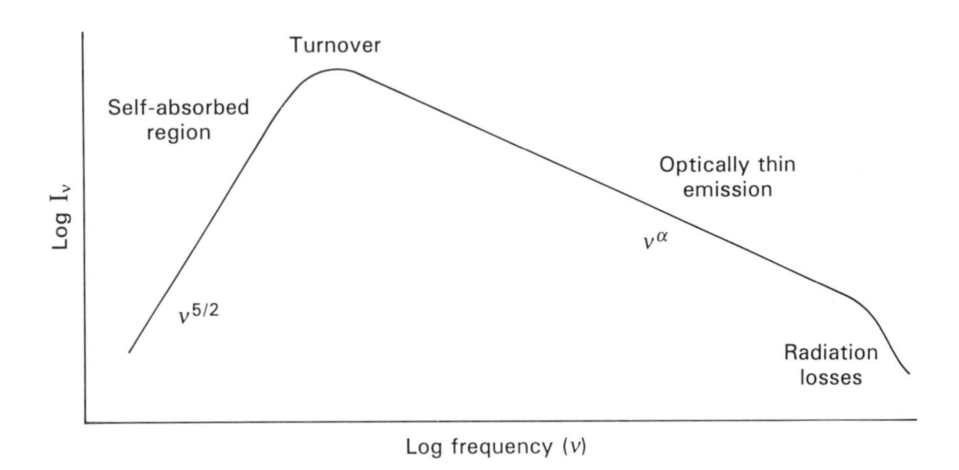

Fig. 4.8. Schematic of a synchrotron spectrum.

A useful relation concerning the turnover frequency, ν_n, is given by:

$$\nu_n = \frac{c}{\Omega}\left(\frac{2\,S_n}{k\,T_b\,\pi\,(1+z)}\right)^{0.5} \qquad (4.39)$$

where Ω is the angular size of the source, T_b is referred to as the brightness temperature of the source as measured by a radio telescope, k is Boltzmann's constant, S_n is the flux density at ν_n and z is the redshift. By making the assumption that the source size (Ω) can be approximated by $\Omega \approx \theta^2$ (where θ is the angular diameter of the source) it can be shown that

$$B = \frac{\pi^3\,\theta^4\,m^3\,\nu^5}{S_\nu^{\,2}\,e\,18} \qquad (4.40)$$

at all self-absorption frequencies. Therefore, from a measurement of the angular size and the flux density of the source at a self-absorbed frequency, the magnetic field strength can be determined.

We can also calculate the lifetime of the synchrotron radiating electrons; this will be invaluable in learning more about the production mechanisms and their locations in terms of AGNs. Consider eqn. 4.29, which gives the energy loss rate of the electrons. Rewriting this and substituting E/m_0c^2 for γ, we obtain

$$\frac{dE}{E^2} = \frac{-\,10^{-14}\,B^2}{(m_0c^2)^2}\,dt\ , \qquad (4.41)$$

which can be integrated to obtain the electron lifetime. Let us integrate between the limits of an initial electron energy of E at a time t = 0, to an energy E/2 at a time t_h, the time for the electron to lose half of its initial energy. After integrating we have

$$\frac{1}{E} = \frac{10^{-14}\, B^2}{(m_0c^2)^2}\, t_h \qquad (4.42)$$

or

$$t_h = \frac{(m_0c^2)^2}{10^{-14}\, B^2\, E} \quad \text{(seconds)}\,. \qquad (4.43)$$

We see that the relativistic electrons have a lifetime dependent on the inverse of the magnetic field squared and the electron energy.

As an example of synchrotron radiative lifetimes let us consider the Crab Nebula, the remains of a supernova that was observed to explode in the year 1054. This is one of the best examples of an astrophysical synchrotron emission spectrum, extending from the radio wavelengths to gamma-rays. The magnetic field strength can be calculated from eqn. 4.40 and gives a value of B ~5×10^{-8} T. The energy of the electrons producing the radio emission can now be determined from eqn. 4.26. This yields a value for E of 5.8×10^8 eV. The radiating lifetime of these electrons is therefore 8.7×10^4 y (from eqn. 4.43).

At the higher electromagnetic frequencies such as in the visible and X-ray wavelengths, the electrons responsible for this emission have much higher energies and correspondingly very much shorter lifetimes. From eqn. 4.26 we can see that the electron energy for X-ray synchrotron emission is related to the energy of radio synchrotron emitting electrons by

$$\left(\frac{E_{X\text{-ray}}}{E_{radio}}\right) = \left(\frac{\nu_{X\text{-ray}}}{\nu_{radio}}\right)^{0.5}, \qquad (4.44)$$

so the energy of the electrons producing X-ray photons at an energy of, say, 20 keV ($\nu = 4.8 \times 10^{18}$ Hz), compared to those producing the radio emission is

$$E_{X\text{-ray}} = E_{radio}\left(\frac{\nu_{X\text{-ray}}}{\nu_{radio}}\right)^{0.5}, \qquad (4.45)$$

which in the case of the Crab Nebula gives $E_{X\text{-ray}} = 4.0 \times 10^{13}$ eV and a radiative lifetime for these electrons of only 1.3 years. But this is much less than the age of the nebula (~950 years) and it immediately tells us that within the Crab Nebula there must be an energy source that is constantly re-energizing the electrons. This was an unsolved puzzle until the discovery of the pulsar at the heart of the Crab Nebula. The energy supply for the synchrotron emitting electrons is being supplied by the gradual spin-down of the rapidly rotating neutron star.

However, there are still difficulties in explaining the X-ray emission coming from the entire nebula because the electrons accelerated by the central pulsar have insufficient time to reach the outer edge of the nebula before radiating all their energy. We now believe the explanation is that the magnetic field in the Crab Nebula must be very lumpy rather than homogeneous. This allows regions of much lower magnetic fields to exist and hence the electrons can travel outwards along such paths with much higher radiative lifetimes. Higher spatial resolution X-ray telescopes have shown that this picture is more or less correct, the emission is far from homogeneous.

Synchrotron radiation for AGNs is confined to the jet sources which we will explore further in chapters 5 and 8. However, we know that in these sources the magnetic field is of order 10^{-4} T and therefore from eqn. 4.26 we see any 10 keV X-ray synchrotron emission, such as from blazars, comes from electrons of energy ~ 10^{12} eV. We will see that these energies are important when we discuss flaring of jet-sources and their gamma-ray emission linked to inverse Compton scattering.

A final comment about synchrotron radiation in terms of observable consequences is that unlike thermal blackbody radiation, it is linearly polarized. Calculations give values of polarizations up to almost 80% for a single electron, but for an ensemble of astrophysically radiating electrons, much lower levels are to be expected in practice. For flaring blazars (which as we saw in section 3.6 are dominated by synchrotron emission) polarization levels of order tens of percent are observed on certain occasions.

4.3.2 Summary of synchrotron radiation

Synchrotron radiation is a common feature of high-energy astrophysics and denotes the presence of relativistic electrons and a magnetic field. Unlike the low-energy classical world, where an electron radiates as a dipole, synchrotron emission from an electron is highly peaked in the forward direction giving a cone of radiation (fig. 4.6(b)) of opening angle inversely proportional to the relativistic factor (γ) of the electrons. The electrons of a given energy radiate at a specific frequency in a particular magnetic field as given by eqn. 4.26. The electrons lose energy as they radiate, given by eqn. 4.29, which leads to a radiative lifetime expressed by eqn. 4.43. The lifetime is dramatically reduced as the magnetic field and electron energy increase.

A power-law electron energy distribution produces a power-law synchrotron radiation emission spectrum and the index, s, of the differential electron energy distribution is linked with the optically thin radiation energy spectral index, α, by a very simple expression (4.38). A synchrotron emission spectrum from such an ensemble of electrons is expected to be power-law in shape (fig. 4.8) and observation of α then allows s to be determined. For many sources α is found to be around -0.7 giving a value of ~ +2.4 for s. At low frequencies the synchrotron radiation spectrum becomes self-absorbed, the spectral slope in the optically thick regime is then $\alpha = 2.5$. At very high frequencies, the spectrum departs from the power-law shape due to the severe electron radiative energy losses.

The observation of synchrotron emission shows that electrons have been accelerated to relativistic energies. The higher the frequency of the observed synchrotron emission, the more energetic the acceleration mechanisms that have operated to produce the higher energy electrons. The similarity of the electron energy index (s) in several astrophysical contexts points to a common mechanism at work. Measurements of synchrotron radiation from sources allow us to make deductions about a number of parameters of the radiating region. The X-ray synchrotron emission observed from some blazars requires electron energies of ~10^{12} eV, which are readily attainable as demonstrated by cosmic rays.

4.3.3 Compton and inverse Compton processes

In laboratory physics, the Compton effect is observed when a high-energy photon interacts with a low-energy electron. The electron energy is boosted and the photon loses energy. In terms of sources and sinks of energy, Compton emission is a sink of energy for the photon field and a source of energy for the electrons. This effect was experimentally discovered in 1923 and is often referred to as Compton scattering, because the photon is effectively scattered, losing energy in the process and hence changing wavelength/frequency.

Figure 4.9 shows a diagrammatic representation of the Compton effect. In following the interactions outlined below, one must be very careful about the frame of reference in

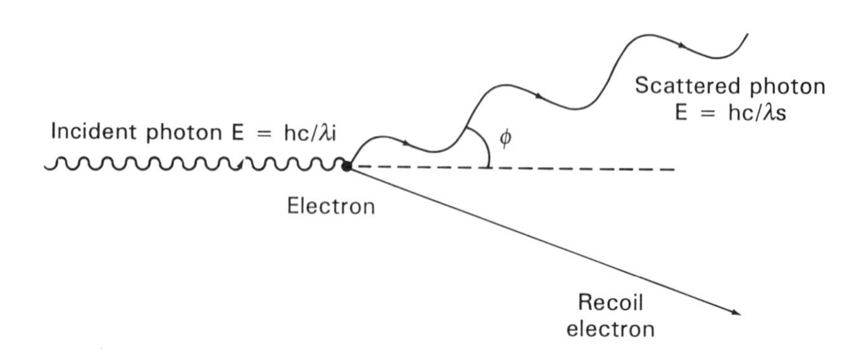

Fig. 4.9. Schematic diagram illustrating Compton scattering.

which the interaction is viewed. It can be shown that the relation between the incident photon wavelength, λ_i, and the scattered wavelength, λ_s, is given by

$$\lambda_s - \lambda_i = \frac{h}{m_o c}(1 - \cos\phi) \quad (4.46)$$

where m_o is the rest mass of the electron, h is Planck's constant and ϕ is the angle through which the photon has been scattered. This equation can be rewritten as

$$\lambda_s - \lambda_i = 2\lambda_C \sin^2\frac{\phi}{2} \quad (4.47)$$

where $\lambda_C = (h/m_0 c)$ is called the Compton wavelength of the electron (from $E = h\nu = hc/\lambda = m_0 c^2$) and has a value of 2.4×10^{-12} m. The lower energy case for Compton scattering is classically termed Thompson scattering, and it should be borne in mind that it is the same fundamental physical interaction process although the energy of the photon remains constant in this case. Compton emission is important in those regions in which electrons are immersed in an intense high-energy photon field. Such conditions occur near the accretion disk of a black hole and the subsequent particle acceleration acts as a cooling process for the photons.

In astrophysical situations, the inverse process is also possible due to the presence of electrons with extreme energies. Not surprisingly this is referred to as inverse Compton

(IC) scattering. Here, a high-energy electron interacts with a photon and the photon energy is boosted at the expense of the electron energy. inverse Compton emission is therefore a source of energy for a photon field and a cooling process for the high-energy electrons.

Compton emission and inverse Compton emission can be treated similarly as long as the correct frame of reference is used. The complex details of the relativistic transformations in going from Compton scattering to the inverse Compton case are beyond the scope of this book and the reader is referred to the specialized texts in section 4.6. Nevertheless, inverse Compton scattering is worthy of closer examination for a number of reasons, in particular the close link with synchrotron emission. We can note that in the case of synchrotron emission a relativistic electron interacts with a magnetic field, whereas for inverse Compton emission, a relativistic electron interacts with a photon field.

In 1865 the Scottish mathematical physicist James Clerk Maxwell elegantly showed that electromagnetic radiation propagates at a constant velocity (the speed of light in vacuum) and is composed of two vector components: the magnetic vector and the electric vector. These imply the presence of two fields, the electric and magnetic fields. A thorough study of light (more precisely e-m radiation) is only possible alongside a detailed understanding of the operation of what are known as Maxwell's equations, which are the basis of undergraduate courses in electromagnetism. Although a thorough grasp of these equations are essential to a student's deep understanding of a central core of physics, this book is not the place to pursue this. Reference works which I have found both understandable and enjoyable to read are provided at the end of the chapter. We shall just say that the inverse Compton process is the electric analogue of the synchrotron process. Because we have dealt with synchrotron emission in some detail we shall list one or two equations central to IC radiation.

The equation governing the rate of energy loss of a relativistic electron interacting with a photon field of energy density, U_{rad}, is given by

$$\frac{dE}{dt} = \frac{4}{3}\sigma_T\, c\, \gamma^2\, U_{rad}\ , \qquad (4.48)$$

where σ_T is the classical, or Thompson interaction cross-section. We note that eqn. (4.48) has precisely the same form as the equation for synchrotron emission (eqn. 4.27), except that the magnetic energy density is replaced by the photon energy density. Again, the spectral shape of the emitted radiation spectrum depends on the electron energy distribution. An important point to note is that in the interaction, the photon number density is conserved and the photons are boosted to higher energies by a factor given by

$$\nu \approx \gamma^2\, \nu_0 \qquad (4.49)$$

where ν_0 is the frequency of the 'seed' radiation photons and ν is the boosted frequency.

As an example, let us take the seed photons as originating from the synchrotron emission of a flat spectrum blazar radiating at a wavelength of ~1 mm (300 GHz) with an electron energy given by $\gamma = 1{,}000$, we find that the boosted photon has a frequency of 3×10^{17} Hz, which lies in the kiloelectron volt X-ray region. With a similar electron energy, optical seed photons would then be boosted to MeV gamma-rays, while X-ray photons could be boosted to explain the TeV gamma-ray emission observed from objects such as Mkn 421. We therefore see that the inverse Compton process provides an obvious source for producing X-ray and gamma-ray photons from a population of relativistic electrons. But remember, there are two key requirements for inverse Compton emission to be observationally important: a population of relativistic electrons and a photon field. Both must be present with significant energy densities. Again, we do not have the space to pursue the calculation of the interaction cross-section but remind the reader that in calculating the luminosity of a source of Compton emission, or any other source of emission, the interaction cross-section is a critical parameter. However, it also turns out that for these particular photon–particle interactions, calculation of the interaction cross-section is far from simple, so we shall merely reiterate the idea that a high-density plasma of relativistic electrons and a strong photon field are a potential source of astrophysical inverse Compton emission.

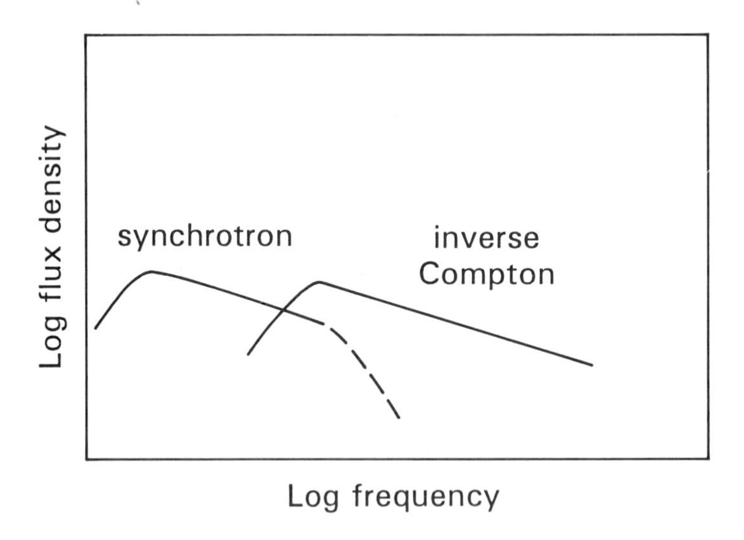

Fig. 4.10. Schematic diagram demonstrating synchrotron-self-Compton emission. Note how the inverse Compton component has the same spectral shape as that of the synchrotron.

This brings us to a key aspect for those AGNs in which synchrotron and inverse Compton emission have complementary roles. Consider a self-absorbed synchrotron source, for example as found in a radio-loud quasar (see chapter 8). At first sight it might be anticipated that the temperature of the electrons in a region of space in which the synchrotron emission is self-absorbed can increase without limit, because of the upward spiral of photon–electron–magnetic field interactions. Let us expand on this. The synchrotron photons scatter the electrons to higher energies, which immediately produce more synchrotron photons by interaction with the magnetic field, and so on. Can this continue indefinitely? We have probably guessed that the answer is no. This upward spiral is eventually halted by the onset of inverse Compton emission which provides an internal cooling mechanism for the electrons. This occurs when the electron and photon energy densities are sufficiently large that the high-energy electrons producing the synchrotron emission now upscatter their own synchrotron photons to higher frequencies. Unlike the low-energy (long-wavelength) photons of the self-absorption synchrotron radiation that are trapped because of the high optical depth, these high-energy inverse Compton photons have a very low optical depth in the medium, and promptly escape. In so doing they provide a direct cooling mechanism for the electrons and an upper limit of the electron temperature is predicted to be $\sim 10^{12}$ K.

In fact this is borne out as shown by radio observations of compact radio sources as we shall see in chapter 8. This process of synchrotron self-Compton (SSC) radiation is anticipated to be a key feature of radio-loud (jet-like) AGNs and as such we need to know what to look for. An inverse Compton spectrum should show the same spectral energy dependence as the synchrotron spectrum from which it derives, but upshifted in frequency as we saw in eqn. 4.49. Schematically this is shown in fig. 4.10. Fig. 5.3 shows a good example of long wavelength synchrotron radiation from the blazar 3C279 and what we believe to be short-wavelength inverse Compton radiation.

4.3.4 Pair production

Pair production is the final high-energy process that we shall consider. This exotic reaction is rare and only important in regions of very high gamma-ray photon energy density. The process consists of a very high-energy photon, interacting with the field of a nucleus (or another photon or a magnetic field) and being converted into an electron-positron pair. It is a spectacular source of particles and a dramatic sink for the radiation field. Schematically it can be thought of as

$$\gamma + (\gamma \text{ or nucleus}) = e^{+} + e^{-} \tag{4.50}$$

Because both energy and momentum must be conserved simultaneously in this process, the reaction involves either another photon or a particle, which is why the second gamma-photon is shown in brackets. The inverse of pair production is electron–positron annihilation with the emission of two photons.

Showing that a photon cannot spontaneously produce an electron–positron pair, but requires the presence of another field to conserve simultaneously both momentum and energy, is a favourite problem posed in an undergraduate physics course. The solution goes as follows. Imagine that the electron and positron of rest mass m_o and energy E_{ep} are produced travelling in the same direction as the incident photon of energy E_γ. Conservation of energy then gives

$$E_\gamma = E_{ep}, \quad \text{i.e.,} \quad h\nu = 2\gamma m_o c^2 . \tag{4.51}$$

But the momentum, p_{ep}, of the electron–positron pair is

$$p_{ep} = 2\gamma m_o v \tag{4.52}$$

whereas the momentum, p_γ, of the incident photon is from eqn. 4.51

$$p_\gamma = \frac{h\nu}{c} = 2\gamma m_o c . \tag{4.53}$$

Equating eqns. 4.52 and 4.53 we see that because v cannot equal c, momentum and energy cannot be conserved simultaneously in free space. The presence of another body or field is required to take up some of the recoil momentum and this can either be the field of a nucleus or another photon. In the case of AGNs, it is usually the former.

Because the electron and positron each have a rest mass energy $m_0c^2 = 0.511$ MeV, this places the interaction mechanism firmly in the low-energy gamma-ray or extreme X-ray regime. For the photon emission case, the photon will be an emission line at the rest energy of the electron/positron, 0.511 MeV. We will discover in section 7.3.4 that this has now been observed from the general direction of the Galactic Centre, showing that positron production processes must be at work somewhere in the region.

In passing, we note that ultra-high-energy gamma-rays interact with the very low energy (10^{-3} eV) but very high-energy density (≈ 0.3 eV m^{-3}) photons of the cosmic background radiation field which permeates the Universe. This limits the distance that these ultra-high-energy gamma rays can travel before being converted into an electron-positron pair. Calculations suggest that gamma-rays of energy in excess of 10^{15} eV should be Galactic in origin.

4.3.5 Summary of Compton processes and pair production

We will see the importance of both Compton and inverse Compton processes in later chapters. Inverse Compton radiation can be thought of as being the electric analogue to its magnetic counterpart, synchrotron radiation. It becomes important for regions where the energy density of high-energy charged particles (which are usually electrons in astrophysical situations) is large, along with the presence of a high-energy density photon field. It is a source of continuum emission that is expected to be observed from the infrared to gamma-ray regions (depending on the photon field that is interacting with the electrons). For highly compact synchrotron sources, such as the self-absorbed cores of the radio jets seen in VLBI studies, its cooling ability provides a natural upper limit to the temperature of a self-absorbed synchrotron region of around 10^{12} K. This has now been verified observationally. Compton processes are important when there is a high photon field and a supply of electrons, such as in the immediate surroundings of the central engine of an AGN.

Pair production is the production of an electron–positron pair from very high-energy photons. It is rare but could be important in the central regions of AGNs where very high photon energies are known to be present. Pair production presents a sink for high-energy photons and a source of high-energy charged particles. The positron emission will usually be followed by the inverse reaction, electron–positron annihilation, which produces line emission at a photon energy of 0.511 MeV. This has now been detected from regions close to the Galactic Centre.

4.4 LINE EMISSION AS A DIAGNOSTIC

4.4.1 Introduction

The basis for line emission was presented in section 2.6, and by now you should be familiar with forbidden and permitted emission processes and understand that line emission and absorption are fundamental processes in physics. They will feature prominently in our investigations of the observable phenomena from active galaxies and the subsequent deductions concerning the energy sources powering them. If you are in doubt, go back and read section 2.6.

A special branch of physics deals with the prediction of the wavelengths of line emission, line strengths and line intensity ratios for differing physical conditions; this is the realm of atomic spectroscopy. It is a highly complex discipline and experiments to determine the accuracy of the model calculations are extremely difficult, due to the very special conditions applying in space that are hard to reproduce in the laboratory. Frequently, lines observed in astronomical spectra are used to test the model predictions and so refine the process further. Much progress has been made in this field during the last twenty years, due in no small part to the drive produced by astronomy and the availability of fast computers.

The great difficulty in applying a simple theory to situations occurring in space is that the conditions in space are far from simple. Indeed, even a relatively local gaseous nebula is an incredibly complex cauldron of interactions. Consider the factors that need to be taken into account in order to produce a satisfactory model of a gas cloud illuminated by a central source. The gas cloud is composed of an ensemble of atoms of differing elemental species, with some particular abundance ratios which are *a priori* not known, and the atoms will be at a temperature or range of temperatures which is also not known. Furthermore, neither the geometry nor density are known (in the sense that the atoms may be in a homogeneous region, a clumpy region or a stratified region) and a magnetic field may be permeating the gas. Finally, the spectral energy distribution and strength of the illuminating photon field, both of which are often unknown, will have a critical bearing on the physical conditions. For all the situations in which we are interested, the photon field is responsible for pumping energy into the system and we will be dealing with photoexcitation and photoionization conditions.

Given the daunting list of unknowns, deducing a perfect model is impossible. Astrophysicists naturally attempt to reduce the number of unknowns and the only sure way of doing this is by observation. This means attempting to observe all the spectral lines that are emitted from the region, preferably with spatial information, along with the underlying continuum that is powering the region. This is not an easy task, but one that has absorbed years of telescope time on the world's major facilities. Excellent progress has been made and we shall review the current state of knowledge when we discuss the emission line regions from the central parts of an AGN in chapter 6.

We should say from the outset that the modelling task is immense, because as we have just seen, there are a huge number of potential interactions to consider. It turns out that although the gas density and geometry are crucial to the fine details of the theoretical modelling, they are very hard to determine observationally. Many models assume a range of abundances and iterate to a range of solutions, finally settling on the one that gives the best fit of output parameters to those observed. The models contain highly sophisticated physical arguments, but the solution to the overall problem is bound by observational limitations rather than theoretical capability. Nevertheless, in nearly all the situations considered, the region is assumed to be in local thermodynamic equilibrium, which is probably an adequate approximation for most cases most of the time.

4.4.2 The centrally photoionized gas model

The simplest generalized astrophysical situation is a photoionized cloud of hydrogen gas surrounding a hot star; this is the relatively well understood 'HII region' treatment. The differences of chemical composition from a pure hydrogen cloud can be taken into account by modifications to the simple model. The HII region model calculations predict an ionized zone bounded by a sharply defined edge called the Strömgren sphere. However, as we shall see in chapter 6, the analogy breaks down for AGNs because the Broad-Line Region (BLR) is far from being a homogeneous zone. Nevertheless, ignoring this complicating factor, we will give a brief overview of the steps taken in deducing the physical parameters pertaining to a region of photoionized gas using classical recombination line theory because this has a direct bearing on the deductions from the observations of the BLR described in chapter 6. The treatment of a centrally photoionized medium is complex and is presented in many textbooks; I have found the description given by Weedman (see his section 4.6) to be excellent and this line of attack will be followed in the overview below.

Starting from the assumption of local thermodynamic equilibrium (LTE) we need to be able to determine the equations of state for the material. These relate conditions of pressure or energy density to the variables of temperature and mass density. This allows us to determine the luminosity in each emission line of the radiating gas. But to do this we need to re-introduce a term which we first met in discussing synchrotron radiation. This is the emission coefficient. For line emission, this is written as j_{mn}, where the subscripts refer to the transition from atomic state m to n. The emission coefficient is in units of watts per cubic metre (W m^{-3}) and is a measure of the power radiated by the line per unit volume of the gas. The total luminosity of this line is just j_{mn} multiplied by the volume, V, of the gas. In active galaxies, unlike galactic regions, the volume of the BLR subtends an angle smaller than can be mapped by current telescopes; differing density structures will not be immediately apparent and so homogeneity and a spherical geometry are first assumed. In what follows it is assumed that all the radiated intensity escapes from the source unattenuated, i.e., there is no absorption by dust. Where dust absorption is *clearly* evident, it is taken into account. Observations of infrared lines have dramatically improved the situation, because they are much less susceptible to dust absorption and can be observed from the innermost regions (see chapter 6).

The emission coefficient is related to the physical parameters of the radiating atoms by

$$j_{mn} = A_{mn} E_{mn} N_m \tag{4.54}$$

where A_{mn} is the spontaneous transition probability (in units of s^{-1}) for the transition from state m to n, $E_{mn}(= h\nu_{mn})$ is the energy of the photon emitted by the electron in this transition, and N_m is the number of atoms per unit volume in the atomic state m. The values of A_{mn} and E_{mn} are known from atomic physics and can be assumed to be constants; the unknown variable is the number of atoms, N, in state m. This is usually called the population number, because it shows how the various states of the atom are populated. This is not at all easy to determine and relies on concepts of atomic physics beyond the scope of this book. Nonetheless, equations which predict the population distribution of the levels can be obtained as a function of the temperature of the ensemble of atoms using Boltzmann's Law. Degeneracy is handled using statistical weights, described further in the specialized texts in section 4.6.

We now need to refer again to Boltzmann's equation (eqn. 4.9) which can be modified to take into account the quantum mechanical description of the atomic levels and can be rewritten in the form

$$\frac{N_m}{N_n} = \frac{g_m}{g_n} e^{-(E_{mn}/kT)} , \qquad (4.55)$$

where g_m and g_n are referred to as the statistical weights of the final states (levels) m and n, respectively, and E_{mn} is the difference in energy between levels m and n. The statistical weights refer to the equivalent number of quantum states allowed in the particular level. We assume local thermodynamic equilibrium (LTE) exists otherwise all the upper levels or all the lower levels would *eventually* become populated and line emission would not be observable. Because line emission is seen to produce the same lines with the same intensity ratios from a wide variety of objects, and in particular for ionized regions within our Galaxy, the equilibrium condition must be a reasonable assumption. The necessary conditions for equilibrium basically reduce to the requirement that the rates of populating and depopulating a given level are equal. Given LTE, we can then use the principle of detailed balance, which states that each microscopic process is balanced by its inverse. This can also be interpreted by saying that in going from a start reaction to an end point, the intermediate pathway is irrelevant. This then allows us to solve the case of ionization equilibrium and we will proceed to give a very simplified example of such a process, which will provide a basis for understanding the observations of the BLR and the introduction of the Standard Model in section 6.2

4.4.3 The Case B approximation

The set of equations which in principle need to be taken into account is intimidating and consists of up to eleven terms describing the number of ways of populating and depopulating levels. Included in these are the population of the level from direct recombinations from the ionized state, radiative transitions from higher levels, stimulated emission and radiative absorption, collisional excitation and de-excitation and others. A key factor in solving such a complex equation is to identify terms that can be ignored in the situation in question. This leads to suitable simplification and a particular example for AGNs is the use of what is termed the Case B condition. This is an approximation applicable for large optical depth conditions for Lyman lines. Every Lyman-line photon is scattered many times in the gas, and ultimately this results in all the Lyman lines being converted into a Lα photon, a Balmer line photon and other photons as well. In principle, the latter can immediately escape while the Lα photon is 'trapped' (by successive scatterings) for a long time and therefore has a much higher probability of being destroyed by any dust present in the medium.

The individual values of N_m or N_n cannot be found when applying Case B conditions to an HII region. However, using two spectral lines allows the ratio to be determined and thence the relative numbers of atoms in the two levels and thus the temperature. For temperatures between 20,000 K and 5,000 K, and for a range of densities, the ratio of the strengths of the Balmer lines, Hα/Hβ, is predicted to lie between 2.75 and 3.0. It is very satisfying to find that the Hα/Hβ line ratios observed in many HII regions lie precisely in this range. The ratio of the strengths of the Hα/Hβ lines is usually referred to as the Balmer decrement. If the observed value of the Balmer decrement turns out to be wildly different from the predicted range then this brings up a number of possibilities. The most obvious is that either the Case B approximation is invalid for the object in question. Alternatively, (or additionally) dust is preferentially absorbing the shorter wavelength Hβ, thereby giving a larger value for the Balmer decrement. It turns out that for AGNs, the value of the Balmer decrement always exceeds the 2.75 to 3.0 range. We will revisit this problem when we discuss the BLR in section 6.2. In passing we can also note that the assumption of the Case B condition implies that the ratio of Lα/Hα should be about 13.

For the purposes of our discussion below, we shall assume that the Case B approximation is valid and we have a dust-free environment. The easily measured Balmer lines can then be used to determine the luminosity of the underlying continuum radiation. This latter point is very important and arises in a situation where the ionized region is said to be radiation bounded. This means that all the ionizing radiation from the central source is absorbed by the atoms of the surrounding gas which is thus optically thick. Hence the ionizing continuum source is hidden from the external observer. However, because of the Case B condition, the number of recombinations per unit volume per second is equal to the number of Lα photons produced per unit volume per second which in turn is equal to the number of Balmer photons produced per unit volume per second. Therefore, measuring the Balmer photons gives the luminosity of the ionizing continuum. However, this does not give the spectral energy distribution of the ionizing continuum.

Line ratios can also be used to calculate the structure, gas mass and size of the emitting gas region to be calculated. These are extremely worthy goals and highlight the importance of line emission in photo-ionized regions. We shall pursue these further in chapter 6 when discussing the BLR of active galaxies.

4.4.4 Use of the forbidden lines

Let us now move on to consider forbidden transitions. These give a key indication of the electron density and to some extent the temperature of the ionized region (a good example is the ratio of the 500.7/436.3 lines of [OIII]). The major difference between the permitted and forbidden lines is that the former arise from recombinations to higher levels followed by cascading downwards, whilst the latter are the result of collisional excitations from the lower states. We remember, from section 2.7, that the term 'forbidden' just means the lines are not seen under normal laboratory conditions. In terms of quantum mechanics, a forbidden line has a small value of the transition probability, A ($A_{mn} = 1/\tau$). This means that the atom will spend a very long time, τ, in the excited state, m, before decaying radiatively to level n. In high-density environments (such as in the laboratory), electron collisional de-excitation will occur before radiative decay and so no line will be observed. In low-density situations, on the other hand, the radiative forbidden decay is seen. Hence the forbidden lines are very sensitive to *density* and it is this that makes them so vital in determining densities in astrophysical plasmas.

The collisionally excited lines act as thermostats for the gas. When the gas heats up there are more collisions, resulting in more atoms in collisionally excited states and this excess energy is radiated away via forbidden lines. Since the region is optically thin to these lines, they can escape directly, thereby cooling the gas in the process. But as the gas cools, the number of collisions decreases, the number of atoms in collisionally excited states decreases, the number of emitted forbidden photons decreases and so the radiation cooling falls. By this thermostat-like process, the kinetic temperature of the free electrons is maintained at an electron equilibrium temperature, whereby the loss of electron energy by collisions matches the forbidden lines escaping the region and the electron kinetic energy is in equilibrium with the ions in the region.

If we consider a number of such regions, then if the input ionizing spectrum and the cooling mechanisms are the same (the latter is equivalent to requiring the chemical abundances to be the same), we would predict that all these regions should show the same electron temperature at equilibrium. For HII regions this is calculated to be around 10,000 K. Again, observations show that this seems to hold for most HII regions, giving confidence that the theory and modelling are on the right track. When differences are seen, it is found that the presence of a higher abundance of the heavy elements leads to greater cooling. This is easily understood in that a small increase in their abundance provides a much greater number of collisionally induced transitions available.

So much for the temperature, but what about the density in the region? We saw above that the forbidden collisionally induced transitions are sensitive to density. Again the situation is treated as an equilibrium process and although it is exceedingly complex, a standard treatment is to work through the solution for a two-level ion of levels m and n (with n higher than m). The equilibrium condition is that the number of transitions into a certain level must equal those out of it and the equilibrium equation for level 'm' is given by

$$\underset{\text{(in)}}{N_n A_{nm} + N_n N_e \sigma_{nm}} = \underset{\text{(out)}}{N_m N_e \sigma_{mn}} \qquad (4.56)$$

and for level 'n' by

$$\underset{\text{(in)}}{N_m N_e \sigma_{mn}} = \underset{\text{(out)}}{N_n A_{nm} + N_n N_e \sigma_{nm}} \qquad (4.57)$$

where σ_{mn} is the collisional cross-section for excitation from level m to level n and σ_{nm} is the de-excitation cross-section from n to m, A is again the transition probability and N_e is the number density of the free electrons. We can immediately see, as might be expected for an equilibrium situation, that eqns. 4.56 and 4.57 are symmetric and therefore cannot be solved for a given value of N. They can be solved for a ratio of N_m/N_n and after a little algebra we obtain

$$\frac{N_n}{N_m} = \frac{\sigma_{mn}}{\sigma_{nm}}\left(\frac{A_{nm}}{N_e\sigma_{nm}} + 1\right)^{-1}. \qquad (4.58)$$

The first term in eqn. 4.58 is by far the most important and depends directly on the electron density. It turns out that the condition for either permitted or forbidden line emission to dominate, depends on whether N_n/N_m is much less or much greater than unity, respectively. In other words, collisional de-excitations are comparable to radiative de-excitations and therefore become important at this point. We can then redefine the electron density in terms of a 'critical density' ($N_{e\,crit}$). This is very important and is defined by

$$N_{e\text{-crit}} = \frac{A_{nm}}{\sigma_{nm}} \tag{4.59}$$

giving

$$\frac{N_n}{N_m} = \frac{\sigma_{mn}}{\sigma_{nm}} \left(\frac{N_{e\text{-}crit}}{N_e} + 1 \right)^{-1} . \tag{4.60}$$

To determine the relative importances of the transitions, we need more than just the populations in the levels, we need the emission coefficient corresponding to the transition. In dealing with levels m and n, clearly the transition will be from n to m in order to produce a photon and so we consider the emission coefficient, j_{nm}, given by

$$j_{nm} = N_n A_{nm} E_{nm} \tag{4.61}$$

where E_{nm} is the energy difference between levels n and m. Substituting for N_n from eqn. 4.60 we obtain

$$j_{nm} = N_m A_{nm} E_{nm} \frac{\sigma_{mn}}{\sigma_{nm}} \left(\frac{N_{e\text{-}crit}}{N_e} + 1 \right)^{-1} . \tag{4.62}$$

If we take a single two-level system, with m being the ground state, then N_m is the density of the ground state. This is related to the density in the entire region, and hence is proportional to the electron density by $N_m = X \; N_e$ where X is some constant of proportionality. Taking the high-density situation of $N_e >> N_{e\text{-}crit}$, then eqn. 4.62 simplifies to

$$j_{21} = X N_e A_{nm} E_{nm} \frac{\sigma_{mn}}{\sigma_{nm}} \tag{4.63}$$

The only variable on the right-hand side of eqn. 4.63 is the electron density N_e. Therefore in the high-density situation the emission in the forbidden lines increases linearly with N_e but for the recombination lines at the same densities, the emission coefficient is proportional to N_e^2. Therefore in the *high-density situation, where the electron density greatly exceeds the critical density, the recombination line emission dominates the forbidden lines.*

Now let us consider the low density region. Here, $N_e << N_{e\text{-}crit}$ and 4.62 reduces to

$$j_{nm} = X N_e A_{nm} E_{nm} \frac{\sigma_{mn}}{\sigma_{nm}} \left(\frac{N_e}{N_{e\text{-}crit} + N_e} \right) \tag{4.64}$$

or

$$j_{nm} = X N_e A_{nm} E_{nm} \frac{\sigma_{mn}}{\sigma_{nm}} \left(\frac{N_e}{N_{e\text{-}crit}} \right) , \tag{4.65}$$

but since from eqn. 4.59 we know that $N_{e\text{-}crit} = \dfrac{A_{nm}}{\sigma_{nm}}$ we have

$$j_{nm} = X N_e^2 E_{nm} \sigma_{mn} , \tag{4.66}$$

and now we see that the forbidden lines also scale as N_e^2 and so will be readily observable.

For a particular transition in an ionized gas that is below the critical density, the ratio of the forbidden line of the transition to the hydrogen lines (for example) remains constant and is independent of the electron density N_e. However, once the critical density has been well surpassed, the ratio of the forbidden line to that of hydrogen falls rapidly with increasing electron density. This neatly explains the observations that in many different objects the ratios of the forbidden lines to hydrogen are the same.

We have identified a powerful potential diagnostic for the electron density in the ionized region. The above example was for the simple case of a two-level ion, whereas astrophysically, there are many transitions to be considered. However, in principle the differing ratios of forbidden lines to hydrogen should give a good handle on the particular electron densities present in the ionized gas. For example, the critical density for the forbidden [OIII] doublet (500.7, 495.9) is about 10^{12} m^{-3}, for semi-forbidden CIII] (190.9) it is around 10^{16} m^{-3}, whilst for CIV (154.9) it is 10^{22} m^{-3}. Another interesting aspect of the [OIII] doublet is that when collisional de-excitations are negligible, the ratio of the line strengths is given by the ratio of the transition probabilities. This is almost exactly equal to 3, and it is indeed observed in practice.

4.4.5 Summary of line diagnostics

Line emission is extremely important. Observations of the lines, which originate from the quantised world of the atomic energy levels of an atom, allow deductions to be made about the physical state of the emitting gas. We saw that the situations in space are very complex, but that suitable approximations can be made that seem to work well for understanding ionized hydrogen (HII) regions. We moved on to consider the details of the radiation transport mechanism for lines and saw that we could define a critical density for the electrons. This then produced a diagnostic for determining the electron density, or at least whether it was in the high- or low-density situation compared to the critical density. In the former, the permitted lines dominate the forbidden line emission, whereas in the latter, the ratio is constant with density and can be relatively easily determined.

We should note that the treatments of photoionized regions undertaken in section 4.4.4 have concentrated on the use of optical lines because they have given a wealth of information and set the scene for our study of the BLR and NLR in chapter 6. However optical line ratios are severely hampered by unknown dust absorption, and near infrared lines are now being used as a diagnostic, especially in those regions where dust absorption is believed to be important. Also, with the advent of X-ray and gamma-ray spectroscopic satellite capability, lines at these energies are becomingly increasingly important as diagnostic tools for the very central regions of an AGN, those regions very close to the accretion disk of the black hole. These X-ray lines originate from the inner shells of 'heavy' elements and will be explored when we study the details of the central engine in section 6.

In section 4.3.4 we discussed the possibility of electron–positron pair production and we also noted that the inverse reaction, the production of an annihilation line at 0.511 MeV, has now been observed from the Galactic Centre region. Other lines are present in the gamma-ray regime due to radioactive decay of elements produced under catastrophic conditions. These are commonly found in supernova remnants and will not feature further in this book.

As far as line widths are concerned, we will just recount that the intrinsic width of a line emitted by a single atom is *extremely* narrow, and is related to the inverse of the transition probability. Line broadening occurs because we observe the lines from a large number of atoms that do not have the same velocity along the line-of-sight. We therefore see a collection of lines of slightly different wavelengths and the resultant Doppler broadening is related to the line-of-sight velocity spread of the atoms.

4.5 SUMMARY

In this chapter we have seen how thermal blackbody emission is produced by an optically thick source and the form of the spectral emission was described. A useful equation, known as the Wien Displacement Law (eqn. 4.8), shows the relation between the wavelength of the maximum of the emission and the absolute temperature of a blackbody. Blackbody radiators are rare and usually there is an emissivity component that modifies the output spectrum in the Rayleigh–Jeans region.

Interactions are crucial in astrophysics, the degree of interaction being dependent on the physical conditions pertaining. These include the energies and energy densities of the particles, photons and electromagnetic fields. The probability of an interaction is related to a quantity that is easy to visualize, the mean free path. Interactions and passage through a medium lead us into the realm of radiation transport, optical depth and opacity. When a region is completely or partially ionized free–free emission (thermal bremsstrahlung) is often important. This also has a characteristic output spectrum.

High-energy interactions of charged particles (electrons) and magnetic or photon fields were next presented in the form of synchrotron and inverse Compton processes. In each case, the electron loss mechanism was described and the form of the emergent photon spectrum shown. Equations for the rate of loss of energy and subsequent electron radiative lifetimes were presented and these will be used in astrophysical situations in later chapters. The production of electron–positron pairs from interactions of high-energy charged particles interacting with a photon were explained and the relevance of this to regimes of very high energy and high energy density was described.

Line emission was reviewed and extended further with a discussion of simple photoionization models for a radiation-bounded system. Forbidden and permitted transitions were explained in some detail and the powerful use of certain lines as density indicators was explained in terms of their critical densities.

Finally, we can remind ourselves that when we observe radiation from a astrophysical source, this represents the cooling of the region. The measure of the amount of cooling, gives the heating energy assuming the region is in a steady-state situation, which is the usual case. We are now in a good position to explore the details of the central engine, the power source of active galaxies.

4.6 FURTHER READING

General and reviews

Aller,L.H., *The physics of thermal gaseous nebulae*, Astrophysics and Space Science Library, Vol. 112, Reidel, 1987. (Meaty reading, excellent coverage.)

Davies,P., Ed., *The new physics*, Cambridge University Press, 1989. (An excellent series of reviews of modern concepts in physics and astrophysics.)

Harwit,M., *Astrophysical Concepts,*. Springer–Verlag, 1988. (An excellent astrophysics text, very thought provoking.)

Longair,M.S., *High Energy Astrophysics. Vol. I Particles, photons and their detection*, Cambridge University Press, 1992. (A superb text book.)

Osterbrock, D.E., *Astrophysics of Gaseous Nebulae and Active Galactic Nuclei,* University Science Books, 1989. (A really excellent treatment of radiation processes including the extension to the central regions of AGN. Told by the master.)

Rybicki,G.B., & Lightman,A.P., *Radiative processes in Astrophysics,* Wiley. (Advanced and not for the faint hearted, but excellent student material.)

Shu,F., *The Physics of Astrophysics, Vol I Radiation, Vol. II Gas Dynamics*, University Science Books, 1991. (Advanced and excellent.)

Swihart,T.L., *Astrophysics and Stellar Astronomy,* Wiley, 1968. (The text book of my undergraduate days, well explained and easy to read.)

Swihart,T.L., *Quantitative astronomy*, Prentice Hall, 1992, (Updated version of above.)

Specialized

Davidson,K., & Netzer,H., 'The emission lines of quasars and similar objects', *Rev.Mod Phys.*, **51**, 715, 1979. (A classic paper, challenging.)

Gear,W.K., 'Continuum emission from active galactic nuclei', in *Millimetre and Submillimetre Astronomy*. Eds., Wolstencroft,R. & Burton,W.B., Kluwer Academic Publishers, p307, 1988. (Excellent treatment of dust emission from AGNs.)

Hilderbrand,R. 'The determination of cloud masses and dust characteristics for submillimetre thermal emission',*Q.J.R.ast.Soc.*, **24**, 267, 1983. (An excellent treatment of thermal emission from dust, a classic paper.)

Spinoglio,L., & Malkan,M.A., 'Infrared line diagnostics of active galactic nuclei', *Astrophys.J.*, **399**, 504, 1992. (A very good review of line diagnostics and an excellent source of further reference material.)

5

Continuum emission from active galaxies

5.1 INTRODUCTION

This chapter could equally well be subtitled 'searching for the central engine'. One of the primary goals of observational AGN study is to identify the major luminosity components and their energy generation mechanism and hence understand the operation of the central engine. This is an ambitious task. Clearly we need observations across the entire e-m spectrum, and because many of these objects are variable, (which in itself provides crucial clues of physical processes and scale sizes) simultaneous, or at least quasi-simultaneous observations are desired. This is an even taller order and even today it has only been achieved for a very limited number of objects.

We would anticipate that the most luminous component is the primary radiation from the central engine and should be the driving force for all other secondary processes, determining the relative luminosities of the various components is therefore crucial. In so doing however, we need to be extremely vigilant about possible absorption affects that could seriously influence our deductions. The primary source could be fractionally (or completely) absorbed and secondary radiation might dominate the luminosity budget. The case of the Far Infrared Galaxies (FIRGs) is a good example. We measure the secondary radiation from the far-infrared luminosity, and so deduce the 'invisible' primary UV luminosity from the hot stars. This is a prime case of reprocessing, and reprocessing will figure prominently from now on. For AGNs, the usual assumption is that the primary engine produces the mid- X-ray to gamma-ray power-law photons, which subsequently interact by various mechanisms to produce much of what we subsequently observe. Therefore, without space-borne telescopes, our goal of identifying the primary radiation from the central engine is hopeless. Where jet emission is apparent (for the radio-loud AGNs) this is also produced directly by the central engine but must be a special case because it is rare. Chapter 4 has stood us in good stead for dissecting the AGN spectra into their individual components and hence determining the emission processes. In all of what follows we shall mostly ignore both the starlight from the underlying galaxy and the emission from regions significantly more extended than the narrow line region. The exceptions to this are the radio lobes and jets and the far-infrared emission.

In our quest for the central engine power source, we are talking about regions of space which are generally far smaller than our telescope beams. It is only for the relatively nearby objects that imaging studies can separate out the various emission components, and even then we do not get down to sizes of the accretion disk or even the molecular torus (see section 6.5). Let us remind ourselves of the scale sizes in question. The diameter of an accretion disk around a 10^8 $M_\odot$ black hole is approximately solar system scale. To resolve this at a modest distance of only 100 Mpc would require an angular resolution of ~ 1 micro-arcsecond. There is no hope of achieving this in the near future. Even at the 'next-door' distance of M31 (only 0.6 Mpc distant) the same accretion disk only subtends an angle of ~ 0.1 milli-arcseconds. Therefore, imaging the actual accretion disk in AGNs will probably remain an unobservable phenomenon. However, the latest HST images have shown that the much more extended circumnuclear disks of gas can already be probed (see plates 4 and 6 for example). We saw in chapter 2 that synthesis radio interferometry is a key tool for obtaining the highest spatial resolution and the king of this technique is centimetre VLBI, with angular resolutions better than 1 milli-arcsecond. Although this sounds very promising for imaging of the accretion disk in M31, we realize that this technique is only applicable to radio-loud phenomena, specifically synchrotron radiation from AGNs, which although fine for the jets, is not expected to be important for the thermal emission of an accretion disk.

Indeed, information on the sizes of the most central regions, specifically the BLR and inwards can really only be obtained through variability studies. The technique of reverberation mapping (seeing how spectral lines respond to a continuum stimulant) is crucially important for probing the structure of the BLR (and NLR). We shall briefly review this topic in section 5.4.4 but leave the details of the BLR to the next chapter where we probe the central engine region in detail. In this chapter we will concentrate on the continuum variability and especially the correlations between the response of one continuum component and another.

Over the past decade it is clear that polarimetry, and especially spectropolarimetry have become increasingly important. We shall leave this topic mostly to the next chapter where we shall show that its value was vividly established with the discovery of a reflected BLR in a type 2 object. This naturally led to eventual acceptance of the putative molecular torus in the central zones and opened up the scene for unification ideas based on obscuration and relative orientation.

We will start by looking at the overall output spectra of AGNs and to do this we require multifrequency observations, preferably simultaneous across all wavebands and over a period of time which is adequate to take into account variability for the source in question. Although coordinated multifrequency observations are a key observational tool, not enough has yet been made of it. (I admit to a personal bias being a strong supporter of such techniques and involved in a number of campaigns.) It is worth stressing that the difficulty of obtaining these broad-band spectral coverages is not necessarily because of technical difficulties, but due to the complexity of the process. A successful campaign generally involves extensive collaborations involving many groups, facilities and nations. The co-ordination of the observations is a nightmare due to the large number of different ground-based and space-borne facilities, all of which have different scheduling committees, observational practices and weather constraints.

Furthermore, determining the true variability forces one to perform the measurements at the different wavelengths essentially simultaneously, and over a sufficiently long period and sufficiently well sampled to understand the variable phenomena at work. Given the overwhelming practical and logistical difficulties this process entails, it is amazing that so many programmes have been so successful. At the time of writing, there are major programmes involved in both reverberation mapping of the BLRs for a small number of AGNs, along with the detection of blazars by the Compton Gamma Ray Observatory (CGRO) and subsequent multiwavelength monitoring. Both of these are undertaken as a collaboration using a wide range of space and ground-based facilities, including my own, the JCMT and UKIRT.

In determining the primary energy mechanism, it turns out that observations in the high-energy part of the e-m spectrum are vital, especially the UV to shorter wavelengths. Unfortunately, we remember that a key region, the EUV, is severely hampered due to absorption by interstellar gas in our Galaxy and the EUV sky is more or less opaque as far as extragalactic studies are concerned. Even with all the modern facilities, the problems in determining the emission processes in AGNs remain high. Often, indeed usually (and especially at UV and shorter wavelengths), there is more than a single radiation component present in the continuum emission at any one wavelength. Disentangling these additive effects compounds the difficulty, but progress has been dramatically improved with higher sensitivity and better spectral resolution X-ray telescopes as we shall see. Variability studies can be a boon in these situations. In some cases line emissions can blend together to form a pseudo-continuum. This is a real problem for the low energy X-ray and UV wavelengths, the latter due to FeII lines.

We will see that apart from the jet component, which when present appears to get out of the zone relatively unscathed, the innermost parts (the central parsec) of an AGN are typified by interaction processes. Absorption of 'primary' (or even secondary) radiation and reprocessing undoubtedly occurs and untangling the interactions is far from trivial. Absorption by gas in the BLR, dust in the putative molecular torus and the ISM of the parent galaxy are obvious sources of interaction (and there are others). The latter pose obvious sources of observational uncertainty when attempting to de-redden the observed emission to deduce the primary energy spectrum. So, given all these what seem to be almost insuperable difficulties, what can we say about the overall continuum emission of AGNs and clues to the fundamental power source?

In our discussion of the continuum emission from AGNs it is useful to separate the radio-loud and radio-quiet categories. The synchrotron emission that dominates the radio regime is seen to extend over at least ten orders of magnitude in frequency for BL Lacs, from the radio to the X-rays. For some of the superluminal AGNs there is now evidence of gamma-ray emission, possibly from inverse Compton scattering. In some senses, synchrotron dominated sources such as BL Lacs are relatively easy to understand in terms of continuum processes as they are 'clean', being dominated by a single production mechanism. Because the synchrotron emission originates in a relativistic jet, we are immediately driven into production mechanisms for the jet. We shall duck the details of this until chapters 7 through 9, and in this chapter we shall tend to concentrate on the radio-quiet objects and the respective continuum production mechanisms.

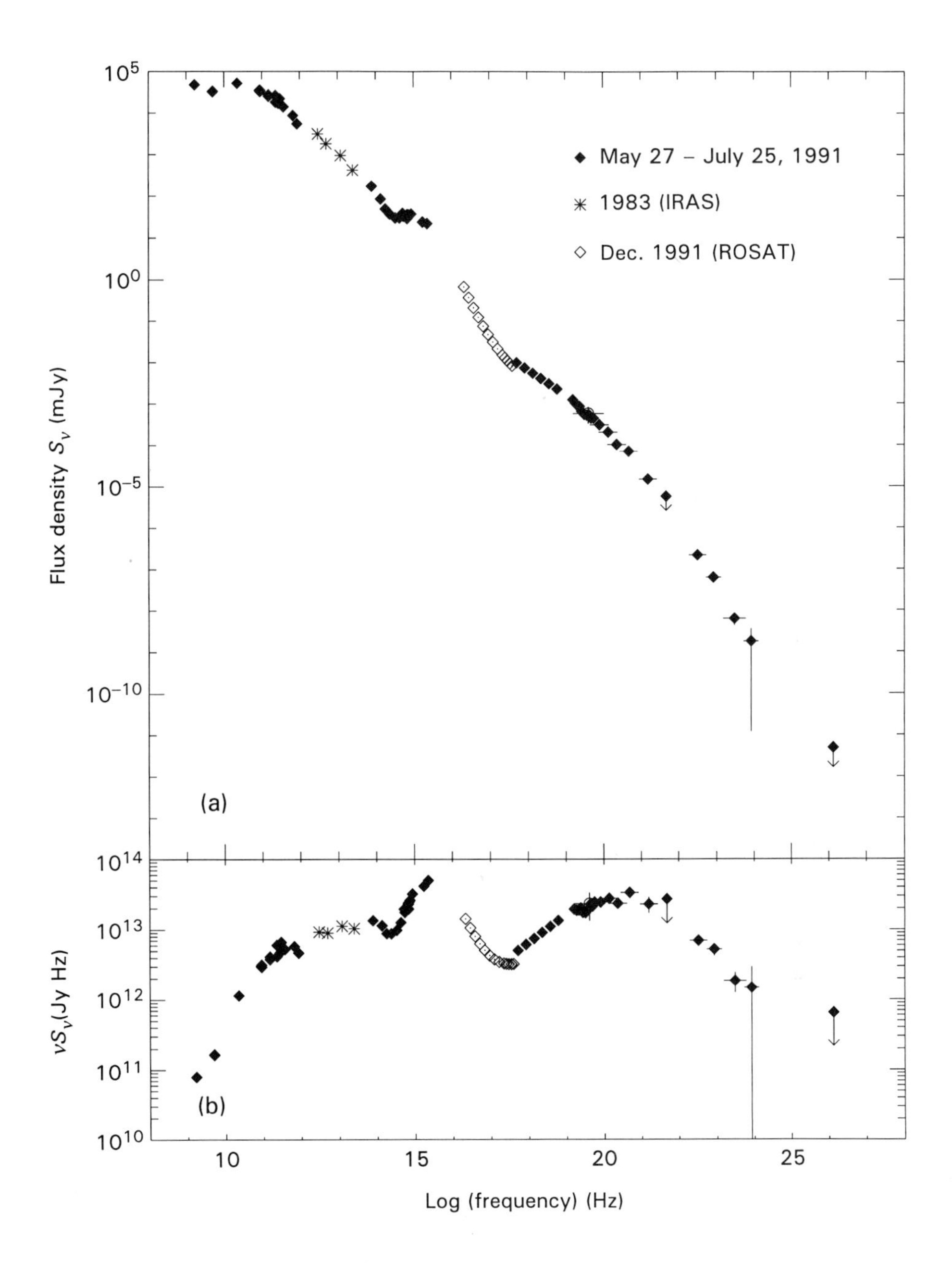

Fig. 5.1. The multiwavelength continuum spectrum of the radio-loud quasar 3C273 plotted in (a) flux units (S_ν) and (b) luminosity (νS_ν). The data span a wavelength range from 11 cm in the radio to almost 1 TeV gamma-rays and the dates of the observations are shown. (Adapted from Lichti, *et al.*, *Astron.Astrophys.*, **298**, 711, 1995, courtesy of Corinna von Montigny.)

Our current ability to determine the continuum energy distribution of AGNs is shown for the radio-loud quasar 3C273 in fig. 5.1(a). The top panel shows the observed flux density (S_ν) plotted as a function of frequency (ν) using logarithmic scales. Such a graph is often referred to as the *spectral energy distribution* (SED) of the source. The plot is a series of observational points at different wavelengths, and when these points are suitably joined together the overall distribution is revealed and hopefully the various physical emission components. Obviously the more points that cover the plot the closer to reality will be the resulting deductions. Where there are large gaps in wavelength, such as in the EUV of fig. 5.1, extrapolation is essentially based on physical arguments and faith but the reader must be aware of potential consequences. Even when good spectral coverage has been obtained, a source can sometimes reveal a smooth continuum spectrum that the astronomer mistakenly identifies as a single radiating mechanism, but is in fact a cosmic conspiracy. The spectrum shown in fig. 5.1 contains one such example that will be discussed later.

Another way of looking at the overall emission of an AGN is to consider the total radiated power (luminosity) per unit frequency plotted against frequency. This is usually plotted as log (νS_ν) versus log (ν). Figure 5.1(a) can be replotted in these units to give fig. 5.1(b). Observationalists tend to favour the SED plot, but for studying the output of the object in order to identify the most powerful luminosity regions (the cooling zones), or in modelling the various radiating components, the luminosity plot is more useful. Let us remember that luminosity refers to a cooling process for the source in question. From this point on we shall make use of a combination of flux and luminosity plots; however, the reader must be careful to note the ordinate label. [I have seen many members of a conference or seminar audience being misled when a mixture of the two plots is used and the difference is either not clearly stated by the speaker, or the legends are too small to be read by anybody other than those in the first row!] For comparison, a source with equal energy at all frequencies has a power-law spectrum of the form $S_\nu \propto \nu^\alpha$ and in this case $\alpha = -1$, so a *log* S_ν versus *log* ν plot has a slope of -1. However, the same emission has a slope of 0 (i.e. is flat) on a *log* νS_ν versus *log* ν plot. This difference is readily seen in fig. 5.1.

From plots of the spectral energy distribution or the luminosity distribution the contributions to the emission from individual mechanisms such as synchrotron, dust emission, starlight, etc. can then be identified. We chose to use a radio-loud object such as 3C273 to demonstrate the energy distribution because these generally give the widest frequency coverage due to the presence of the jet and its associated effects. Also, apart from the classical BL Lacs, radio-loud AGNs possess all the attributes of radio-quiet quasars such as the NLR, BLR and big blue bump. I will adopt the view that to a first approximation a radio-loud quasar is a radio-quiet quasar with a jet. 3C273 also happens to be very luminous and relatively close-by for a quasar, making high signal-to-noise observations possible across the entire e-m spectrum. Figure 5.2 gives a good illustration of the difference between radio-loud and radio-quiet AGNs in terms of a νS_ν plot and some of the separate components are identified.

Although a treatment of the continuum energy distribution of AGNs could be discussed in terms of categories of sources, we will discuss wavelength regimes and physical radiating mechanisms. It is often the case that within a particular wavelength regime a single radiating mechanism dominates making life that much simpler. Good examples of this are synchrotron radiation in the radio, and thermal emission from cool dust in the far infrared. This clearly helps in identifying and separating out the major physical processes occurring in AGNs. Where we can unambiguously identify the physical process and source or location of the emission mechanism in the AGN we will, (in the case of 3C273, at least four continuum emission processes can be readily identified), where this is indeterminate, we shall draw attention to competing views and leave further discussion to later chapters.

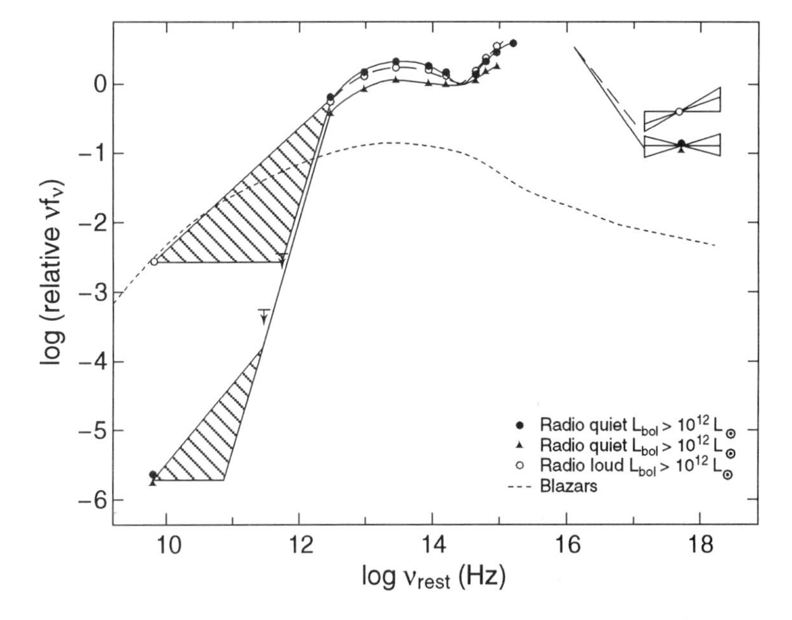

Fig. 5.2. A plot of the mean continuum emission from radio-quiet and radio-loud AGNs, the shaded portion in the radio shows the range of emission observed. The pure synchrotron component can be identified with the blazar line. Note the very large difference between the radio-loud and radio-quiet AGNs in the radio, but the close similarity of the $L > 10^{12} L_\odot$ objects in the infrared through UV. This strengthens the view that the physical processes responsible for the infrared through UV emission are basically the same in the two types of object, but the presence of the jet causes additional emission components to be seen. (From Sanders, *et al. Astrophys.J.*, **347**, 29, 1989 courtesy of Dave Sanders.)

In section 5.5 we will take our first step in constructing a coherent picture of the emission mechanisms for AGNs (the big picture) using the evidence from continuum, imaging and variability data as appropriate. It is impossible to review adequately all the information in the literature, but a lengthy list for further reading is presented in section 5.6. This is split into topic areas so that readers can select those most appropriate to their interests or requirements.

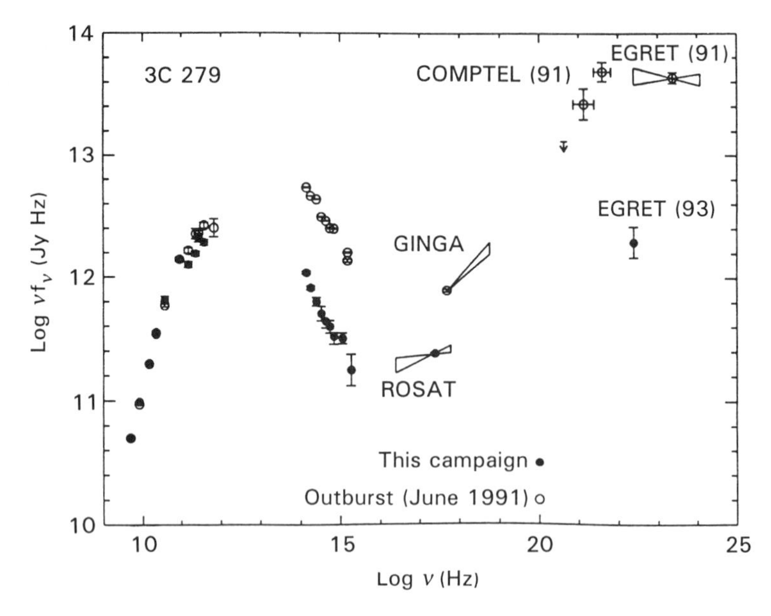

Fig. 5.3. A log (νS_ν) versus log (ν) luminosity spectral energy distribution plot for the blazar 3C279 at two epochs. The two main luminosity components are synchrotron emission and what is believed to be inverse Compton emission. Note that the spectrum appears to be simpler that that for 3C273 probably due to the greater dominance of the synchrotron emission. (From Maraschi *et al. Astrophys.J.*, **435**, L91, 1994.)

5.2 THE CONTINUUM ENERGY DISTRIBUTION

5.2.1 Radio

In many respects this is the simplest of the wavelength regimes to consider because as we noted above, when radio emission is important in active galaxies in terms of luminosity, it is invariably synchrotron radiation that dominates. Obvious candidates exhibiting this phenomenon are radio galaxies, OVV quasars and BL Lacs. We shall reserve the discussion of the emission from the very extended lobes of radio galaxies to chapter 8 and concentrate here on the synchrotron emission emanating from the central core of the galaxy.

We observe two different types of spectrum from radio-loud AGNs. These are referred to as either 'steep spectrum sources' where the emission is dominated by synchrotron emission from the radio lobes, or 'flat spectrum sources' in which the compact core dominates. Examples of synchrotron spectra are shown in figs. 3.9 and 4.8. We note that their typical characteristics are a self-absorption turnover (section 4.3.1) occurring somewhere in the radio region, and an optically thin power-law (with a spectral index of $\alpha \sim -0.7$ on a *log* S_ν versus *log* ν plot), declining through the infrared and optical regimes. Polarization and variability data all show that this emission is due to synchrotron radiation.

For a small number of radio sources, the OVV quasars and classical BL Lacs (i.e. the blazars), the radio spectrum has a much flatter spectral index and extends to the millimetre and occasionally the submillimetre regime. The synchrotron emission in these flat spectrum radio sources originates from relativistic jets blasting out from the central engine (chapter 7). Note that 3C279 is a flat spectrum radio source but this is not readily apparent when plotted on a luminosity distribution (as in fig. 5.3).

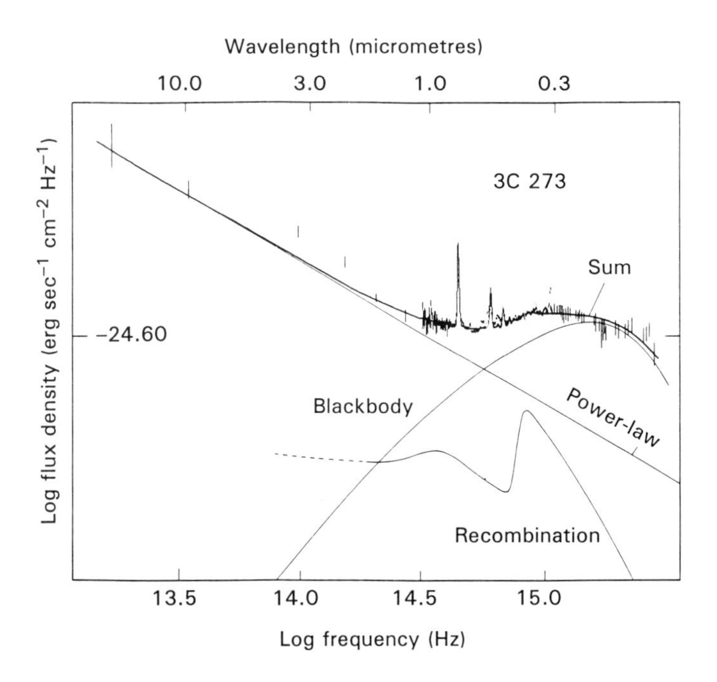

Fig. 5.4. Infrared through UV continuum of the quasar 3C273 showing the big blue bump in the UV. The overall spectrum has been fitted by a synchrotron power-law and the thermal emission of a hot accretion disk. The power-law is now known to have an additional component in the IR. (From the classic paper of Malkan & Sargent *Astrophys.J.*, **254**, 22, 1982, which opened up the field of spectral energy deconvolution).

A number of flat spectrum radio sources have been observed simultaneously at several radio wavelengths and the increased wavelength sampling reveals that the apparent 'flat' spectrum is composed of a series of superimposed individual self-absorbed synchrotron sources, each appearing as a local maximum on the spectral energy distribution. This is the most famous of the so-called 'cosmic conspiracies' and is shown in fig. 5.5 for the source NRAO140. The spectrum extends to even shorter radio wavelengths of around 1 mm where the synchrotron spectrum declines to the IR with an optically thin power-law.

In some sources (especially the BL Lacs), the synchrotron radiation outshines the emission from the host galaxy and the featureless continuum (no blue bump) extends to the X-ray or even γ-ray region (see fig. 3.9). Flat spectrum quasars tend to show strong emission in the optical from the BLR and also the big blue bump (fig. 5.4). Identification of strong synchrotron radiation is a key result which clearly demonstrates that some acceleration mechanism is at work to produce the relativistic electrons responsible for the emission.

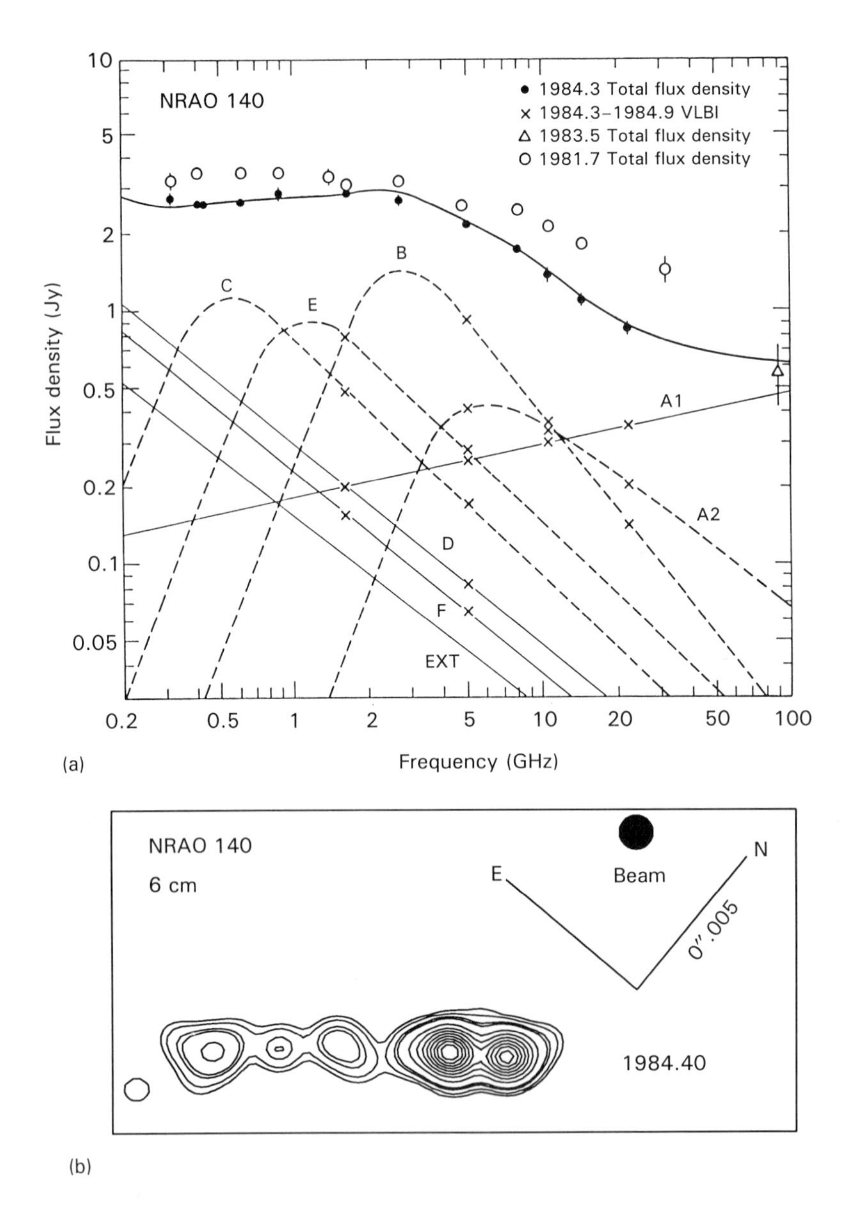

Fig. 5.5. (a) The radio spectrum of the quasar NRAO140 showing how the individual self-absorbed synchrotron components (A2, B, C & E) merge together at low spectral resolution to form the so-called flat spectrum. The core is component A1, while D and F are not detected by the VLBI array. EXT represents the extended arcsecond scale emission. (b) Single epoch VLBI map of NRAO140. (Adapted from Alan Marscher, *Astrophys.J.*, **334**, 552, 1988.)

We should briefly note in passing that the presence of radio synchrotron emission from a galaxy does not automatically mean that it is an active galaxy. Synchrotron emission can be generated by other mechanisms apart from a central engine. In our Galaxy and other spiral galaxies, we see synchrotron radiation from cosmic ray electrons interacting with the diffuse galactic magnetic field. However, in this case the emission originates from the entire galaxy and is weak in terms of its overall luminosity. Supernova remnants in the disks of galaxies are well-known sources of synchrotron emission and when galaxies are probed to very faint flux levels, individual remnants can be identified. Radio emission can also be generated by free–free emission from supernova remnants and from star formation regions and again this is not a pointer to AGN activity.

For our purposes, strong radio emission from an AGN signifies the presence of a relativistic jet, and we will leave the discussion of this to chapter 8. We can summarize by noting that for the flat-spectrum radio sources, the highest frequency synchrotron component has its turnover at around 1 mm wavelength. This is attributed to synchrotron radiation from relativistic electrons from the innermost part of the jet. Therefore variability studies at millimetre to infrared regions are crucial in order to determine the physical processes underway close to the central engine. From the discussion on synchrotron radiation in section 4.3.1, we saw that determination of the turnover frequency allows us to make estimates of the strength of the magnetic field. For a number of AGNs, magnetic field values of order 10^{-5} T (0.1 gauss) appear to be typical of the fields within the radiating part of the jet.

5.2.2 Submillimetre—far-infrared—mid-infrared—near infrared

As we are now aware, the vast majority of AGNs are very weak radio emitters and are classed as radio-quiet, being much brighter at optical wavelengths. For nearly all AGNs, the flux density increases with increasing wavelength from the optical through to the infrared. Furthermore, for the still relatively small number of objects that have been detected by the IRAS satellite and ground-based telescopes, this increase continues through 10 μm to ~100 μm, beyond which it declines steeply to the very weak flux levels of the radio regime. For many objects in the Universe, the submillimetre—far-infrared—mid-infrared regime is typically dominated by thermal emission from heated dust. Is this the case for AGNs?

As well as thermal radiation from dust, self-absorbed synchrotron radiation can produce a spectrum that peaks in the far-infrared. How can we discriminate between the two processes? Chapter 4 has already provided us with the tools we require. We note that the overall spectral shapes are different, the polarization properties are expected to be very different and any variability should also occur on widely differing timescales. Let us tackle each in turn starting with the spectrum.

As we saw in the previous chapter, a self-absorbed synchrotron spectrum has a spectral slope of $\alpha = +2.5$, whereas the Rayleigh-Jeans component of thermal emission from dust should have spectrum steeper than $\alpha = +2$ (the blackbody value), the precise value depending on the emissivity of the dust. Observations of dust in our Galaxy and other galaxies, give emissivities in the range of 1-2, so we expect that the dust spectral index should lie in the range $3 \leq \alpha \leq 4$. Unfortunately, as we saw in chapter 2, the determination of the submillimetre through infrared spectral energy distribution is still not in very good shape due to the relatively small number of sources that can be observed across the entire region.

Polarization studies (albeit on a small number of sources) provided no support for a synchrotron hypothesis, all the data being consistent with thermal emission. Furthermore, data from variability studies argue strongly against a synchrotron explanation, there being no evidence for rapid variability and the data are again consistent with thermal emission from dust. Nevertheless, if only a single radio-quiet quasar or Seyfert with strong far-infrared luminosity revealed significant 100 μm variability on a timescale of around a week, then all bets would be off as this would be very hard to explain in terms of dust emission which should come from a region much larger than indicated by this timescale.

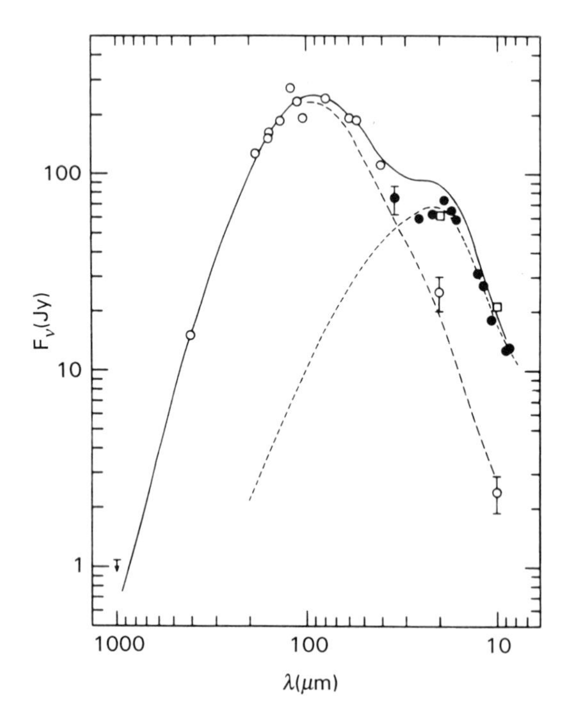

Fig. 5.6. Submillimetre to infrared spectral energy distribution of the Seyfert 2 galaxy NGC1068 showing the prominent far-infrared dust emission from a 3 kpc circumnuclear disk in addition to the nucleus. (From Telesco *et al.*, *Astrophys.J.*, **282**, 427, 1984.)

Determining the submillimetre spectral energy distribution slope is the answer, but this is still difficult. The submillimetre through infrared SED in the early 1980s was well determined for only about three galaxies, and just one of these was an AGN, the Seyfert galaxy NGC1068. The SED of NGC1068 is shown in fig. 5.6. Although IRAS detected a large number of Far Infrared Galaxies (FIRGs), the number of radio-quiet AGNs detected remained small. The problem of proving conclusively that the far-infrared emission from AGNs is due to thermal emission from dust has been at the heart of much of the work in the submillimetre and infrared regimes over a number of years.

However, the simple test of the spectral index measurement was somewhat blurred due to the work of ingenious theoreticians who managed to produce steeper self-absorbed synchrotron slopes ($\alpha \leq 2.9$) by modifying the electron energy distribution. Nevertheless, not to be outdone, detailed observations using the JCMT, IRAM and the IRAS database strongly suggested (albeit for a very small sample of the brightest sources—quasars or Markarian Seyfert 1s) that the long wavelength spectral slope is around $\alpha = 3.8$. This is highly suggestive of thermal emission from heated dust and virtually rules out a synchrotron process (see fig. 5.7). We will assume from now on that it is indeed dust emission.

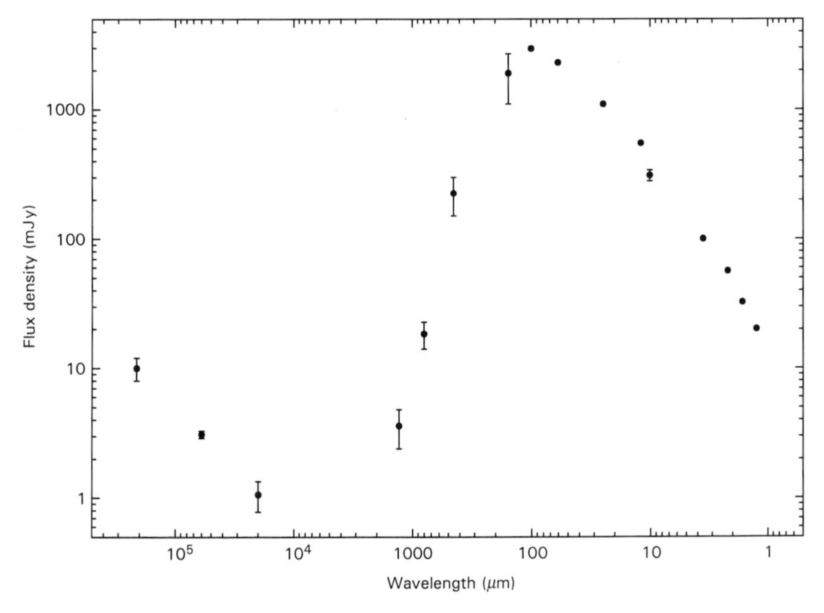

Fig. 5.7. The radio to optical SED of the radio-quiet quasar PG0050+124 (1Zw1). The spectral index, α, in the submillimetre exceeds 4.0, consistent with thermal emission from dust and the data constrain the dust temperature to lie between 39 K and 55 K. (Adapted from Hughes, Robson, Dunlop & Gear, *Mon.Not.R.Astron.Soc.*, **263**, 607, 1993, courtesy David Hughes.)

Much better determination of the complete IR to far-infrared spectrum is still required and two new facilities will provide the key to this, the ISO satellite and SCUBA on the JCMT. Recent millimetre and submillimetre observations have detected continuum emission from a distant radio-quiet quasar with a redshift of $z \sim 4$. If this is also due to dust emission, then not only is the luminosity extremely high, but there are cosmological implications for forming such large quantities of heavy elements at these very early epochs. We will pursue this further in chapter 9; suffice it to say that my bet is firmly on the side of dust emission.

This further suggests that extensive dust emission is a normal component of AGNs at most cosmological epochs. This will have some level of impact on observations of

AGNs, sometimes seriously. The effects of absorption by dust and potential orientation effects of the galaxies with respect to us will seriously need to be considered in forming the 'big picture'. We can easily imagine that the presence of dust in a torus or sphere will cause severe problems for optical and UV observations because of extinction. On the other hand, the dust emission is a bonus in that for heavily enshrouded AGNs (buried) the re-radiated far-IR at least luminosity tells us the minimum AGN luminosity (assuming that the heating is due to the AGN.)

But what about the dust itself? The temperature lies in the range 20 to 80 K, but with some debate over the precise values. The mass of the dust can also be estimated, although there are a number of free parameters that must be modelled and a unique solution is not always possible. However, data from my own observing programmes leads to dust masses exceeding 10^{10} $M_{\odot}$ and extending over a region of radius around 1 kpc.

The near to mid infrared energy distribution of AGNs has been relatively simple to determine since the late 1970s and resembles a power-law in shape. Power-laws with negative slopes tend to be indicative of synchrotron or non-thermal emission from a power-law electron distribution and this IR spectral shape was interpreted by some as being indicative of non-thermal processes at work. This was strengthened when the first studies of near infrared and X-ray data from AGNs appeared to show a strong correlation between the two and indeed an extrapolation of the IR spectrum often joined smoothly to the X-ray spectrum (a large extrapolation in frequency space). This latter aspect suggested an intrinsic IR to X-ray power-law which had a non-thermal origin. In this case the far-infrared turnover of the radio-quiet AGNs must be the self-absorbed synchrotron peak. However, more detailed studies using carefully selected samples (such as that of the Palomar-Green radio-quiet quasar survey) obtained no such correlation between the IR and the X-ray fluxes. The earlier apparent correlation is now probably explained in terms of a redshift selection effect. Gradually the IR to X-ray power-law faded from the scene, but the fact remains that the IR-mid IR has a general power-law shape. Further study in fact showed the power-law was not perfect and indeed, there seemed to be subtle differences in the shape of the spectrum between quasars and Seyfert 2s. In chapter 4 we saw that the thermal emission from a centrally illuminated source can easily produce a non-blackbody spectrum on the Wien side of the peak, and the emission from dust at a range of temperatures can be shown to resemble a power-law over a wide range of wavelengths. This is the answer that we shall adopt, all the mid IR-far-submillimetre emission is from re-radiating dust.

As we move through the near infrared to the optical regime, the spectrum for radio-quiet AGNs is typically representative of a transition region between the longer wavelength thermal emission from dust and the optical emission from the blue bump or starlight. However, there is a peculiar and unexplained feature that seems to be relatively common. This is a mysterious near infrared bump located between 4 μm and 5 μm. It is most frequently referred to as the 3.5 micron bump and it might be the result of very hot dust lying close to the central engine and on the point of evaporating. It is just discernible in figs. 5.1 and 5.4. Observations reveal this emission component to be non-variable and non-polarized. By analogy with our discussion of the 100–μm emission, this is again strongly suggestive of dust emission.

A further clue to the origin of this bump came from observations of the radio-loud quasar 3C273. It had long been believed that the centimetre to infrared spectrum was due to synchrotron radiation, but a long-standing puzzle was that although the centimetre to millimetre emission was polarized at the few percent level, in its non-flaring state the near infrared emission from 3C273 had a polarization consistent with zero. In 1986 however, the millimetre through 10 μm energy distribution had fallen to a very low level, a level never before observed with such sensitive equipment. It then became clear that the spectrum of 3C273 was composed of at least two components: a variable millimetre to mid-IR synchrotron component, plus a steady infrared (1–3 μm) component. For some unexplained reason, nature conspired that 3C273 has a radio synchrotron spectrum which apparently joins smoothly to the near IR bump at ~ 5 μm and when sparsely sampled, gives the illusion of a single smooth (synchrotron) spectrum. Furthermore, the near infrared bump was also found to be non-polarized.

The picture is now clear. When the synchrotron emission increases during a flare, the entire infrared through millimetre flux increases due to synchrotron radiation alone. This has been tested and reveals the expected infrared polarization. As the electrons radiate and lose energy, the spectrum declines until the near IR bump becomes comparable to the synchrotron emission. With the further decline of the synchrotron emission, the near IR (non-polarized) bump emerges and is clearly revealed.

So what is the source of this IR component in 3C273? With a self- absorption turnover at a wavelength of 4 μm, the magnetic field is estimated to be a rather implausibly high value of almost 10^{-2} T and the source size would be of the order of a light-day or less. Both of these suggest rapid variability and high levels of polarization, but neither are observed. An entirely different possibility is free–free emission. There are models in the literature which predict that the broad-line region has a very high plasma density, of order 10^{17} electrons m^{-3}, and that this would give rise to free–free emission which could produce a very similar infrared bump to what is observed. Although this remains an intriguing possibility, it is not supported by most pundits. The favourite explanation as we saw above is of thermal emission from dust on the point of evaporating.

To conclude this section we can summarize by noting that many of the radio-quiet AGNs radiate almost 30% of their total luminosities in the far-infrared. For the most extreme cases this figure can increase to almost 90% of the bolometric luminosity of the galaxy. Thermal emission from dust is understood to be responsible for this emission which dominates the submillimetre through IR regions. The dust mass may be as high as 10^{10} $M_{\odot}$ or, as we shall see later, even greater. A very high redshift quasar has been found to radiate in the submillimetre and this suggests that dust emission from AGNs is common at all epochs. This can have dramatic effects on observations as the optical through shorter wavelengths are very susceptible to extinction due to dust. The properties of AGNs might be seriously compromised and orientation effects may play a significant role.

5.2.3 Optical and UV

Determining the infrared continuum energy distribution of AGNs is relatively straightforward because the ratio of line to continuum emission is low and the infrared bandwidth is large (a factor of 5 at least). The situation changes in the optical. Not only is the spectral wavelength coverage reduced by a factor of about 2 for the standard optical regime, but the lines from the BLRs in the type 1 AGNs pose a hazard for continuum determination. This is because the line emission is both strong and broad, bringing about the possibility for line shoulders to blend together, making determination of the continuum somewhat difficult. In what follows we shall ignore the smooth and featureless synchrotron spectra of the BL Lac objects, but the other blazar sources, the flat radio spectrum OVV quasars with their BLRs and blue bumps, will be discussed.

An interesting aspect of many radio-quiet AGNs is that their continuum luminosities show a pronounced minimum at around 1 μm in their local rest frames, the luminosity increasing towards both longer and shorter wavelengths. In terms of the energy spectra, this manifests itself as a 'flat' spectrum in the optical–UV rather than the steep spectrum of the BL Lacs (see fig. 5.4). This seems to be a general characteristic of AGNs because for those sources where spectral steepening has been observed in the UV, it can readily be explained by extinction from intervening intergalactic absorption. This minimum does not seem to be dependent on the luminosity of the AGN, therefore it must represent a fundamental parameter of a distinct physical process or processes. We have already met the long wavelength process, thermal emission from dust. At temperatures close to 2,000 K, the dust evaporates and so this leads to a pronounced fall in the infrared emission from dust. The next question is to investigate the optical–UV side of this minimum.

But before commencing our discussion of the thermal emission from the putative accretion disk of the central engine, let us not forget that the dominant source of emission from normal galaxies in the IR–optical–UV is that from the photospheres of stars. For the powerful active galaxies starlight can be safely ignored. As we descend the hierarchy of activity, starlight becomes more important as the ratio of luminosity in the central spike (representing the central engine and close environments) to parent galaxy falls.

Returning to investigations of the overall optical–UV continuum emission from AGNs, further complicating factors are associated with line emission. Emission from Balmer radiation is a good example of this and occurs in the rest wavelength range of 380-270 nm. This arises from complex atomic energy states in the pumped UV field of an AGN and results in a cascade of electrons moving through their respective energy levels, releasing predominantly Balmer photons (as explained in sections 2.6.2 and 4.4). Another source of emission, whose importance has only been recognized over the last 15 years, is the blending of lines of singly ionized iron (FeII). When observed at low wavelength resolution, this blend forms a continuum-like feature predominantly in the 180–350 nm region (but also in the 440–480 nm and 500–550 nm regions). Along with the Balmer radiation, this forms a pronounced humping in the spectrum. This is sometimes referred to as the 'little blue bump' or the '300 nm' bump. However, even when this has been taken into account, another component still appears to be present. This increases in strength to the shortest UV wavelengths and is the famous *big blue bump* or just blue bump for short.

Greg Shields was the first to suggest that the flat optical–UV continuum of quasars might be due to a hot accretion disk and the idea was soon refined by Matt Malkan. He used a sample of quasars and Seyferts and, in a classic example of spectral energy deconvolution, found that after removing the underlying continuum due to other processes an excess emission component remained. Malkan and others believed that this was the long wavelength thermal emission from an accretion disk of temperature between 25,000 K and 35,000 K. In fact this does not prove the existence of an accretion disk. The observations are consistent with radiation from an opaque thermal emitter, radiating up to $\sim 10^{38}$ W from a size of around a few AU or so, but it does not have to be an accretion disk. For modelling purposes energy units favoured by X-ray astronomers are often used and in this case the emission is said to be of energy around 3 eV.

The big blue bump will therefore extend into the UV, EUV and soft X-ray regions. (Remember that some of the hottest blue stars, such as Rigel (β Ori), have effective temperatures of this order and are very strong UV radiators. The optical emission from Rigel that we see with our eyes is only the long-wavelength Rayleigh–Jeans part of the thermal emission. We never see the true brilliance of this star.) One might typically expect an accretion disk to radiate like a very hot star with temperatures ranging from 30,000 K of the outer photosphere, to over 10^5 K from the much smaller inner core. A rough guide to which wavelength range this covers is obtained by assuming a blackbody radiator approximation and the application of the Wien Displacement Law (eqn. 4.8). This reveals that the peak of the accretion disk emission should lie in the range 100 nm to 30 nm, which as we now know lies smack in the middle of the EUV region. Hence the importance of observations in this regime is readily apparent. However, the above calculation was based on the assumption that the accretion disk radiates like a blackbody. In reality this is not expected to be the case and complex models of accretion disk structure and radiation produce output energy distributions that deviate significantly from that of a blackbody.

Nevertheless, a significant problem remains in that it is difficult to fit the observed continuum energy distributions with the spectra predicted by accretion disk models (see chapter 6). The difficulty for the accretion disk pundits is further exacerbated by variability studies of the big blue bump. There is a distinct lack of correlation between the spectral shape of the big blue bump and both its luminosity and the luminosity of the AGN in question. This is an unexpected result if it is the emission from an accretion disk. Because of this and other difficulties (which will be revealed by variability studies—see later), the hot accretion disk theory explanation for the big blue bump and flat optical–UV energy distribution has competition. As we saw from section 4.2.5, free–free emission produces a naturally 'flat' spectrum and there are good reasons to suppose that free–free emission from a highly ionized plasma in the vicinity of the central engine should be present. Might this be a better explanation for the optical–UV continuum?

One might think that the answer should be easy to determine as the optically thin free–free spectrum is quite different from a greybody thermal spectrum. Alas, the answer is far from simple as we shall see when we consider the details of the central engine and accretion disks in chapters 6 and 7. One test to distinguish free–free from a 'cold' thermal emitter is to look for evidence of absorption in the continuum spectrum, either from hydrogen gas or cold atomic material, the latter giving rise to bound–free absorption edges. In principle this is a clear indicator that the emission is from neutral (or at least not highly ionized) gas, but again, this does not necessarily have to be in the form of an accretion disk. To observe such absorption effects we need to move to even shorter wavelengths, the X-ray domain, which also leads us into the realms of reprocessing the hard X-ray energy into lower-energy emission.

5.2.4 EUV

The EUV or *extreme ultraviolet* regime extends to higher energies than the short wavelength limit of ~100 nm of the IUE satellite and the HST. It spans the 100 nm (12.4 eV) to ~12.4 nm (0.1 keV) range. IUE observations showed that the de-reddened spectrum of some AGNs continued to rise to shorter wavelengths, into what is

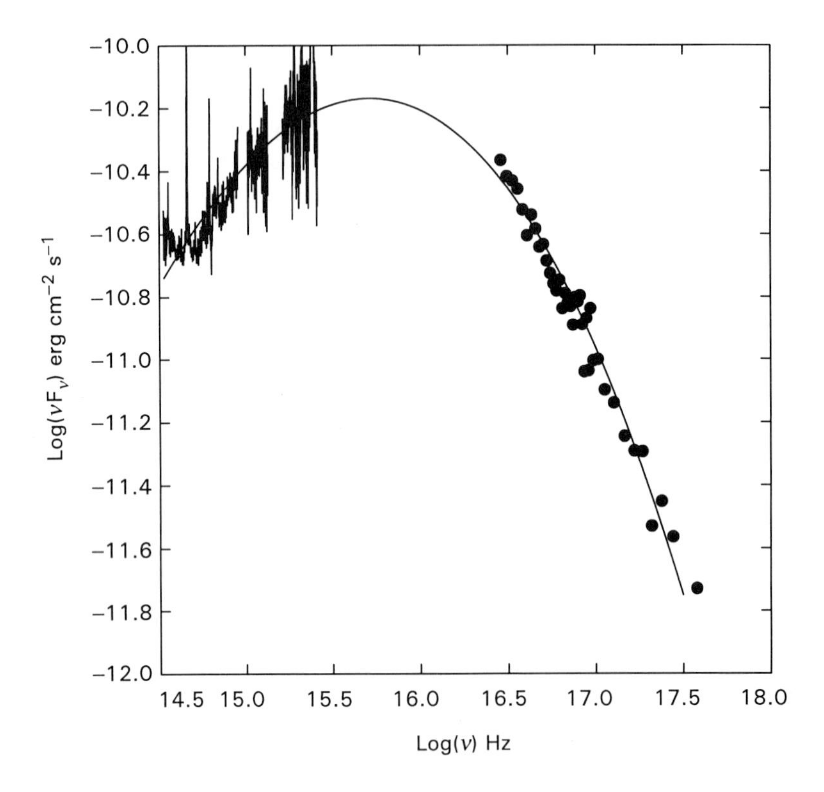

Fig. 5.8. The infrared (948 nm) to X-ray (~1 keV) energy distribution of the Seyfert 1 galaxy Mkn478. The strong emission lines in the optical and UV are clearly seen, along with the big blue bump. Although the EUV region is inaccessible, it is tempting to connect the X-ray emission directly with the UV. The so-called soft X-ray excess is not obvious from the figure because the medium and high-energy X-ray data are not displayed (no simultaneous data) although the low energy X-ray emission clearly deviates from a power-law. Note that this is a log(νF_ν) plot similar to fig. 5.2. (Courtesy of Prab Gondhalekar, see 'Reverberation mapping of the broad-line region in active galactic nuclei', *Ast.Soc.Pacific.Conf.Series*, 69, 1994.)

sometimes referred to as the vacuum ultraviolet. Furthermore, as we shall see in the next section, the X-ray emission from these sources was found to be decreasing with increasing photon energy (decreasing wavelength). The obvious conclusion was that the emission from these AGNs must peak somewhere in the EUV.

The major difficulty of observation (apart from the technology of the spacecraft and detectors) is that photons in this wavelength region have high interaction cross-sections to cold gas in our Galaxy. This causes very high absorption. The detection of extragalactic objects is severely hampered, and indeed rendered nigh on impossible apart from directions away from the plane of the Milky Way where the column length of absorbing gas is reduced. Even so, only the brightest of extragalactic objects are detectable. The most successful telescopes operating in this regime are the EUVE satellite launched in 1993 and the wide field camera aboard ROSAT. For AGNs the studies are totally dominated by observations at the lowest X-ray energies, namely from 0.1 keV to ~1 keV. Indeed, observations in this regime was one of the driving forces for the ROSAT mission (see section 2.4.5 and the next section).

We shall have to live with the difficulty of not knowing the spectral shape in the EUV (looking at a high redshift quasar does not help in this case as other problems then come into effect). The best that can be done is to extrapolate from the UV and lowest-energy X-ray regimes. However, investigating the correlations between the UV and X-ray emission provides an important handle on what might be happening in this regime. We shall explore this in section 5.4.

5.2.5 X- and gamma-ray

These two regimes are treated together because with the improvement in observing platforms it is becoming increasingly difficult, and indeed somewhat artificial to separate the high-energy X-ray from the low-energy gamma-ray region, although in terms of emission processes, there are a number at work in this entire regime.

The X-ray regime has brought a wealth of precious data for active galaxy study. Using the sophisticated instruments on recent satellites, a detailed picture of emissions from the innermost parts of the central engine is now slowly emerging, but the observations are still not at the required level of accuracy to constrain the various different models. Radiation in the X-ray regime is one of the most common aspects of AGNs. Indeed it is frequently a significant (if not major) contributor to the bolometric luminosity and it is the only region in which large amplitude temporal variations of less than one day are seen for radio-quiet objects. Furthermore, the discovery of a significant number of new AGNs in the ROSAT X-ray survey now ranks with those from the optical.

Over the past few years, the ever-improving X-ray observations mean that we are now close to grasping the big picture, but there are still important details to be completed. Nevertheless, it is likely that the precise details of the physical processes of the central engines of AGNs will derive from advances yet to be made in this particular region. We should note that determining the precise spectral shape in the critically important 0.1 keV to 20 keV regime is complicated by the contribution from a number of different radiating components but which are observationally difficult to disentangle. X-ray instruments tend to have either have wide wavelength (energy) coverage with low spectral resolution, or low wavelength coverage with high spectral resolution. The ASCA satellite is an example of the latter and although brilliant for detecting line emission and linewidths, is relatively inferior for determining the spectral shape in the 2-10 keV regime.

ROSAT, which has carried out the most sensitive survey in this region, discovered emission from many hundreds of quasars and Seyferts and a strong 'soft X-ray excess' has been found in almost a third of the X-ray selected AGNs. These are mostly Seyferts but some quasars also fall into this category. One of the most important conclusions from the 0.1 to 10 keV studies of radio-quiet AGNs is that there is a large degree of uniformity in spectral energy distribution, despite a range of several orders of magnitude in luminosity. To first order the overall spectra are relatively well fitted by power-laws (but remember the story of the infrared). However, there are important deviations as we shall see. In discussing spectral slopes, X-ray astronomers use photon energy terms, where the spectrum is characterized by the form $F(E) = C\,E^{-\Gamma}$ in units of photons cm^{-2} s^{-1} keV^{-1} (with apologies for the lack of SI units). In terms of the energy spectral index used in this book ($S_\nu \propto \nu^{\alpha}$), the indexes are related by $\alpha = -(\Gamma - 1)$.

The 'soft excess' is an excess of X-rays with energies below a few keV, and is an excess with respect to an extension to lower energies of the general 2–20 keV power-law. However, determining both the shape of the excess and even the power-law is complicated. As we noted in the previous section when discussing EUV observations, absorption by cold gas in the interstellar medium in our Galaxy becomes severe for the lowest energy X-ray range, being a major obstacle below 0.3 keV. One of the points to come out of the studies on X-ray absorption column lengths is that there seems to be little evidence for very high levels of intrinsic absorption by cold gas in the type 1 AGNs observed so far. Limits on the absorbing column density are now very stringent and typically lie in the range $\sim 10^{20}$ cm^{-2} (this indicates the equivalent number of absorbing atoms in a column of cross sectional area 1 cm^2 along the line-of-sight to the object; the Galactic value of the column density $N_H = 1.65 \times 10^{20}$ atoms of hydrogen cm^{-2}.). Where absorption is found, the levels are either consistent with that produced in our Galaxy or only about a factor of ten larger. This latter value might point to absorption by broad-line clouds rather than the general interstellar medium of the AGNs. For the type 2 and low luminosity AGNs however, there is good evidence of absorption, and values as high as $\sim 10^{24}$ cm^{-2} are found.

At this point we should just note that although X-ray telescopes lack the spatial resolution to make detailed imaging possible on all but the closest of the active galaxies, ROSAT data already show that the X-ray emission is extended in a number of sources. NGC1068 is one of these and we now know that the X-ray emission originates from a size of the order of the blue disk (~ 1 kpc) superimposed on an unresolved nuclear core. This is compatible with a model of a point-like central engine surrounded by a starburst disk. The starburst galaxy M82 also shows very extended X-ray emission. In this case it is most likely from hot gas that is bound to the body of the galaxy and demonstrates what a starburst but 'non-active' galaxy looks like in X-rays.

After taking absorption into account the corrected spectral slope of this excess emission appears to have a mean value of at least −1.0 and often steeper. It has also been found that the luminosity and steepness of the low energy excess are correlated for many AGNs. The correlation is in the sense that the higher-luminosity galaxies have steeper low-energy X-ray spectra. It is also the case that the lower-luminosity Seyfert 2 galaxies tend to have, on average, flatter spectra. Additionally, there appears to be a correlation in both quasars and Seyferts between the strength of the optical FeII lines and the steepness of the low-energy X-ray excess. However, it is not clear whether selection effects are playing a significant part in producing this correlation. Indeed, the correlation is in the opposite sense to what was expected from theoretical considerations, where the X-rays were anticipated to be one of the sources contributing to the line emitting region and hence producing the optical FeII emission. In this case, the flatter X-ray spectrum would be expected to produce a higher X-ray luminosity that would in turn generate more FeII. The apparent contradiction of this prediction is a puzzle that needs solving, either by finding a hidden selection effect or a better understanding of the production of the FeII emission.

We have previously stated that the usual interpretation of the soft X-ray excess is an extension of the big blue bump. In this case it is thermal emission from either the accretion disk, or gas surrounding the central engine. However, we repeat that this appealing interpretation has a number of problems and the current view is coming around to the idea that the optical-UV big blue bump is more likely to be reprocessed material by the accretion disk (see next chapter and chapter 9). We will see below that variability arguments support the decoupling of the soft X-ray and big blue bump, strongly indicating that the soft X-ray excess is a separate component originating from much hotter gas (but still $< 10^6$ K). The picture is still far from being clear and we shall investigate the implications of these observations in section 5.4 and later chapters.

Although the power-law form of the X-ray continuum appears to be similar for quasars, Seyfert 1s and 2s, as we noted earlier it is found that nearly all the type 2 objects show significant absorption by cold material. This has the effect of making the Seyfert 2s appear generally weaker in X-ray luminosity than Seyfert 1s. When this is taken into account, the X-ray luminosity of the Seyfert 2s is increased and often exceeds that of the lowest luminosity Seyfert 1s. This has two implications: firstly that the X-ray luminosity is similar between the two types and secondly that the low-energy X-ray excess in Seyfert 2s is not precisely the same as for Seyfert 1s. It cannot be a continuation of the higher-energy power-law (which would be completely absorbed due to the large column density of absorbing gas), and so it must be yet another distinct component. To date, the favoured explanation is that these low-energy X-ray photons are *scattered* into our line of sight by some reflecting material, perhaps the surface of the accretion disk which is irradiated by primary X-rays above the central engine (see chapters 6 and 9), whereas the higher energy power-law photons are observed directly. This is then an example of reprocessing.

This is a critically important deduction and the concept of reflection explaining a number of components in the continua of AGNs has become very attractive over recent years. It is further supported by specific observations of X-ray line emission from the K (and sometimes L) shells of iron (fig. 5.9). These lines are different for different types of object (obscured and unobscured), but both point towards reflection being a primary source of the emission. We will return in detail to this important clue when we look at the central engine (chapter 7) and the unification models (section 9.3).

For the 2-20 keV range, the spectra give reasonable fits to a power-law and overall this has a photon index ~1.7 (and spectral index ~ −0.7), but determining this figure is beset with problems due to contributing components such as absorption and emission from atomic gas along with possible Compton scattering. Figure 5.9 shows what has been achieved in terms of spectral deconvolution. We should note that as the X-ray energy increases, substantially fewer sources are sufficiently bright to provide good detections. Hence, conclusions based on the results for the X-ray bright populations of the rather limited (hard X-ray) flux-limited sample may be invalid for the general AGN population.

The much higher signal-to-noise data from the more recent X-ray telescopes have revealed a definite flattening of the power-law beyond ~10 keV. This is believed to be due to two distinct components being present: an intrinsic (primary) steep spectrum that predominates in the 2–10 keV regime, superimposed upon which is a flatter spectrum that extends to higher energies. It is believed that the latter may be the result of some form of reprocessing by Compton scattering (the reflection model) very close to the central source (fig. 5.9 and section 9.3). When this scattering is taken into account, the intrinsic power-law is found to have a slope of around −1.0, although for one source recently studied by the ASCA satellite, the 'bare power-law' slope turns out to be the same as the original value of $\alpha \sim -0.7$. Furthermore, the few lower X-ray luminosity Seyfert 2s that have been observed have generally steeper spectra in this regime. Perhaps this means that the power-law is not present or is different. With more high quality data, the suspicion is beginning to arise that the optical classification of type 1 and 2 (as applied to Seyferts) may not be telling us the entire picture in terms of the X-ray emission and we may have to resort to different descriptions for AGNs in terms of the physical emission processes. Clearly the picture is complex and not yet solved but we will return to this in chapters 7 and 9. Suffice to say at this point that if the power-law is the primary radiation from the central engine, then the slope of this spectrum has critical implications on its production mechanism.

Moving to the very high X-ray energies (E ~ 50 keV), breaks in the power-law producing spectral steepening are apparent. However, only the brightest of the sources have so far been observed in this regime and so some caution must be given to generalizing this result. On the other hand, a limit to the X-ray emission from sources is imposed by the diffuse X-ray background radiation. This is not cosmological in origin (like the 2.7 K cosmic background radiation), but is the result of emission from a large number of individual galaxies. The latest ROSAT survey data show that the brighter of these objects are quasars, but when much fainter levels of X-ray emission are probed, weak emission line galaxies at redshifts between $z \sim 0.2$ to ~ 0.5 take over. A simple calculation reveals that spectral steepening of the high-energy X-ray spectrum must take place to ensure that the flux in the X-ray background is not exceeded.

The gamma-ray region has only come into serious contention for AGN study over the past few years, and before the launch of the Compton Gamma Ray Observatory (CGRO) the picture was bleak. Until then, only one AGN was known to be a high-energy γ–ray emitter. This was the quasar 3C273 (detected by the COS-B satellite in the 1970s) while at the lower energies of ~ 1 MeV, three other AGNs (NGC4151, Centaurus A and MCG-8-11) had been detected. As an aside, it was the *non*-detection of certain radio-loud flat spectrum quasars by COS-B that provided additional firm evidence for superluminal motion. This was an important conclusion as it was independent of the apparent motions of the moving blobs as observed by VLBI as we shall see later in section 8.3.

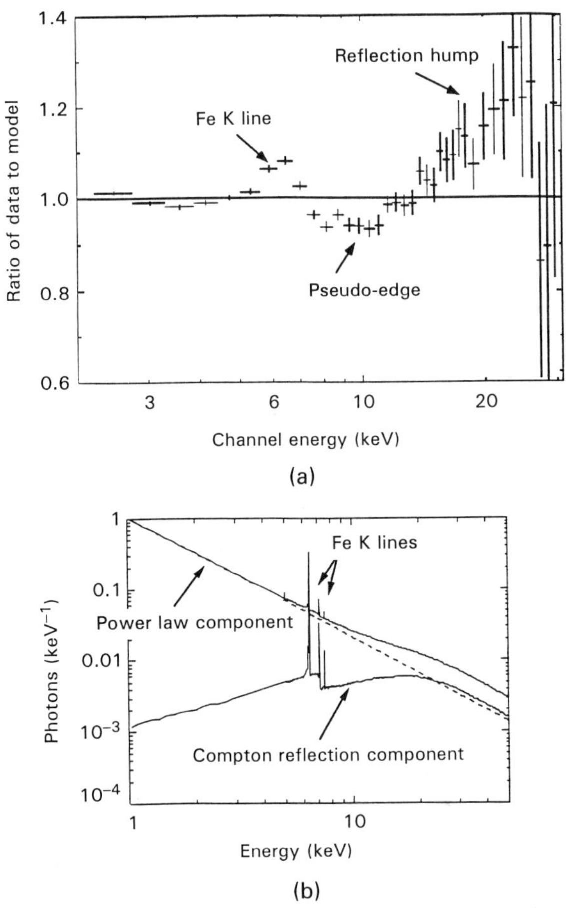

Fig. 5.9. (a) Observations from Ginga ratioed against a power-law showing deviations at ~ 6 keV, 8 keV and E > 10 keV. These can be interpreted in the reflection model as being due to the iron K-α line in emission, an edge absorption feature and the high-energy Compton reflected (scattered) component respectively. (b) A theoretical model from George and Fabian. (From Mushotzky, Done and Pounds, *Ann.Revs.Astron.Astrophys.*, **31**, 717, 1993.)

The launch of the CGRO in 1990 provided a welcome boost for γ–ray astronomy, and by mid-1991 the OVV quasar 3C279 was detected by the EGRET experiment. At the time, and perhaps fortuitously, 3C279 was undergoing a major synchrotron flare. Since then, observations of AGNs have been one of the major areas of study for the satellite and almost forty have now been detected. All of these demonstrate blazar characteristics and nearly a third show superluminal motion. (The other sources have not been observed by VLBI on a sufficient number of epochs to determine whether they are superluminal and so it is quite possible that superluminal motion is a requirement for observable γ-ray emission.)

These sources range in redshift from $z \sim 0.03$ to $z > 2$ and have γ-ray luminosities typically in excess of 10^{41} W (assuming isotropic emission). This is a very high luminosity and hints (but by itself does not prove) that beaming is probable. As we saw in section 3.6.1, the source luminosity is reduced by the ratio of the solid angle of the beam to the 4π solid angle of isotropic emission, and we shall see in section 8.3.3, that this is further drastically reduced by relativistic boosting. Measuring the spectral index of the emission is difficult due to the relatively poor spectral sampling of the CGRO detectors. Nevertheless, a power-law slope with a spectral index ranging from −1.7 to −2.4 (in terms of the energy spectrum) seems to cover those sources observed to date.

Currently, the most popular theory for the origin of the gamma-ray flux is first order inverse Compton scattering. However, the source of the seed photons is not yet clear. One obvious possibility is the synchrotron photons produced by the electrons in the relativistic jet of the quasar (fig. 5.3) and we shall explore this further in section 5.4.

At the extreme gamma-ray energies of 10^{12} eV (1000 GeV = 1 TeV), Mkn 421 and even more recently, Mkn 501, have been detected by the ground-based Whipple Gamma-Ray Observatory in Arizona. Mkn 421 is a nearby (100 Mpc) elliptical galaxy which shows blazar characteristics and it has also been detected by the CGRO. For isotropic emission, its TeV gamma-ray luminosity is 10^{36} W. The production of these gamma-rays by the interactions of highly accelerated charged particles is far from simple and is one of the exciting areas of current theoretical work.

5.3 IMAGING STUDIES

5.3.1 Introduction

Now that we have covered the major features of the continuum energy distributions of AGNs, we can turn our attention to imaging studies. Here we immediately find that the problem of spatial resolution is severe. As we saw in the introduction to this chapter, even the 0.05″ resolution of the HST is inadequate to directly image an AGN accretion disk. On the other hand, the HST has enabled the disk of gas surrounding a probable supermassive black hole to be detected in the galaxy M87 (plate 4). VLBI gets closest in probing the central regions of AGN but of course these observations are only possible for the jets of radio-loud objects and this discussion is postponed to chapter 8.

5.3.2 Optical and infrared imaging of host galaxies

Optical and infrared imaging with CCDs has allowed breakthroughs to be made in the study of AGNs, producing information not possible before the introduction of these devices. The first ‘fuzz’ around quasars was detected in 1973, but the major advances came in the mid-1980s. These studies are incredibly difficult due to the high contrast between the extremely bright central spike of the AGN and the much lower surface brightness of the surrounding galaxy. A number of imaging studies investigating the hosts of Seyferts and quasars have now been carried out both in the optical and infrared, although the number of sources observed remains at less than ~100.

A general conclusion from these studies is that quasar hosts often show very disturbed morphology, sometimes with multiple 'nuclei' in the parent galaxy. Furthermore, the parent galaxy is often seen to be surrounded by blobs of emission. These could be companion galaxies representing a rich environment (such as a cluster of galaxies) or just foreground objects. The precise picture is inconclusive as these companion 'blobs' are still much too faint to be spectroscopically measured but on the whole, the incidence of the brighter of the surrounding blobs is consistent with them being companion galaxies. The overall view seems to be that quasars are disturbed galaxies and probably surrounded by a number of companion galaxies. This strongly suggests that interactions play a key role in the quasar phenomenon. Very recently, HST images of quasars, Seyfert 1 and Seyfert 2 (although with buried BLRs) have supported this view (e.g. plate 7). Many of the AGNs imaged are highly disturbed, further supporting the merger picture.

In assessing the big picture we would like to know whether AGNs are preferred in certain morphological types of galaxies, and if there is a link between luminosity and the luminosity of the parent galaxy. Chapter 3 gave us an overview of our current knowledge, which we saw was very limited for the quasar population. We have just seen that the imaging studies reveal that the host galaxies for quasars are often disturbed, thus assigning a morphological type, spiral or elliptical, is very difficult without incredibly hard work and statistically meaningful samples and rigorous test criteria and control samples.

The first optical CCD imaging studies in the 1980s tended to show that host galaxies of radio-quiet quasars appeared bluer than those of radio-loud quasars, although the samples were small as were the differences in colours. This supported the widely held belief at the time that radio-loud quasars were located in giant elliptical galaxies whereas radio-quiet quasars were in spirals. The driving force for this had come from two observations: all powerful radio galaxies were giant ellipticals and many Seyferts were clearly spirals. The link between radio galaxies and radio-loud quasars was then made. However all was not so simple. Further studies indicated that the radio-loud quasar hosts were both brighter and bluer than those for the radio-quiet quasars. This was odd because if radio-quiet quasars were in spirals, these should be bluer than the radio-louds (because elliptical galaxies are on the whole red objects being noticeably deficient in blue light).

We now realize that a number of selection effects were present in many of the early samples, resulting in lower luminosity radio-quiet quasars being selected for study because these quasars were closer. Indeed, the picture was far from clear after almost a decade of study. More carefully selected samples and longer wavelength (*i*-band) optical CCD studies then produced a significantly different picture. These showed that elliptical galaxies could not only provide the host for the radio-loud population, but also for a substantial fraction of the radio-quiet population as well. Also, although in agreement with the earlier studies that the radio-quiet quasar hosts are on the whole fainter than their radio-loud counterparts, the difference had now fallen to less than a factor of two.

Infrared imaging turned out to be an even better tool for examining the host galaxy of powerful AGNs. This is because the contrast between the central AGN spike and the surrounding starlight is less than in the optical. (This is one of the very rare cases where the aim is to reduce the contrast in the observation, since the very bright central spike has to be removed in order to study the fainter underlying component.) The results of the infrared studies (see fig. 5.10 for example) fitted in well with those from the *i*-band work. The differences in infrared luminosity between radio-quiet and radio-loud host galaxies were now negligible.

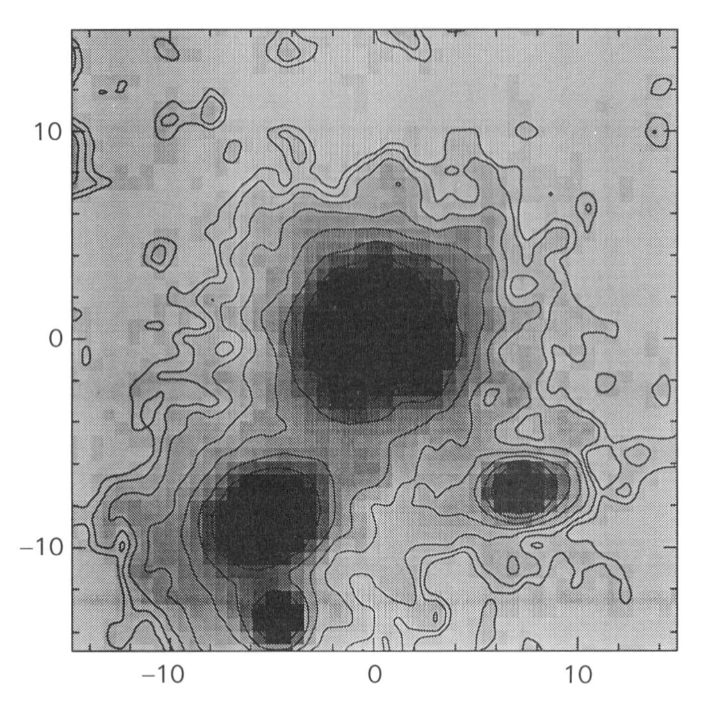

Fig. 5.10. An infrared (2.2μm) 30" x 30" image of the radio-quiet quasar 0923+201. The bright spots below the core (0,0) lie at the same redshift and are probably the remnant nuclei of an interacting system, or close companion galaxies. (Courtesy of Geoff Taylor and adapted from Dunlop, Taylor, Hughes & Robson, *Mon.Not.R.Astron.Soc.*, **264**, 455,1993.)

An important result of the infrared studies is that the host galaxy luminosities of radio-loud and radio-quiet quasars, along with powerful radio galaxies, could only be fitted by galaxies at the very high end of the galaxy luminosity function. Furthermore, because the most powerful elliptical galaxies are much more luminous than the most powerful spirals, this strongly suggests that most quasars are found in elliptical galaxies. Specifically, quasars are only found in galaxies whose infrared (K-band) absolute magnitude is brighter than ~ −25. This is consistent with earlier studies which indicated that galaxies with absolute V magnitudes fainter than ~ −21 had difficulty in making quasars (see fig. 5.11). Indeed, this luminosity division found by Véron-Cetty and Woltjer in 1990 became one of the standard criteria for differentiating quasars from Seyfert 1s.

However, very recent HST data have cast doubt on the above interpretation, and in fact thrown the studies of quasar hosts into confusion. John Bahcall and colleagues found that quasars appear to reside in galaxies with a wide range of luminosity, and generally much less than indicated by the studies above. Very surprisingly, two of the quasars seemed to have host galaxies almost four times fainter than the median luminosity for Seyfert galaxies and they were unable to even detect a host galaxy for some quasars. These should have been readily detectable if it they were luminous galaxies. There are reasons why the HST might not be sensitive to very low-level emission. It is still a small telescope and the exposures are not all that deep and obtaining a sufficiently precise point-spread-function is far from simply. Even orientation of the host galaxy could increase the contrast between the saturated central spike and the galaxy. Nevertheless, the results are very unexpected and while the numbers are still very small (~10 objects), there is a sufficient range of luminosity to make them spectacular. It is highly unlikely that there are any hidden selection effects plaguing the ground-based programmes because the IR studies have used very carefully selected samples, (but as I am deeply involved in these studies I have to admit to a personal bias). Perhaps the HST is giving a false picture for this type of study (and these are difficult observations even for the HST), we do not yet know. As they say in the papers, watch this space.

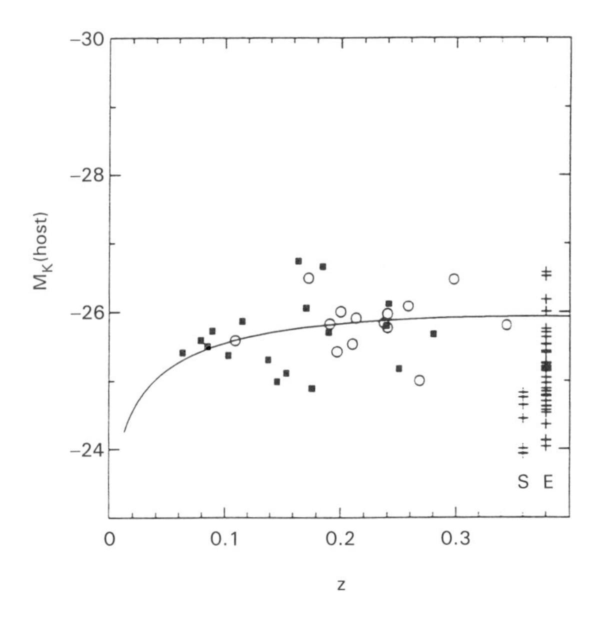

Fig. 5.11. Infrared absolute magnitude versus redshift for radio-quiet (circles) and radio-loud (filled squares) quasars. Also plotted vertically is the upper range of the luminosity function for spiral (S) and elliptical (E) galaxies. (From Dunlop, Taylor, Hughes & Robson *Mon.Not.R. Astron.Soc.*, **264**, 455, 1993, courtesy of James Dunlop.)

Notwithstanding the above controversy, we can also tackle the investigation of spiral versus elliptical hosts by another method, that of galaxy profile fitting. Ellipticals have a light distribution that reflects their basic shape (elliptical or spheroidal), whereas spiral galaxies display a disk-like distribution. Naturally for spirals one must be aware of orientation which can range from edge-on to face-on. From the latest data of our infrared imaging studies (which are about to go to press) we find a general conclusion (which is statistically highly significant) that all radio-loud quasars lie in luminous elliptical galaxies, radio-quiet quasars lie in both luminous spiral and elliptical galaxies (albeit with a preference for the former), while radio galaxies are ellipticals. There is no evidence to support the bifurcation hypothesis that radio-quiet quasars are spirals and radio-louds ellipticals.

The fact that spirals can form both radio-loud and radio-quiet quasars is intriguing and we will return to this in the final chapter. The conclusion about luminosity stands, quasars are only found in the most luminous galaxies. Also, in terms of the size of the host galaxy, one fact is common for all quasars so far measured from our studies: the host galaxies are large. The infrared sample found that all the galaxies had minor axes exceeding 40 kpc. Therefore it appears that at least luminosity determines whether a host galaxy can form an AGN, only the most powerful can form quasars, and morphology has something to do with whether they are radio-loud or quiet. All radio-loud quasars seem to be found only in ellipticals but radio-quiet quasars are found in both elliptical and spiral galaxies. (Note, when we talk of luminous galaxies we mean the luminosity of the galaxy due to its ensemble of stars.) We further note that as we move down the AGN luminosity ladder, then the lower luminosity AGNs (the Seyferts etc.,) are found in lower luminosity galaxies, many of which are spiral. We will return to this question of what makes an AGN in chapter 8.

5.3.3 Imaging of the central engine

Turning now to the AGN itself, one of the major goals of the HST programme was imaging the central engine. The first results were disappointing, the imaging quality being severely degraded by the well-known mirror defect. Not to be daunted, work continued. The reward came in an exciting image of the nuclear regions of the second-brightest radio galaxy in the nearby Virgo cluster, the galaxy NGC4261, also known as 3C270. This is shown in plate 6 and strongly hints at the presence of a disk of gas and dust. However, even this exciting discovery was eclipsed by some of the first images from the refurbished HST. The target of these studies was the favourite galaxy M87, for which earlier studies had pointed towards the presence of a black hole.

The HST Wide Field Planetary Camera 2 images and accompanying spectroscopic data of the central regions of M87 taken early in 1994 were a major breakthrough in the field of active galaxy study. Two images were obtained through a 2.8 nm wide narrow-band filter centred at 659 nm (that included the redshifted lines of Hα and [NII]) along with a further two continuum images through a 48.7 nm wide filter centred at 545.4 nm. The WFPC2 plate scale is 0.0455″ per pixel, giving a spatial resolution of 3.3 pc at the assumed distance to M87 of 15 Mpc. The narrow-band image showing the ionized hydrogen and nitrogen gas (plate 4) clearly reveals the presence of a small disk of gas surrounding the nucleus. Note the scale of the figure, with the bar denoting 1″. Closer inspection of the raw data gives the distinct appearance of a spiral shape to the gas in the disk, or at least mini, spiral-arm-like extensions. The disk is elliptical in shape with dimensions of 140 pc by 40 pc, and it is believed to be inclined to the line of sight by around 42°. The orientation and inclination of the disk are consistent with the radio and optical jet being emitted close to normal to the disk (at least within 10°).

The associated spectroscopic observations probed the kinematics of the disk, which enabled the gravitational potential of the central regions to be determined. As we shall see in section 7.2.4, these data strongly suggest the presence of a supermassive black hole of mass $\sim 2\times10^9$ $M_\odot$.

5.3.4 The extended narrow-line region and first steps in anisotropic emission

While ground-based optical telescopes cannot match the dramatic resolution of the refurbished HST, improving technology (such as adaptive optics) has resulted in better quality imaging from the latest generation of facilities. A large number of studies on spectral imaging of the central regions of the nearer AGNs have been undertaken. These are especially suited to investigating the Narrow-Line Region. One of the most exciting discoveries to come out of these studies is the conclusive evidence for a non-isotropic central radiation field. The observational technique is the same as discussed in the previous section for the HST observations of M87 and for some of the optical imaging studies of the host galaxies of quasars. As well as obtaining continuum images (showing the starlight for example), very narrow-band filters are selected to isolate a particular emission line for a particular galaxy (at a given redshift). This is placed in front of a panoramic (CCD) detector to obtain an image of the line emission in the galaxy.

An alternative technique is to use a very special instrument, a two-dimensional (imaging) Fabry-Perot interferometer. A well-known example is called TAURUS. This produces a mega-cube of data for what seems to be endless analysis. In a particular scientific paper, one of the pundits of this technique uses the tongue-in-cheek explanation 'with a magic wand the raw data-cube can be transformed into the—final answer'. The reader is then referred to a very detailed paper on the complexities of the analysis process. This technique is not for the faint-hearted. However, a number of stalwarts have persevered with this mind-blowing, data-analysis marathon to produce spectacular results. Although many objects have been investigated, the two relatively nearby Seyferts, NGC4151 and NGC1068 are favourite targets and are the best studied examples.

The key results from these various studies (supported by recent HST observations) is the presence of high excitation emission-line gas extending up to 20 kpc from the nucleus, and showing a distinctly non-isotropic distribution. This is referred to as the extended narrow line region (ENLR). We shall not dwell on the details of this region, but rather shall use it as a tool to highlight the anisotropic emission. We must be clear that the ENLR is not just an extension of the NLR; it is a separate component and the gas has a very small line width (typically less than 50 km s^{-1}). The velocity field is consistent with the gas rotating with the bulk of the galaxy. Furthermore, this emission is asymmetric, usually appearing very elongated, with a preferred extension along the direction of the central radio emission, often referred to as 'cone-shaped'.

The now accepted explanation for this lack of isotropy is that the central source of the ionizing continuum ultraviolet radiation is somehow propagated into the bulk of the galaxy anisotropically, probably by selective absorption by dust surrounding the central engine. Further support for this hypothesis comes from studies of individual galaxies, where the distribution of spatially resolved emission lines shows a roughly conical appearance. In a few cases, a clear bi-conical distribution is apparent, morphologically resembling the bipolar outflow pattern commonly found in the early phases of star formation. Using the highest possible angular resolutions on nights of excellent seeing, ground-based observations showed that the anisotropy and ionizing cones permeated to the NLR itself. One of the key targets of this study is the enigmatic Seyfert galaxy NGC4151. We shall return to the study of this object and the HST evidence of anisotropic emission in chapter 9, but in section 6.5 we shall take up the idea of an obscuring torus when we investigate the central engine itself.

5.4 VARIABILITY STUDIES

5.4.1 Introduction

Variability studies are immensely important but due to the fierce competition for telescope time and the difficulty of scheduling long-term variability studies, insufficient use has been made of this technique. The very fact that a significant fraction of the luminosity from an AGN can change on timescales of minutes (X-ray emission from Seyfert galaxies) or many hours (emission from quasars), enables firm conclusions to be drawn on the size of the emitting region. (The only time in which this simple measure fails to supply the source size is when we are dealing with beaming and relativistic motion; we then need to correct for the relativistic time dilation and orientation effects.)

In chapter 1 we saw that one of the first observational results for quasars revealed that their optical emission varied on timescales of less than a year. This highlighted the problem of the compactness of the energy source. The importance and impact of variability studies have not lessened since. Indeed, with the observation that many active nuclei show X-ray variability on timescales of 100 seconds or less (representing a distance scale of about one-millionth of a parsec—or just over twenty times the diameter of the Sun), the impact is arguably even greater. We have met some of the examples of variability, or lack of variability earlier in this chapter. The absence of evidence supporting significant variability in the infrared emission of radio-quiet AGNs on timescales of less than a year fits in well with the concept that the near and far-infrared emission originates in radiating dust of a size of the order of a parsec.

Variability studies come in a number of forms: monitoring the long-term behaviour of a particular AGN or a sample of AGNs; looking for short-timescale and dramatic activity; and multi-wavelength monitoring searching for links between variability at different wavelengths. As we shall see, these may be the only way of unravelling the complexities of the emission mechanisms and testing theoretical models for AGNs. In spite of this obvious importance, there is still a dearth of information on the long-term behaviour of active galaxies, apart from a small number of relatively well studied examples. These are generally bright at most wavelengths and mostly radio-loud. We shall make use of some of these 'case studies' to further our understanding of AGNs. Also in this section we will introduce and touch on the time-delayed response of the emission lines to variations in the central engine. This technique is called reverberation mapping, because it maps the location of the lines that react to the injection of energy by flaring of the central engine.

5.4.2 Multiwavelength monitoring

Multiwavelength studies serve two purposes. Firstly, by making a series of observations across the entire electromagnetic spectrum the luminosity of the object can be determined along with some possibility of identifying major emission processes (see figs 5.1–5.3). Secondly, monitoring the time evolution of these processes gives a much better handle on their inter-relations.

Being bright at all wavelengths, the quasar 3C273 is one of the best sources studied by multiwavelength monitoring. We saw in section 5.2 that monitoring led to the discovery of the non-synchrotron nature of the near infrared component and solved a long-standing puzzle of the polarization properties. Furthermore, by combining data from radio through to X-rays over a period of years astronomers have been able to show that there are *at least* four distinct radiating processes at work in this quasar. Amazingly, they all have about the same luminosity. This may be a pure coincidence, or it may be the result of close coupling between the components producing approximate equipartition of the energy fields. Figures 5.1 and 5.2 show the results from such studies and synchrotron, near IR dust, optical–UV–big blue bump–EUV–X-ray excess and inverse Compton (perhaps) gamma-ray components are seen. However, further detailed studies showed that the situation is even more complex. ROSAT has so far failed to find a correlation between the low- and medium-energy X-ray bands, demonstrating that there must be two distinct electron populations responsible for these emissions, and furthermore, that they arise from two physically distinct regions.

This nicely leads us into the next step, the search for temporal correlations between components. The aim of these searches is to identify clear links between the components, hence separating primary from secondary processes and obtaining a clearer picture of the location of the mechanisms with respect to the central engine. The starting point is a lightcurve. The subsequent analysis is beset by problems for a number of reasons. When looking at a complex lightcurve, the eyeball tends to pick out periodicities, as do some of the special mathematical analysis tools. (Indeed, the literature is sprinkled with claimed periodicities, most of which have faded into oblivion with the addition of more data or the use of more sophisticated analysis tools.)

Correlating data from lightcurves at different wavelengths is even harder. Because of the vagaries of telescope scheduling and/or weather, the data at each wavelength are usually irregularly sampled. To sample the faster of the variable sources, observations are required every few days over many months in order to have sufficient and sufficiently well sampled data that correlation analysis can be meaningful. It does not require many days of bad weather for a single telescope to spoil a perfect record, and so many telescopes tend to be used to provide some level of redundancy. Piecing all this data together is highly complex and the subsequent analysis techniques are beyond the scope of this book. The reference to Edelson and Krolick in section 5.6 gives guidance for the keen student.

Figures 5.12 and 5.13 show a set of infrared to radio lightcurves and the resulting correlations from my studies on 3C273. The ordinate of fig. 5.13 is the degree to which the two wavelengths are correlated as a function of time. The peak is the best correlation, and the corresponding time is that by which one of the wavelengths lags (or leads) the other. In all the cases shown in fig. 5.13, the longer wavelength lags the

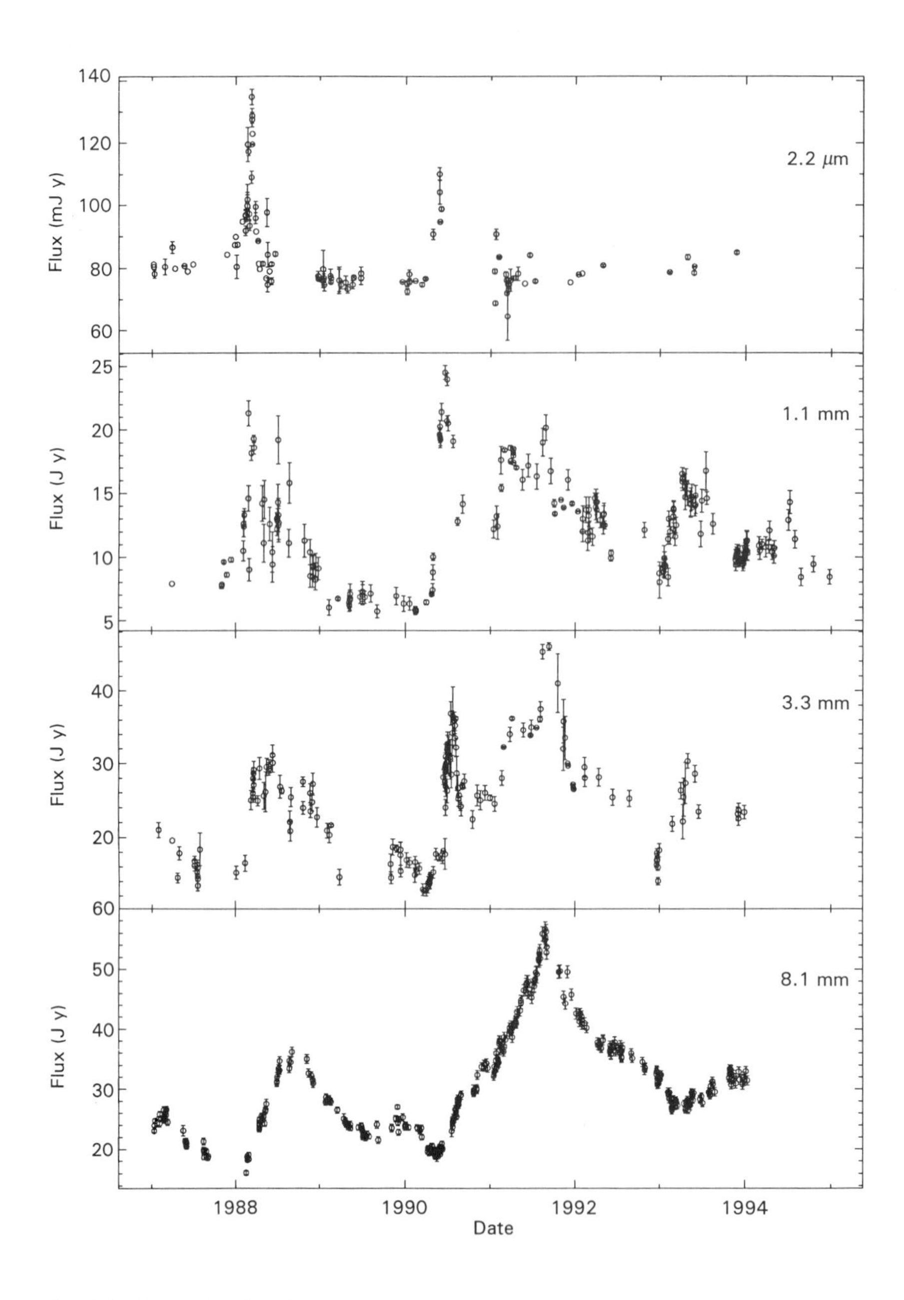

Fig. 5.12. Infrared (K-band) to ~ cm radio lightcurves for 3C273 from 1987 to 1995. Although the sampling is much denser, it is clear that the 8 mm data have a much longer time constant than the higher-frequency data.

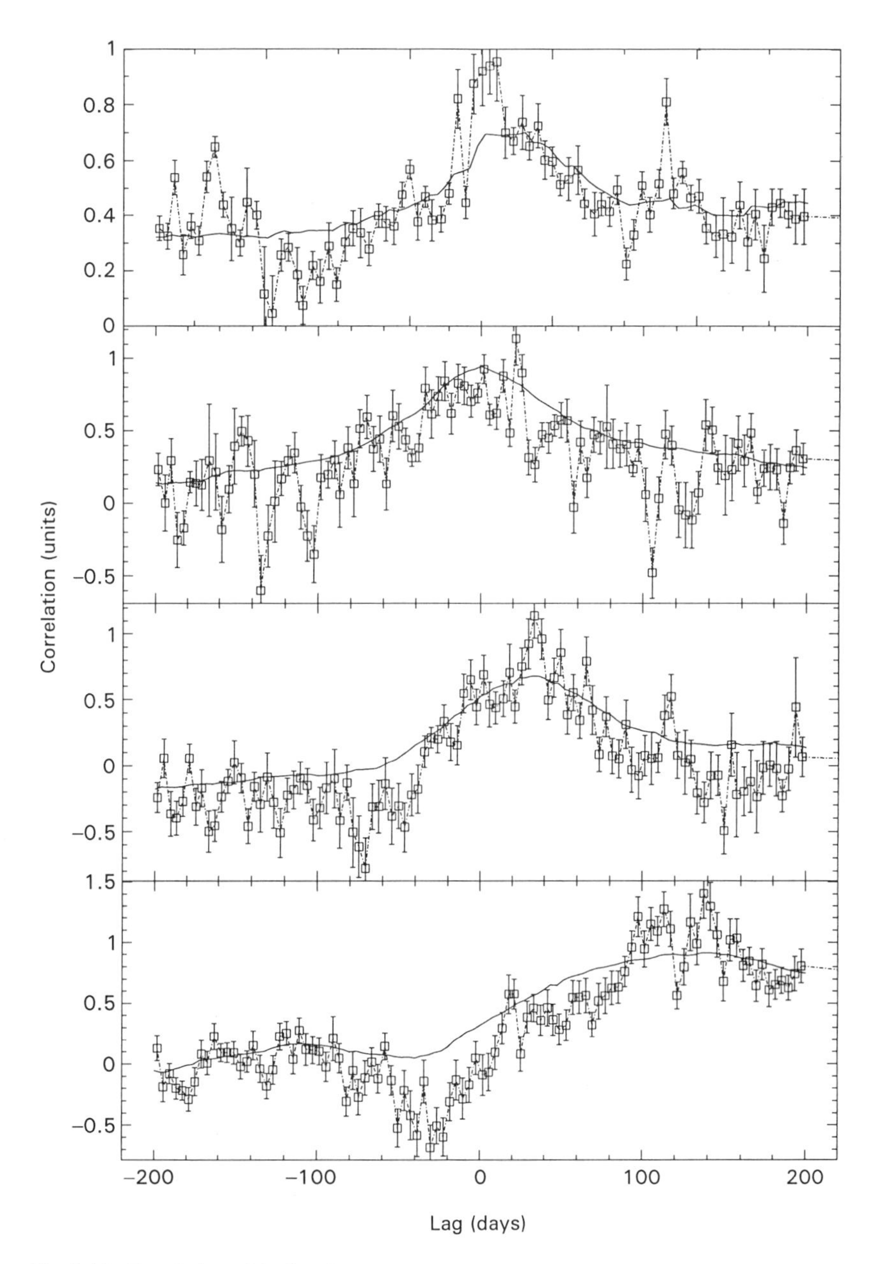

Fig. 5.13. Correlation of the lags between the flux variations from 3C273 from the data in fig. 5.12. The ordinate indicates the degree of correlation while the abscissa shows the lag in days between the two wavelengths. From the top are plotted 2.2 microns versus 1.1 mm showing zero lag; 1.1 mm versus 3.3 mm also showing zero delay; 1.1 mm versus 3.3 mm showing a lag of around 30 days and 1.1 mm versus 8.1 mm showing a lag of more than 100 days. This tells us that the prompt synchrotron emission is the infrared to 2 mm component, and at longer wavelengths we are seeing the propagation effects within the relativistic jet.

shorter wavelength and the picture is well described by a non-varying near infrared component (probably due to heated dust), along with a highly variable synchrotron jet. In terms of the other components, as we mentioned above, the correlations have so far failed to show conclusive direct links apart from the UV and the radio. We now know that 3C273 is, unlike 3C279 which seems to be a much 'cleaner' synchrotron-self Compton blazar, one of the most complex of sources to tackle and in many ways, is not an ideal object of study.

Because they show the most variability, the blazars are the best studied AGNs. This is an obvious selection effect. In terms of the blazar picture, some studies have apparently shown a link between increased optical emission (perhaps from the accretion disk or processes intimately linked to the central engine) and a subsequent synchrotron flare. However, the precise details have yet to be fully worked out and more data are required for confirmation.

Considering the variability at radio wavelengths, the rise and decay of flares reveal smaller timescales and usually higher amplitudes at the shorter wavelengths. This fits in well with a picture whereby the relativistic electrons are injected or accelerated in the throat of the jet. This gives rise to the prompt emission of synchrotron radiation in the UV–optical–millimetre region, perhaps accompanied by X-ray inverse Compton emission. As the electrons propagate along the jet, shocks and opacity effects transfer this energy to lower energies and the flare moves to the longer radio wavelengths as it loses energy through radiation losses. The details of the jet mechanisms are too complex for this text and are far from being well understood, but suitable references are provided in section 5.6. The lightcurve of fig. 5.13 serves to demonstrate the above propagation ideas.

There is some evidence for the central engine stimulus, as observed by the UV, to lead the radio synchrotron by the order of a few months to a year. This timescale then represents the time for the new electrons to be injected into the base of the jet (wherever it is in relation to the black hole) and to form a shock which subsequently radiates the radio synchrotron (see chapter 8). The length scale for this is around a parsec, and this is governed by the speed of the components in the jet, not the speed of light.

X-ray emission through inverse Compton scattering of the radio–IR synchrotron photons by the relativistic electrons (see section 4.3) should result in a very close temporal link between the radio–IR seed photons and the resulting X-ray emission. For a small number of well-studied blazars, including BL Lac itself, this correlation seems to hold well. For these sources there is no evidence that the X-ray emission precedes the IR-radio emission, at least on a timescale of days.

The discovery of gamma-ray emission from blazars triggered a number of intense coordinated campaigns of space- and ground-based telescopes. It is worth stressing the importance of variability at these extreme energies because although the database for variability is currently small, the tremendous energy carried by a single photon has dramatic implications for energy generation mechanisms. Examples of extreme variability include a 50% change at 1 MeV in the nearby radio galaxy Centaurus A (NGC5128) on a timescale of a week, while amazingly the TeV output of Mkn 421 is also variable on the same timescales, or even shorter. This very rapid variability timescale places constraints on possible emission models as we will see later.

However, the most dramatic variability results from the EGRET detector on the CGRO are for the blazar 3C279. The overall luminosity and gamma-ray variability for 3C279 are shown in fig. 5.3. A factor of five variability of the gamma-ray emission is found on a timescale of weeks and closer inspection reveals that lower amplitude variability on timescales as short as 1 day is significant. As more data become available, variability by factors of 3 on timescales of about a week is now being seen in other sources. This could indicate that this level of variability is typical for all these blazar-type gamma-ray sources.

The burning question is what is the production mechanism for these gamma-rays and from where do they originate in the AGN. The possibility of inverse Compton emission was pointed out in section 4.3.3, but in this case we need to ask from where do the seed photons originate. As we shall see in the following chapters, there is expected to be an intense photon field immediately surrounding the central engine. Therefore models of the inverse Compton gamma-ray emission from the blazar sources fall into two categories: those where the seed photons derive from the relativistic electrons in the jet itself (jet-produced), and those where they derive from seed photons external to the jet (perhaps produced by a photosphere of the accretion disk). Correlations and lags between different wavelength regimes are crucial for discriminating between the various models for the energy production very close to the central engine. An increasing number of coordinated multifrequency observations are now being done in this field, driven by the high-energy observations available from the CGRO. We shall return to this topic in chapters 7-9 but for 3C279, an inverse Compton process looks a reasonable bet because of the very similar spectral energy distributions of the synchrotron and X-gamma photon emission. But as yet the source of the seed photons remains unclear.

An important result has recently been obtained from multiwavelength monitoring of the BL Lac object Mkn 421. The TeV emission was seen to flare with a doubling timescale of order 2 days, and a subsequent cooling of order 12 days, albeit with little evidence of variability as seen by the EGRET detector at ~100 MeV. Furthermore, although X-ray observations by the ASCA satellite revealed a flare coincident within a day of the Whipple TeV measurements, there was little evidence for flaring in the radio through UV part of the synchrotron emission. We know that for Mkn 421, the X-ray emission seems to be a continuation of the radio through optical synchrotron, and so we are left with the puzzle of why only the X-rays flared but not the lower energy synchrotron emission and why the ~100 MeV gamma-rays did not show a strong flare. About the only way out of this is to suggest that the flare is due to an increase in the number of high-energy electrons in the jet, perhaps because the electron cut-off energy suddenly moved to higher energies. This effectively produces a new population of high-energy electrons that would radiate specifically at X-ray frequencies as we saw in section 4.3.1. These synchrotron photons could then be inverse Compton scattered to explain the TeV gamma-rays (section 4.3.3). The short timescale for the variability of this TeV flare poses severe problems for alternative models relying on external photon injection from the corona of the black hole.

We will now move on to the radio-quiet AGNs and present selected results from multiwavelength monitoring studies on one or two of the best-studied sources. In passing we shall again remind the reader that these results may, in the fullness of time, turn out not to be representative of the norm of the AGN populations. One interesting aspect which should be pointed out is that Seyfert 2 galaxies show little (if any) evidence of significant (Δflux / flux ~ 1) variability on timescale of around a year or less. This is true in the far-infrared, infrared, optical, UV and even X-ray regimes. So when we talk about significant variability in objects then invariably we are referring to type 1 objects.

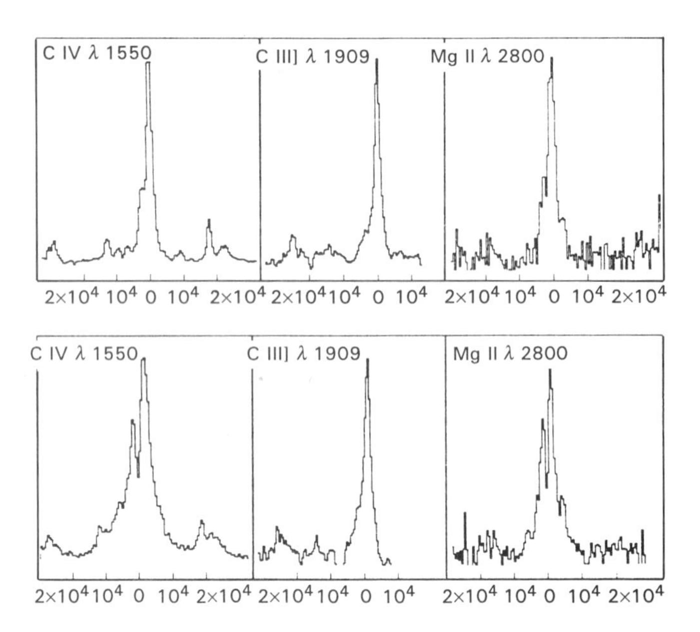

Fig. 5.14. The dramatic change in BLR line shapes from the Seyfert 1 galaxy NGC4151. The top row is when the UV continuum is low while the bottom row is when the continuum is bright and shows that although CIV and MgII are much broader, the CIII] has hardly changed. This was a classic IUE paper showing some of the first evidence leading to stratification of the BLR. (From Ulrich,M.H. *et al., Mon.Not.R.Astron.Soc.*, **206**, 221, 1984.)

Historically, one of the best monitored objects is the Seyfert 1 galaxy NGC4151 and a number of detailed campaigns involving the IUE satellite and ground-based telescopes were undertaken in the 1970 and 1980s. The data from these studies (e.g. fig. 5.14) provided a breakthrough in AGN research in that it triggered a clear change in thinking regarding the classification scheme for Seyfert 1 and 2s. The results from the campaigns demonstrated that over a period of many years, the Seyfert galaxy NGC4151 changed from showing the classical spectral characteristics of a Seyfert 1 galaxy, to those more resembling a Seyfert 1.8 galaxy, and then back again. (Remember, the drivers for the classification are the emission lines as described in section 3.2.) This apparent metamorphosis was accompanied by a change in the strength of the UV continuum which was believed to be representative of the ionizing radiation field.

At a stroke the classification scheme was seen to be reflecting conditions of the central engine. A change to the central engine luminosity changed the classification of the AGN. This showed conclusively that the classification scheme needed to be viewed in a much broader context. In many ways, this new thinking started the ball rolling towards unification scenarios. The work on NGC4151 and the trigger for the unification concepts of today are a fitting testimony and tribute to the career of Dr Mike Penston, who was intimately involved in these studies and who sadly died in his middle years in 1992.

The BL Lac PKS2155–304 has been a favourite target for multiwavelength studies over a number of years. Intensive IR–optical–UV monitoring, including polarization, has produced a complex picture. Perhaps surprisingly, although the optical, UV and soft X-ray fluxes show good correlation, the spectral slopes between them do not demonstrate a correlation with the fluxes. The polarization results are also not easy to understand in terms of simple models. It has recently been shown that the X-ray flux leads the UV by a few hours and that the general variability occurs on timescales of < 1 day. The best interpretation of the observations is that the variability in the IR–optical region is more likely due to synchrotron radiation from a changing jet geometry, or perhaps a changing bulk Lorentz factor (see chapter 8).

Another favourite target for these studies is the Seyfert 1 galaxy NGC5548. Carefully coordinated observations by the Ginga and the IUE satellites revealed the link between the X-ray and UV emission. Eleven series of observations were taken during 1989 and 1990, with the UV and X-ray observations being separated by never more than 1 day. These showed extremely good correlations between the 2–10 keV power-law X-ray emission and the 135 nm UV continuum over the period in question. Furthermore, they failed to reveal any lag between the two wavelengths greater than the temporal resolution of the observations of 6 days. This was even more impressive as during this two year observation period the source varied by a factor of 2.6. Similar, although not as persuasive correlations also exist for NGC4151, NGC 4593 and Fairall 9.

As we saw earlier, one possible mechanism for the production of the X-rays is inverse Compton scattering of the UV photons from the electrons in a very hot thermal plasma. In this case, the UV is the primary energy driving source and would be expected to lead the X-ray emission. However, the picture is actually more complex because the UV could be reprocessed hard X-rays and so saying which is the driving mechanism becomes very difficult to untangle. For example, the hard X-rays might be the driver, being absorbed by dense material, which is heated and subsequently re-radiates at UV wavelengths. In this case the X-rays should lead the UV. So far evidence for the X-rays leading the UV has only come from the studies of PKS2155–304. On the other hand, further strong suggestive evidence of this picture is that the X-ray emission can vary significantly on the timescale of hours or less, whereas the UV seems to be a much more damped system. This suggests that the X-rays *are* the driver, and originate in a more compact system. This is also reinforced by the observation of a 30% increase in the X-ray emission from NGC5548 in the space of 5 hours (unfortunately at a time when there were no UV observations).

Overall, there is a general favouring of the X-rays being the driving source, with the UV then reacting to the energy change. This also suggests that the X-rays are absorbed by the accretion disk, which then re-radiates in the UV–optical along with a hard (and reflected) X-ray component (see section 5.2.5). We shall follow up these detailed correlation studies (including the use of optical and X-ray line emission) in the following chapters because they relate specifically to the modelling of the central engine and the geometry of the accretion disk. Extensive campaigns (described below in section 5.4.4) on two Seyferts also showed that the optical continuum responds essentially simultaneously (within the day or so time resolution of the studies) with the UV continuum. This is a key conclusion because it means that the entire optical–UV continuum (including the blue bump) is from the same emission mechanism. This in fact poses major problems for the accretion disk origin as we shall see in chapter 6.5 and 7. Suffice to say that the temporal studies and the shape of the UV–optical continuum set restrictions on the geometry of the disk.

Very recent work on the Seyfert 1 galaxy NGC5548 has produced some startling evidence where the soft X-ray excess component flared by a factor of three over a few days but this was not accompanied by any change in the power-law X-ray intensity or shape. This means that the hard X-ray power-law cannot be the driver for the soft X-ray, these must come from a separate component, which as we noted in section 5.2.5 could be a region of hot gas, such as material falling onto the accretion disk. We will pursue this in subsequent chapters.

Moving to the infrared, the situation is much clearer. Early work in optical-infrared variability provided strong support for the infrared emission in the Seyfert 1 galaxy IIIZw2 (Mkn 1501, PG0007+106) being due to re-radiating dust. The data are in full agreement with a model whereby the dust is located ~1 pc from the central engine and beyond the BLR. More recent and very detailed studies of two other objects, the Seyfert 1 galaxy Fairall 9 and the radio-quiet quasar PG1202+281, have both provided even stronger evidence for thermal emission from dust being the source of the IR emission. For Fairall 9, the infrared flux lags the UV by ~250 days at 1 μm, ~365 days at 2 μm, and over 400 days at 3 μm. This points to a radius for the dust region of around 1 l.y. (0.3 pc) and gives a temperature of around 1,700 K, which is right on the point of evaporation. The situation for PG1202+281 is very similar, the 2 μm flux responding to an optical–UV flare after ~250 days and the data being consistent with the longer infrared wavelengths responding on even longer timescales. Therefore, although based on a small sample (of two), the picture is consistent with a dusty region surrounding the central engine at a distance that is outside the BLR, but nevertheless within ~1 pc. This is another powerful step in assembling the 'big picture'.

5.4.3 Short-timescale X-ray variability

The detection of very short-timescale X-ray variability in AGNs was one of the highlights of X-ray observations of the early 1980s. Significant luminosity changes on timescales as short as minutes severely restrict the size of the emitting region. In the studies conducted so far, there seems to be no preferred timescale for the X-ray variability. In these studies, reference is often made to the 'power' spectrum of the variability. This reveals whether there is a characteristic timescale of events and how the energy is distributed in different kinds of variability—short timescale to long timescale. The power spectrum is crudely the product of the frequency of a particular event and the energy change in the event. The answer from the X-ray observations of AGNs is that low-amplitude flickering is an almost permanent feature, but major changes, such as a flare with a large change in luminosity (like a doubling) occur much less frequently. This leads to the idea of a chaotic emission process rather than a highly ordered system. This will be pursued when we take a detailed look at the accretion disk and black hole super-corona in later chapters.

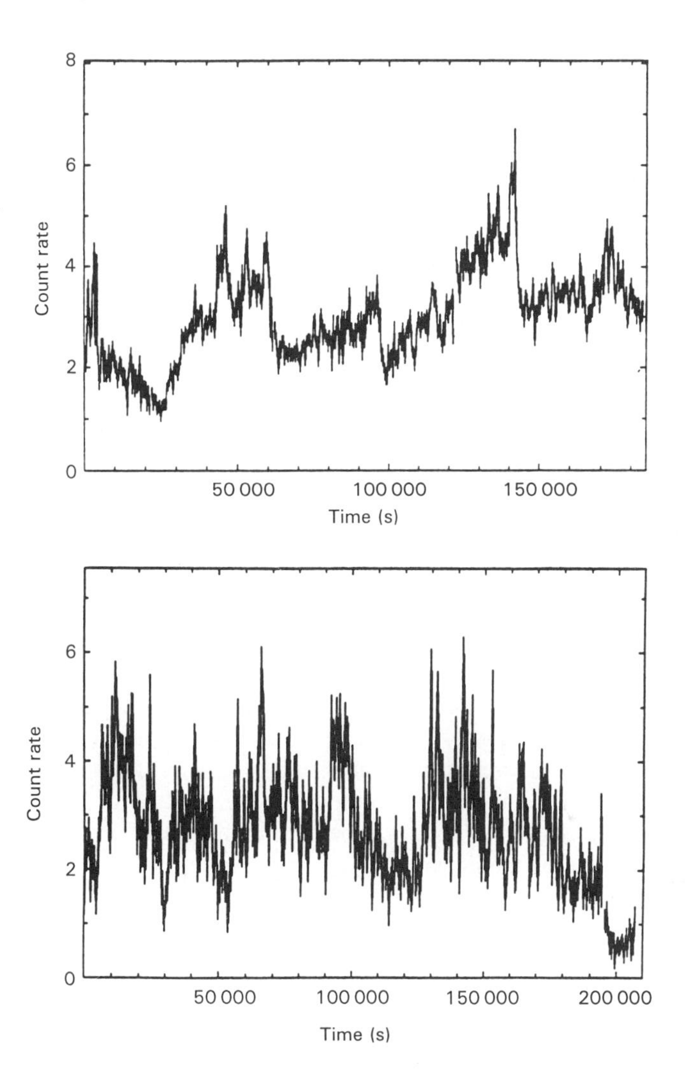

Fig. 5.15. Short-timescale X-ray variability from EXOSAT long exposures on the Seyfert 1 galaxies (top) MCG 6-30-15 at 2–10 keV energies, and (bottom) NGC 4051 at 0.05–1.5 keV energies. Note that 50,000 seconds is 13.9 hours. (Courtesy of Ian McHardy and adapted from '*X-ray variability of active galactic nuclei*', *Proc 23rd ESLAB Symposium on X-ray astronomy*, ESA SP-296, 1989.)

At first sight there appears to be no obvious general correlation between the X-ray variability and spectral slope. In some sources the overall X-ray slope steepens as the source becomes brighter, while for others, the opposite is the case. For the medium-energy X-ray range (a few keV), the slope of the spectrum does *not* appear to change as the X-ray energy brightens and fades. This argues strongly against models in which the UV is upscattered to the X-ray via the inverse Compton process from a variable temperature source of thermal photons. Further work on this aspect is crucial. Observations of variable sources such as dwarf novae and symbiotic stars have demonstrated the rapid increase in X-ray emission as energy (perhaps from magnetic field lines) is suddenly dumped into a region to create an energetic flare. These models are being extended to the AGN core region.

This lack of a characteristic timescale limits the ability to confine the X-ray production to a specific region. For example, a characteristic scale for the accretion disk might be expected to emerge. Furthermore, although the data have been searched for periodicities due to rotation effects, all the claims have subsequently faded away. The lack of any observed predominant timescale and the fact that random variability from 'shot-noise' arguments also fail to explain the observed variability of the 'power' spectra remain one of the most puzzling aspects. This serves to remind us that explaining the X-ray emission is far from a simple process. Perhaps this is the key conclusion: it is unlikely that a single and simple process is responsible for the entire X-ray emission. We shall bear this in mind for future discussions in later chapters .

5.4.4 Reverberation mapping

As we home in on the central parsec of active galaxies, reverberation mapping takes on a crucial role. Reverberation mapping refers to measuring the response of line emission to a change in the continuum as well as the changes in the lines' profiles. Different lines in the BLR and NLR are selected for study, and large amounts of time (but hardly ever enough) on many telescopes are devoted to intensive monitoring campaigns. These are very difficult studies to undertake. In addition to the coordinating and weather problems (where, as we saw above, even gaps of days can be crucial) the measurements of the selected lines and the continuum in the output spectra must be of high signal-to-noise ratio to allow the correlation techniques to be meaningful. An overview of the extensive campaigns of coordinated monitoring was given in section 5.4.2 above. Here we shall briefly review the same studies but look at how the emission lines respond to changes in the continuum flux.

We saw that for Fairall 9 the IR lagged the UV by about a year and the same coordinated studies showed that the emission lines in the BLR responded on a timescale of ~155 days. This locates the BLR clouds ~0.13 pc from the central engine, well inside the dust zone. The most extensive and intensive continuum and line monitoring campaigns have been undertaken for two Seyfert 1 galaxies, NGC5548 and NGC3783. These studies concentrated on sampling the optical–UV continuum and a number of lines in the BLR. We will report on the detailed results of these studies, and their implications for the geometry and size of the BLR, in section 6.3. At this juncture we will note that the response timescales for the broad lines in these objects are less than 20 days, very much less than for Fairall 9. With a corresponding scale-size of < 0.2 pc, we see that this really begins to sample the regions close to the central engine.

5.5 SUMMARY AND FIRST STEPS TO THE BIG PICTURE

In chapter 1, we saw that the bolometric luminosity of a radiating body is a key parameter. We have now seen in which regions of the spectrum the various bodies radiate most strongly and by which physical processes. Recognizing that the problems with the EUV are still with us, we can nevertheless make a list of bolometric luminosities and even tabulate the luminosities for the various wavebands for categories of AGN. Whatever the method of ranking, quasars dominate the luminosity league table. These are the most powerful phenomena in the Universe (we will assume that ultraluminous IR galaxies are buried quasars—see chapters 7 and 9). However, it is also apparent that there is overlap in luminosity across a wide range of category.

We note that the deduced radio luminosities for the jet sources are probably strongly enhanced because of the beamed emission. The correlation studies support the view that over much, if not all the spectrum, these sources (blazars) are dominated by synchrotron radiation from a relativistic jet. The BL Lacs show only synchrotron emission, the BLR region is either absent or more probably very weak in relation to the strongly beamed component. The same must apply to the big blue bump. For the OVV quasars on the other hand, the big blue bump is sufficiently powerful to outshine the synchrotron process in the optical–UV and they display a prominent broad-line region. In fact they are just like radio quiet quasars with the addition of a powerful synchrotron emitting jet.

A number of blazars are now known to be gamma-ray emitters and, for the best-studied, inverse Compton emission appears to be a reasonable explanation. However, whether the seed photons come from the synchrotron radiation in the jet, or from another source (such as a hot photosphere), remains unanswered. We shall pursue this further in chapters 7 and 9. In any event, the TeV detections from Mkn 421 and Mkn 501 shows that very powerful mechanisms for particle acceleration must be at work. At the moment, the rapid variability and correlation of the TeV and keV emissions suggests that the flaring is due to an increase in the number of high-energy electrons in the relativistic jet, which radiate synchrotron X-ray photons which are subsequently upscattered to give inverse Compton emission at the TeV energies. Note that this explanation derives from the observation of a single flare in one source.

For the radio-quiet AGNs, we can now construct spectral energy distribution plots for a very large number of objects of all categories. This allows statistically significant conclusions to be drawn for carefully selected samples. It has been noticed that for a very large number of radio-quiet AGNs a significant dip in the luminosity spectrum is seen at $\sim 1\ \mu m$, even in objects differing in bolometric luminosity by factors of over 10^3. This would seem to be a common factor and the most straightforward explanation is that this region is the transition between two common radiating mechanisms found in AGNs.

The long-wavelength emission is almost certainly due to re-radiating dust in the central zone of the AGN. The near infrared component (the 3 μm bump) most likely being dust on the point of evaporation and presumably originating in the inner zones of the molecular torus (see section 6.5). The temporal monitoring of the differing continuum components in Fairall 9 shows that this zone lies just outside the BLR, at a radius of ~ 1 pc. We will see that this fits in well with our big picture. We note in passing that the far-infrared dust emission is a secondary process, resulting from absorption of

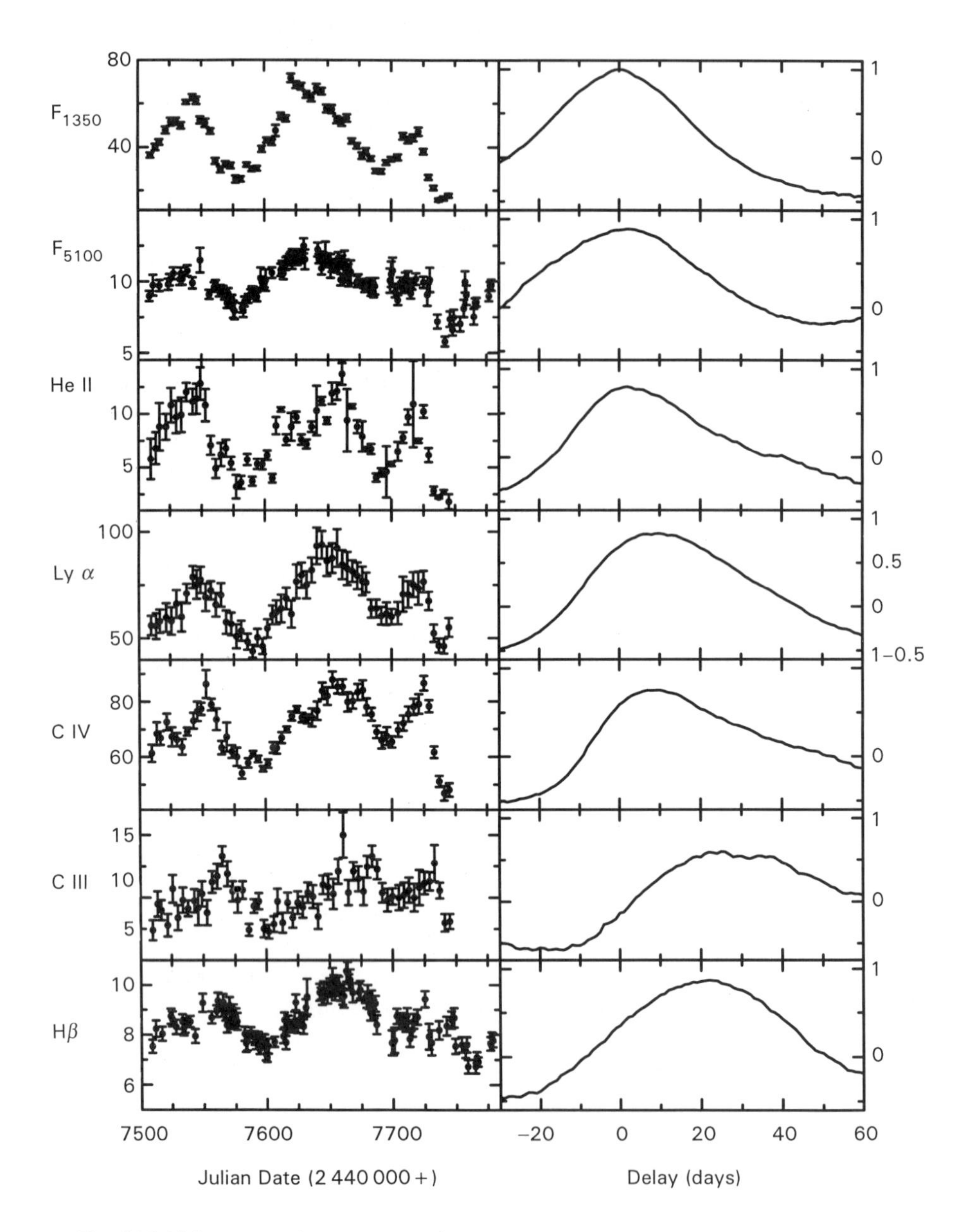

Fig. 5.16. Lightcurves and cross-correlation functions for the Seyfert 1 galaxy NGC5548. The cross-correlations are computed from each of the emission line lightcurves with that of the UV continuum (top left). This shows the pinnacle of what is currently achievable with reverberation mapping techniques and contains a wealth of information. (Courtesy of Brad Peterson and adapted from Peterson, '*Reverberation mapping of the broad-line region in active galactic nuclei*', Ast.Soc.Pacific.Conference Series, 69, 1994.)

energy from the central engine. As such it tells us about the minimum luminosity of the central engine. If the dust is extensive and lies in a spherical zone, then at UV through optical wavelengths it will be able to hide completely the details of the central engine due to severe absorption. Alternatively, the dust might be in the form of a disk, such as a starburst disk of dimensions of order a kiloparsec. This produces the far-infrared–submillimetre emission which may or may not be directly powered by the central engine. Closer to the central engine, the near infrared emission is from dust on the point of evaporation and this probably comes from the molecular torus, the inner face of which (or at least the innermost dust component) lies around 1 pc from the ionizing continuum of the central source.

The emission shortward of about 1 μm is from the long-wavelength end of the big blue bump (and perhaps a contribution from starlight in the parent galaxy for the weaker AGNs). The big blue bump is often assumed to extend to, and join with the soft X-ray excess and in this picture is a single emission component. We stress that this is neither proven, nor universally agreed and in one or two cases there is powerful evidence to suggest that the soft X-ray emission is *not* linked to the big blue bump. Indeed, although the idea that this component is thermal emission from a hot accretion disk surrounding the black hole is attractive, we shall see in the next two chapters that this is rapidly losing ground to other models.

We saw that there is a wealth of data to be obtained in the X-ray region (perhaps too much for the sensitivity of the facilities at the current time). We have gathered by now that this is one of the most important regions for AGN investigations because it arises closest to the central engine and provides the most direct information about the emission processes taking place.

The appearance of a 'soft X-ray excess' seems to be a common feature of AGNs of both type 1 and 2, but the poor spectral resolution means that disentangling the line emission from the continuum is very difficult. The general belief is that the excess is due to radiation from a very hot gas; however, the location of this gas is not clear. The emission could come from the inner edge of the accretion disk, but there are problems with this interpretation and we will resume this discussion after the geometry of the accretion disks have been presented in the next chapter. Much better low-energy X-ray observations are needed to clarify the observational picture further, while regretting the lack of EUV capability.

At higher energies the X-ray emission is complicated but seems to be reasonably well fitted in a variety of objects by a combination of a power-law primary photon spectrum modified by absorption and emission due to 'cold' (almost neutral) gas, along with reflection by Compton scattering off warm (partly ionized) material which produces a flatter spectrum extending to higher energies. The importance of radiating halos and reflection processes will be taken up further in later chapters. Although it is supposed that the power-law is the primary radiating power, from what this derives is far from clear apart from the assumption of a supermassive black hole. These will be discussed further in section 6.6. At even higher X-ray energies ($E \sim 50$ keV and above) breaks in the power-law are apparent in the brighter sources that have so far been observed.

The correlation of the X-ray and longer wavelengths leads to the strong suspicion that the X-rays are the primary radiation compared to the UV–optical, which are reprocessed X-rays. The hard X-ray continuum spectrum is the prime contender and how this reacts with the surrounding medium via absorption and reflection is a vital area of study. More data are needed to clarify the picture. The very short-timescale X-ray variability also points to this region being intimately linked to the central engine, representing very high luminosities emanating from very small regions of space. The observed variability timescales of minutes indicates regions of dimensions of a few tenths of an Astronomical Unit. Furthermore, the chaotic nature of the behaviour is suggestive of a large number of X-ray emitting clouds, of varying scale-size, perhaps contained in a large halo or corona surrounding the central engine.

We will see in the next chapter that there is strong support for a molecular torus. Indeed, the HST imaging of the core of M87 points to a dusty, ionized central zone, extending outwards from around 1 pc to a hundred parsecs or so. In M87 this has the appearance of a possible spiral structure, and we can speculate that this might indicate material being drawn in from the inner parts of the galaxy to fuel the central engine. Currently this is sheer speculation, which we shall pursue in chapters 7 and 9.

The optical and infrared imaging studies show that quasars tend to be found in disturbed galaxies and are frequently surrounded by what are believed to be companions. The possibility of interactions or mergers triggering AGN activity is appealing. This also tends to suggest that because galaxies were closer together in the past, then quasar activity should have been higher at earlier epochs in the Universe. This is indeed borne out by observations.

A key result from the infrared imaging studies is that the old idea of radio-loud quasars being found in elliptical galaxies and radio-quiet quasars in spirals is not tenable. There is now good evidence from ground-based studies to suggest that galaxy luminosity, mass and size are the dominant parameters and that below a certain threshold, galaxies do not form quasars. However, very recent HST observations do not support this conclusion, pointing to the fact that quasars can be found in a wide range of host galaxy mass and luminosity, much weaker than suggested above. Leaving this controversial point aside, the latest data on infrared imaging shows that radio-loud quasars are only found in elliptical galaxies, radio galaxies are also only found in elliptical galaxies, but radio-quiet quasars can be found in either ellipticals or spirals, and with a preference for the latter. We know that as we move down the radio-quiet AGN luminosity league, Seyferts and LINERs are often found in spirals. So a question we need to investigate is what prevents spirals from becoming radio-loud quasars. We know that some Seyferts show weak radio jet-like emission on the parsec scale but this never manages to become a fully fledged kpc scale jet. We will return to this in chapter 9.

So the stage is now set. We see that the improvement in observational techniques has allowed a much clearer picture to be obtained. The imaging capability of the HST, together with the more sensitive and higher spectral resolution X-ray observations, have allowed the central engine to be studied in much greater detail. This in turn has shown the need for more, and even better observations. The major puzzle thrown up by the HST observations of host galaxies of quasars will need to be solved. And so the quest goes on. In chapter 6, we shall explore the central regions of an AGN and consider the primary energy generation mechanisms.

5.6 FURTHER READING

General and reviews

Bregman,J.N., 'Continuum emission from AGNs', *Astron. & Astrophys.Rev.*, **2**, 125, 1990. (An excellent and comprehensive review article—see also his contribution in Courvoisier & Blecha.)

Courvoisier,T.J-L., & Blecha,A., *Multi-wavelength continuum emission of AGN,* IUA symposium 159, Kluwer, 1994. (An up to date series of papers on multiwavelength observations, the latest news, essential reading.)

Courvoisier,T.J.-L., & Robson,E.I., 'The quasar 3C273', *Scientific American*, p50, June 1991. (My favourite quasar.)

Gondhalekar,P.M., Horne,K., & Peterson,B.M., Eds., *Reverberation mapping of the broad-line region in active galactic nuclei,* Astronomical Society of the Pacific Series, vol. 40, 1994. (All there is to know about reverberation mapping.)

Miller,R., & Wiita,P.J., Eds., *Variability in active galactic nuclei,* Cambridge University Press, 1991.

Mushotzky, R.F., Done, C., & Pounds, K.A., 'X-ray spectra and time variability of active galactic nuclei', *Ann.Rev.Astron.Astrophys.*, **31**, 717, 1993. (Required reading.)

Peterson, B.M., 'Reverberation mapping in active galactic nuclei', *Pub.Astronom. Soc.Pac.*, **105**, 247, 1993. (Excellent review.)

Specialized

Continuum energy distribution

Band,D.L., *et al.*, 'Iron K-shell emission from NGC1068', *Astrophys.J.*, **362**, 90, 1990.

Barvainis,R., 'On the optical-to-far-infrared spectral energy distribution of radio-quiet quasars', *Astrophys.J.*, **353**, 419, 1990.

Binette, L., Robinson,A. & Courvoisier, T.J.-L., 'The ionizing continua of active galactic nuclei: Are power-laws really necessary ?', *Astronomy.Astrophys.*, **194**, 65, 1988.

Brunner, H. *et al.*, 'X-ray spectra of a complete sample of extragalactic core-dominated radio sources', *Astron.Astrophys.*, **287**, 436, 1994.

Chini, R. & Krugel, E., 'Dust at high z', *Astron.Astrophys.*, **288**, L33, 1994.

Ciliegi, P., Bassini, L. & Caroli, E., 'On the X-ray spectra of BL Lacertae objects', *Astrophys.J.*, **439**, 80, 1995.

Courvoisier,T.J-L. *et al.*, 'Multi-wavelength observations of 3C273 II, 1986-1988', *Astron.Astrophys.*, **234**, 73, 1990.

Courvoisier, T.J.-L., 'Multiwavelength properties of active galactic nuclei', *Astrophys.J.Suppl.*, **92**, 579, 1994.

Dunlop,J.S. *et al.*, 'Detection of a large mass of dust in a radio galaxy at redshift z = 3.8', *Nature*, **370**, 347, 1994.

Edelson,R., & Malkan,M., 'Spectral energy distributions of active galactic nuclei between 0.1 and 10 microns', *Astrophys.J.*, **308**, 59, 1986.

Guilbert,P.W., & Rees,M.J., "Cold' material in non-thermal sources', *Mon.Not.R. Astron.Soc.*, **233**, 475, 1988.

Gear,W.K., *et al.*. 'A comparison of the radio-submillimetre spectra of BL Lacertae objects and flat-spectrum radio quasars', *Mon.Not.R.Astron.Soc.*, **267**, 167, 1994.

Hughes,D.H., Robson,E.I., Dunlop,J.S. & Gear,W.K., 'Thermal dust emission from quasars I, Submillimetre spectral indices of radio-quiet quasars', *Mon.Not.R. Astron.Soc.*, **263**, 607, 1993.

Kruper, J.S., Urry, C.M., & Canizares., C.R., 'Soft X-ray properties of Seyfert Galaxies. I. Spectra', *Astrophys.J.Suppl.*, **74**, 347, 1990.

Lawrence,A., *et al.*, 'Millimetre measurements of hard X-ray selected Active Galaxies : implications for the nature of the continuous spectrum', *Mon.Not. R.Astron.Soc.*, **248**, 91, 1991.

Lichti, G.G., *et al.*, 'Simultaneous and quasi-simultaneous observations of the continuum emission of the quasar 3C 273 from radio to γ-ray energies', *Astron.Astrophys.*, **278**, 711, 1995.

Lightman,A.P., & White,T., 'Effects of cold matter in active galactic nuclei: a broad hump in the X-ray spectra', *Astrophys.J.*, **335**, 57, 1988.

Litchfield,S., Robson,E.I., & Stevens,J.A., 'Infrared photometry of Blazars: 8 years of observations', *Mon.Not.R.Astron.Soc.*, **270**, 341, 1994.

Macomb,D.J., *et al.*, 'Multiwavelength observations of Markarian 421 during a TeV/X-ray flare', *Astrophys.J.*, **449**, L99, 1995.

Malkan, M.A., *et al.*, 'The ultraviolet excess of Seyfert 1 Galaxies and quasars', *Astrophys.J.*, **254**, 22, 1982.

Maraschi, L., *et al.*, 'The 1993 multiwavelength campaign on 3C 279: The radio to gamma-ray energy distribution in low state', *Astrophys.J.*, **435**, L91, 1994.

Marscher,A.P., 'Contemporaneous X-ray and VLBI observations of the quasar NRAO140', *Astrophys.J.*, **334**, 552, 1988.

Nandra, K., *et al.*, 'Detection of iron features in the X-ray spectrum of the Seyfert I Galaxy MCG-6-30-15', *Mon.Not.R.Astr.Soc.*, **236**, 39, 1989.

Nandra,K., & Pounds,K.A., 'Highly ionized gas in the nucleus of the active galaxy MCG-6-30-15', *Nature*, **359**, 215, 1992.

Netzer, H., Turner,T.J., & George,I.M., 'X-ray color analysis of the spectra of active galactic nuclei', *Astrophys.J.*, **435**, 106, 1994.

Owen, F.N., & Laing,R.A., 'CCD surface photometry of radio galaxies - I. FR class I and II sources', *Mon.Not.R.Astron.Soc.*, **238**, 357, 1989.

Puetter, R.C., *et al.*, 'Optical and Infrared spectrophometry of the quasi-stellar objects: The spectra of 14 QSOs', *Astrophys.J.*, **243,** 345, 1981.

Punch, M., *et al.*, 'Detection of TeV photons from the active galaxy Markarian 421', *Nature*, **358**, 477, 1992.

Robson,E.I., 'Millimetre and submillimetre observations of Blazars', in *Submillimetre Astronomy,* Eds. Watt,G.D., & Webster,A.S., Kluwer, p215, 1990.

Robson,E.I., Hughes,D., & Gear,W.K., 'The importance of submillimetre observations of radio-quiet and radio-loud quasars', in *From ground-based to space-borne submm astronomy,* ESA **SP-314**, p111, 1990.

Robson,E.I., *et al.*, 'The infrared -millimetre-centimeter flaring behaviour of the quasar 3C273', *Mon.Not.R.Astron.Soc.*, **262**, 249, 1993.

Robson,E.I., *et al.*, 'A new infrared spectral component of the quasar 3C273', *Nature*, **323**, 134, 1986 .

Sanders,D.B., *et al.*, 'Continuum energy distributions of quasars: slopes and origins', *Astrophys.J.*, **347**, 29, 1989.

Stevens,J.A., *et al.*, 'Multifrequency observations of Blazars V, Long term millimetre, submillimetre and infrared monitoring', *Astrophys.J.*, **437**, 91, 1994.

Von Montigny, C., *et al.*, 'High energy gamma ray emission from active galaxies: Egret observations and their implications', *Astrophys.J.*, **440**, 525, 1995.

Véron-Cetty,M.P., & Woltjer,L., 'Galaxies around luminous quasars', *Astron. Astrophys.*, **236**, 69, 1990.

Weaver,K.A., *et al.*, 'An ASCA X-ray spectrum of a prototype bare Seyfert 1 nucleus in MCG –2-58-22', *Astrophys.J.*, **451**, 147, 1995.

Imaging studies

Abraham, R.G., McHardy, I.M., & Crawford, C.S., 'Optical imaging of the BL Lac host galaxies', *Mon.Not R.Astron.Soc.*, **252**, 482, 1991.

Bahcall,J.N., Kirhakos,S., & Schneider,D.P., 'HST images of nearby luminous quasars', *Astrophys.J.*, **435**, L11, 1994.

Bahcall,J.N., Kirhakos,S., & Schneider,D.P., 'PKS 2349-014: a luminous quasar with thin whisps, a large off-center nebulosity and a close companion galaxy', *Astrophys.J.*, **447**, L1, 1995.

Dunlop, J.S., Tayler,G.L., Hughes,D.H., & Robson,E.I. 'Infrared imaging of the host galaxies of radio-loud and radio-quiet quasars', *Mon.Not.R.Astron.Soc.*, **264**, 455,1993.

Falomo, R., Pesce, J.E., & Treves, A., 'Host galaxy and environment of the BL Lactertae object PKS 0548-322: Observations with subarcsecond resolution', *Astrophys.J.*, **438**, L9, 1995.

Hutchings,J.B., & Neff,S.G., 'Optical imaging of QSOs with 0.5 arcsec resolution', *Astronom.J.*, **104**, 1, 1992.

Hutchings,J.B., *et al.*, 'HST imaging of quasi-stellar objects with WFPCII', *Astrophys.J.*, **429**, L1, 1994.

Hutchings, J.B., 'Hosts of possibilities', *Nature*, **373**, 118, 1995.

McHardy, I.M., *et al.*, 'Hubble Space Telescope observations of the BL Lac object PKS 1413 + 135 : the host galaxy revealed', *Mon.Not.R.Astron.Soc.*, **268**, 681, 1994.

McLeod, K.K., & Rieke, G.H., 'Near-infrared imaging of low-redshift quasar host galaxies', *Astrophys.J.*, **420**, 58, 1994.

Mcleod, K.K., & Rieke, G.H., 'Near-IR imaging of low-redshift quasar host galaxies. II. High luminosity quasars', *Astrophys.J.*, **420**, 58, 1994.

Pesce, J.E., Falomo,R., & Treves, A., 'Imaging and spectroscopy of galaxies in the fields of five BL Lacertae objects', *Astron.J.*, **107**, 494, 1994.

Zhou,S., Wynn-Williams,C.G., & Sanders,D.B., 'Imaging of infrared luminous galaxies at 3.4 microns', *Astrophys.J.*, **409**, 149, 1993.

Variability studies

Clavel, J., *et al.*, 'Correlated hard X-ray and Ultraviolet variability in NGC5548', *Astrophys.J.*, **393**, 113, 1992.

Courvoisier,T.J-L., *et al.*, 'Rapid infrared and optical variability in the bright quasar 3C273', *Nature*, **335**, 330, 1988.

Dietrich, M., *et al.*, 'Steps toward determination of the size and structure of the broad-line region in active galactic nuclei. IV. Intensity variations of the optical emission lines of NGC5548', *Astrophys.J.*, **408**, 416, 1993.

Edelson, R., *et al.*, 'Multiwavelength monitoring of the LB Lacertae object PKS 2155-304. IV. multiwavelength analysis', *Astrophys.J.*, **438**, 120, 1995.

Edelson,R., & Krolick,J.H., 'The discrete correlation function: a new method for analyzing unevenly sampled data', *Astrophys.J.*, **333**, 646, 1988.

Macomb,D.J., *et al.* 'Multiwavelength observations of Markarian 421 during a TeV/X-ray flare', *Astrophys.J.*, **449**, L99, 1995.

McHardy,I.M., 'X-ray variability of active galactic nuclei', Proc 23rd ESLAB Symposium on X-ray astronomy', ESA SP-296, 1989.

Peterson, B.M., *et al.*, 'Steps toward determination of the size and structure of the broad-line region in active galactic nuclei. VII variability of the optical spectrum of NGC 5548 over 4 years', *Astrophys.J.*, **425**, 622, 1994.

Reichert,G.A., *et al.*, 'Steps toward determination of the size and structure of the broad-line region in active galactic nuclei. V. variability of the ultraviolet continuum and emission lines of NGC3783', *Astrophys.J.*, **425**, 582, 1994.

Ulrich,M.H., *et al.* 'detailed observations of NGC4151 with IUE - III variability of the strong emission lines from 1978 February to 1980 May', *Mon.Not.R.Astron.Soc.*, **206**, 221, 1984.

Ulrich,M.H., *et al.* 'Narrow and variable lines in the ultraviolet spectrum of the Seyfert galaxy NGC4151', *Nature*, **313**, 747, 1985.

6

The central kiloparsec

6.1 INTRODUCTION

This chapter plunges into the heart of the activity of active galaxies, the physical processes occurring in the central kiloparsec. We already know that observational evidence points to the region being photoionized by a very hot and intensely luminous central source. This is responsible for the highly ionized emission lines of the broad-line region (BLR), the narrow-line region (NLR) and we presume the UV to the X-ray continuum emission. The BLR and NLR were briefly described in chapters 2 and 4 and in this chapter we will investigate them in much greater detail, specifically saying how they can be used to tell us about the central engine. Extensive imaging spectroscopic observations have enabled astronomers to construct and refine models describing the physical conditions of the ionized gas in the central kiloparsec. These models describe the geometry, ionization parameters and density of the gas. These studies have led to what is known in the field as the 'Standard Model' describing the ionization of the central zone. We will describe the Standard Model in some detail, highlighting its successes and noting its shortcomings. The outer boundary of the central ionization zone is generally defined by the NLR. This extends to a radial distance of about 1 kpc from the centre.

6.2 THE BROAD-LINE REGION (BLR)

6.2.1 Introduction

One of the most important observational aspects of the powerful AGNs are the Broad- and Narrow-Line Regions. Unfortunately, although this topic is well described in a number of highly specialised texts, very few non-specialist books give a simple description of the physical processes taking place within these regions. This is not surprising as we saw in chapter 4 that these processes are very complex. With this in mind we shall tread a careful path and describe the observations and resulting physical arguments as simply as possible, yet in sufficient detail to be meaningful to the student. We will utilize many of the aspects of line emission discussed in sections 2.7 and expanded in section 4.4. We will then apply the relevant theories to the investigations of the innermost emission line zone of an active galactic nucleus, the BLR.

We saw in section 3.2 that the BLR is associated with broad and strong emission lines with velocity widths up to 10,000 km s^{-1}. As we saw in section 4.4, model calculations describing the physics of the Broad- and the Narrow-Line Regions must eventually be able to reproduce the observational data such as shown in figs. 3.6 and 6.1.

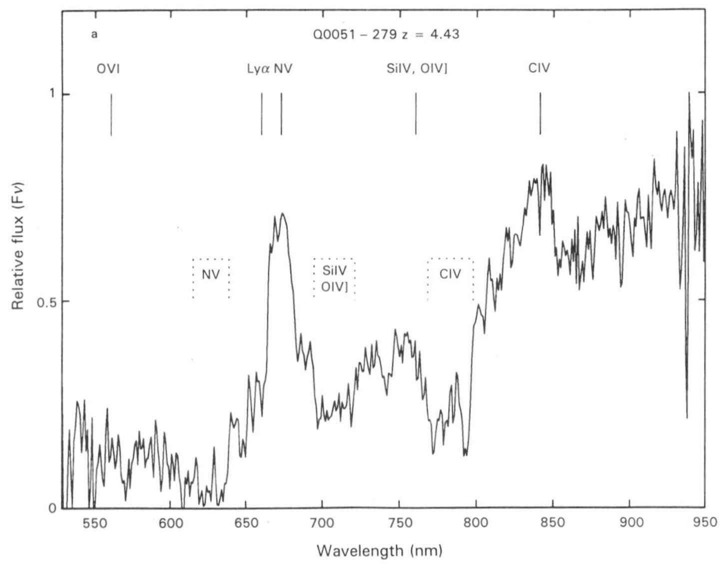

Fig 6.1. Emission spectrum of the quasar 0051-279. Although this is much fainter than the spectrum of the quasar shown in fig. 3.6, the improvement in technology allows identification of the emission lines and the redshift of z = 4.43 to be determined. This spectrum was obtained using the 4 m optical telescope at Kitt Peak. (NOAO).

Let us begin by considering the processes by which astrophysicists commence their quest for understanding the physical conditions within the BLR. Assuming that nothing is *a priori* known about the source, the first task is to determine the redshift of the quasar in order to calculate line luminosities, etc. This requires identification of a pattern of lines corresponding to a particular element. Remarkable progress has been made in this quest, identifying redshift 4 to 5 quasars with a high degree of confidence. For medium- to high-redshift quasars, the very broad lines of hydrogen Lyman alpha (Lα, level 2-1, λ_0= 121.6 nm), CIV (λ_0= 154.9 nm) and semi-forbidden CIII] (λ_0=190.9 nm) are often identified first as these are usually the strongest and most prominent lines in the *visible* spectrum, having been redshifted from their rest wavelengths. Although strong, hydrogen beta (Hβ, 4-2, λ_0= 486.1 nm) and especially hydrogen alpha (Hα, 3-2, λ_0= 656.3 nm) are often redshifted out of the visible band to the near infrared for all but the low-redshift quasars. We will see later that another important line of hydrogen is now frequently measured, the infrared line of Paschen alpha (Pα, 4-3, λ_0= 1.87 μm). In the case of a low-redshift quasar, the forbidden transition of O[III] (λ_0= 500.7 nm) may also be very prominent in the visible spectrum. Note, for high-redshift objects such as in fig. 6.1, the spectrum is very much weaker than in fig. 3.6 but the improvement in technology has allowed adequate signal-to-noise to be obtained so that line identification and hence redshift can readily be determined.

Before moving farther into the topic of emission lines, let us remind ourselves that there are two kinds of transitions: permitted and forbidden. The hydrogen lines above are all permitted lines resulting from recombinations from photoionized states. When discussing the BLR region in AGNs, particularly in photoionization modelling, astronomers often refer to the hydrogen lines as the permitted lines while the rest of the lines are all assumed to be forbidden. This is not strictly accurate because there are other permitted lines observed apart from hydrogen.

wavelength (nm)	element	mean observed relative intensity
103.4	OVI	20
121.6	HI (Lα)	100
124.0	NV	20
140.7	OIV]	10
154.9	CIV	50
164.0	HeII	5
190.9	CIII]	20
279.8	MgII	20
486.1	HI (Hβ)	20
656.3	HI (Hα)	100
1875.1	HI (Pα)	6

Table 6.1. The principal emission lines from the BLR and illustrative values of their relative intensities. Other lines with intensities less than 5% come from NIV] 148.8, OIII] 166.3, HeII 468.6 and HeI 587.6

The determination of the redshift enables the remaining spectral lines to be identified. This produces a list of elements with their corresponding line strengths, linewidths and excitation/ionization states. The lines are separated into the broad and narrow-line regions and a representative list for the stronger lines from the BLR is given in Table 6.1. The abundances of atomic species and ionization states can then be determined. This is far from simple for the weaker, or unexpected lines, but as we will see later is a necessary step as these lines are important diagnostic tools.

It turns out that similar lists had long been known to astronomers studying photoionized emission line nebulae such as HII regions and planetary nebulae in our Galaxy. (Note that the latter also additionally show prominent emission from [OIII].) However, although we believe the basic physics is generally the same in AGNs, the details of the picture are harder to determine. As well as having a much more energetic ionization source, AGNs are very much more complex entities. However, from the outset there was always the firm belief on the part of the workers in the field that the observed emission lines would eventually enable a 'complete picture' of the central region of the AGN to be obtained, including the luminosity of the central ionizing source. This faith was well placed because notable success had been achieved for Galactic HII regions. Their gas densities, chemical composition and luminosities were successfully derived from their observed line spectra by pioneering workers such as Don

Osterbrock and Lyman Spitzer (see section 6.8). In principle, given this success, it should be relatively straightforward to undertake detailed calculations of atomic transition probabilities and radiation transfer (including the effect of any dust in the region) to derive the physical parameters of the emitting gas from the observed spectrum and thence to the source of the energy powering the emission. Alas, reality is far from this simple.

Although we defined the lines in the BLR to be those with the very high-velocity wings of FWHM up to 10,000 km s^{-1} and occasionally even up to ~15,000 km s^{-1}, these are extreme values. A more typical value is ~5,000 km s^{-1}. In the discussion of the classifications of Seyfert galaxies in chapter 3 we saw that improved spectroscopic capability revealed that the Balmer line profiles (for example) often show both a broad and narrow component. These are manifested as a very broad pedestal on top of which is superimposed a much narrower core (e.g. fig. 9.5). The wide wings of the pedestal that define the BLR and the assumption is made that all lines with the same FWHM wings come from the same region of space.

How big is the BLR? No BLR has yet been directly imaged; however, we saw in section 5.4 that an estimation of the scale size is possible through variability studies. Variations in the line shapes and strengths have been observed on timescales of less than a year. We therefore know that at least some of the lines in the BLR come from regions of space of the order of 1 light year or less. If we assume an order-of-magnitude size of 1 pc for the radius of the BLR, then at a relatively nearby distance of 10 Mpc it will subtend an angle of only 50 milli-arcseconds. We immediately see that progress in understanding the BLR derives essentially from observational spectroscopy and theoretical modelling.

The observed emission lines lead to estimated values of density as high as 10^{16} m^{-3} and effective temperatures in the range ~ 10^4 K. We should note that this is not a very high temperature in terms of high-energy phenomena, being only the photospheric temperature of a star such as Vega. We can now commence our modelling treatment by comparing some of the most prominent lines and see where this leads.

6.2.2 The problem of the Balmer decrement

In section 4.4 we saw that for photoionized zones the Balmer decrement (commonly referred to as the ratio of the strengths of the Hα/Hβ lines) should have a ratio lying in the range 2.75 to 3. We saw that this is supported by observations of well-studied Galactic HII regions and so the expectation might be that AGNs should also fall in line. However, this is not the case. The observed Balmer decrement nearly always exceeds a value of 3, often by large amounts. Clearly this must be telling us something important. There are a number of possible explanations, the main ones being that either the Case B approximation is invalid, or that dust is preferentially absorbing Hβ.

In principle we can observationally test for the latter point by comparing the flux ratios of Pα and Hβ. These both arise from the same atomic level (n = 4) and so any differences between observation and theory should not be due to any temperature or density effects changing the population of the levels. On the other hand Hβ would be very sensitive to dust absorption compared to the infrared Pα line. Such observational studies were a prime target for the new generation of sensitive spectrometers operating on infrared telescopes at high, dry observatory sites. However, progress was much slower than anticipated. In practice it turns out that there are two further problems in obtaining an unambiguous answer to the puzzle of the Balmer decrement.

The first comes down to the very small angular extent of the BLR compared to the beamsizes of the first spectrometers. These are generally a few arcseconds in diameter, and at best one arcsecond, making it difficult to obtain a high contrast between the lines and the continuum. The observations failed to reveal a high degree of reddening for many objects (although see the discussion on unification in chapter 9, particularly for objects such as Arp 220), suggesting that although dust may be causing some absorption it is not the cause of the Balmer decrement discrepancy. Furthermore, dust grains can only exist up to a maximum temperature of ~1,700 K before they evaporate. This immediately poses the question as to whether dust grains could co-exist with gas in the UV-energetic environment of the BLR. Most experts in the field now consider this unlikely, and believe that any dust will lie outside the main BLR zone. Therefore we are left with the possibility that the observed absorption is due to dust grains along the line of sight to the observer but external to the BLR. This is an important point because the geometry should be orientation-dependent, to which we will return later and discuss in detail in chapter 9.

The second problem of the Balmer decrement calculation is that the Pα/Hβ ratio turns out to be model-dependent in terms of the geometry of the BLR. We shall return to this later. A further possible mechanism to increase the Balmer decrement is to selectively over-populate level 3 with respect to level 4. Although this will obviously produce an enhanced Balmer decrement, before getting too excited we must ask whether the potential ways of achieving this in the BLR are realistic. One mechanism is by over-populating the lower levels through collisional excitation by free electrons in a dense plasma. Although collisional de-excitation can then take place, radiative decay leads to enhancement of Hα/Hβ. We remember that electron collisional excitation was ignored in the Case B situation in that the appropriate term was equated to zero.

Finally, yet another method of explaining an increase in the Balmer decrement seems to be favoured by theoreticians. This is to evoke a mechanism termed Balmer line self-absorption. Here, level 2 is over-populated in the medium and the Balmer lines are thereby self-absorbed, producing an extremely complicated chain of interactions in the plasma eventually leading to an over-population of level 3 with respect to level 4.

The upshot of all this is that using the Balmer decrement as a test of recombination line theory and equilibrium Case B conditions fails spectacularly. So what else can be used to test the hypothesis that Case B conditions apply? It also predicts that the ratio of Lα:Hα should be about 13, which is supported by observations of HII regions. However, measurements of this ratio in a wide variety of AGNs produce results much closer to unity, an order of magnitude away from the 'correct' result. So what do astrophysicists resort to in order to explain this major embarrassment?

One answer is to look to more complicated models (see later). These become necessary mainly because of the belief that dust absorption is not going to save the day. Although we will not challenge this belief, it should be remembered that Lα is extremely susceptible to absorption by any dust present in the medium through which the photons travel. As we saw in section 4.4.3, for a centrally photoionized region, the Lα is 'trapped'. Remember, this occurs because most of the hydrogen is in the ground state and Lα photons only travel a very short distance before they excite another hydrogen atom. The Lα generated by the subsequent radiative de-excitation is emitted isotropically with respect to the exciting photon. Therefore the Lα has to random-walk its way out of the region. This involves a huge increase in the path-length of the Lα photons through the medium, making the possibility of absorption by even minimal amounts of dust very high. As an example, a reduction of a factor of five in the Lα:Hα ratio requires less than a magnitude of extinction, and this is not easy to distinguish accurately in AGNs by observation.

A further approach is to use a much higher-energy continuum spectrum for the ionizing source. Although this lowers the Balmer decrement back towards the expected value, it is not in agreement with the observed continuum spectra of a number of objects. However, all is not yet lost because although this high-energy continuum should reveal significant UV and soft X-ray emission, we saw in chapter 5 that the high UV to low-energy X-ray continuum emission regime is not that well determined due to the EUV 'gap'. We know that high-energy UV-X-ray emission has been detected from many AGNs, but whether it arises from the ionizing continuum or is the result of secondary processes such as scattering is still not absolutely clear (but see later and chapter 9).

It is also worthy of note to point out that if electron collisional excitation were at work, then as we saw above, this would lead to the lower levels being preferentially populated. But we would then expect the ratio of Lα/Hα to increase, rather than decrease. On the other hand, the Balmer line self-absorption produces effects that are in the correct direction; Lα is reduced with respect to Hα, thereby providing another reason for the enthusiasm supporting this explanation.

Given all these problems relating to Lα and Hβ, one might begin to wonder why bother going to all the trouble to measure them. The answer is that in spite of the difficulties of interpretation, they still remain one of the best tools for providing a framework describing the physical processes within the central regions of an AGN. We end this section by noting that if only Lα is observable, resulting predictions are prone to a large uncertainty.

6.2.3 The ionization parameter

An alternative method of determining the luminosity of the ionizing continuum is by making use of different lines and the *ionization parameter*. This can be defined in a number of ways but we will use the most common form which is given below. The ionization parameter is basically the ratio of the ionizing photons to free electrons, i.e.

$$U = \frac{Q}{4\,\pi\,r^2\,c\,N_e}, \qquad (6.1)$$

where r is the distance from the source of ionizing photons to the gas atoms, c is the velocity of light, N_e is the free electron number density and Q is the ionization rate given by

$$Q = \int_{\nu=\nu 1}^{\infty} \frac{L_\nu}{h\nu} \, d\nu \, , \tag{6.2}$$

with L_ν being the luminosity of the source at a frequency ν and the integral being taken over all frequencies *above* that of the ionization frequency. We see that U depends on the source luminosity, the distance of the cloud from the source, and the free electron number density, N_e. The luminosity is the product of the strength of the ionizing continuum and the shape of the spectrum. If the continuum flux spectrum has an index flatter than −1, then the luminosity is dominated by the highest-frequency photons (see below) and it is these photons that provide the most ionization. In modelling the ionization parameter, a range of values are allowed for the continuum shape due to the observational uncertainty in determining the precise value. The ability to vary this critical parameter has been a great boon to modellers. Somewhat with tongue in cheek it is also true to say that it has maintained the number of free parameters that have allowed model fits to remain valid as better observational data further constrain the available parameter space. To pin down the shape and strength of the ionizing continuum is one of the most important series of observations to be made. The eventual success will severely limit the freedom of theoretical models of the BLR. Based on a number of studies, the current estimates of the ionization parameter tend to favour values around $U \sim 2 \times 10^{-3}$.

We will now see how, in principle, we can deduce the physical picture of the ionized region from the very detailed model calculations. These are both complex and numerous and require large computer codes solving many simultaneous equations of reactions in order to determine most of the required parameters. The computational models boil down to either concentrating on the atomic physics and the energy balance within the region, or concentrating on a more exact treatment of the extremely complex radiative transfer equation. In so doing we shall consider the so-called 'standard model' that has produced much of the impetus for the work in this area.

6.3 THE STANDARD MODEL AND SUPPORTING EVIDENCE

6.3.1 Introduction

The standard model was introduced in the late 1970s and many of those involved had previously been studying planetary nebulae in our Galaxy. These objects seemed to be the physical systems that most closely resembled the emission line regions of AGNs. Some of the limitations of the modelling process were revealed in section 6.2 and the standard model is now known to be only a zeroth order approximation. Nevertheless, it has been a powerful tool in allowing a basic understanding of the physical processes occurring within the central kiloparsec of an AGN. Although there have been many modifications to the model, these have mostly been of a minor nature compared with the overall framework, which has remained to this day.

Many books and conference proceedings have been devoted to the topic of the 'standard model' and the central zone of an active galaxy. Unfortunately the modelling results are still far from what astronomers desire and in some respects this topic has faded from fashion in favour of accretion disks and gas flow studies. In this section we will review the main aspects of the standard model, realize its limitations, but applaud the achievements. There is no doubt that testing the predictions of the standard model has spurred more detailed and penetrating observations to be undertaken. Theoreticians have been goaded to even greater heights of imagination to produce new offshoots to the model when predictions are not borne out by the new observations. Not to be outdone, the computational modellers have strained the available computing power to the limit in order to produce synthetic spectra with which to confront the observational data. Again we see the power of scientific progress at work.

6.3.2 The framework of the standard model

The starting point for the standard model is the assumption that the BLR is driven by photoionization from a hot and highly luminous central object. This ionizing continuum source is significantly more energetic than starlight in terms of higher-energy photons. We can feel secure with this assumption because the observations demonstrate that for the AGNs that show a prominent BLR, the wide variety of ionization stages present are much greater than can be produced by normal starlight continuum. The evidence for photoionization providing the energy source comes from three main aspects: (i) observations of forbidden lines yield electron temperatures in the range 15,000 K to 20,000 K, which is characteristic of a gas energized by photoionization; (ii) the luminosity in the permitted lines is well correlated with the luminosity of the underlying continuum; (iii) the intensity ratios of lines are more indicative of photoionization than of shock excitation.

Determining accurately the physical parameters of the region from the observed line spectra is incredibly difficult. The problems associated with this goal were described in the previous section. The standard model has been improved since its introduction through better understanding of atomic cross-sections and more detailed line trapping-mechanisms. The ability to include dust to provide a degree of relative reddening of the lines has also helped. These modifications, however, have not dramatically changed the predicted intensities of the strongest lines. This should not be too surprising as it is these lines that provide the main cooling for the hot gas and as such they are not very dependent on the intricate details of the model. Indeed, it was these lines that were observed in all the early spectra and so the main framework of the model was built around them. The weaker lines, on the other hand, are much more model-sensitive. It is in the prediction of the strengths, numbers and ratios of the weaker lines that most changes have occurred since the early model calculations of workers such as Davidson, Netzer, MacAlpine and Baldwin, subsequently followed by Kwan and Krolik and others in the 1980s.

The central continuum source ionizes a series of components within the surrounding gas, which are in turn confined by some mechanism. This could be the gas pressure of a hot surrounding medium (see, e.g., Davidson and Netzer, 1979) and we will return to this important point later. Within this framework the model must be able to reproduce the emergent observed spectrum (taking into account any extraneous factors such as dust between the BLR and the observer). It should be able to deduce the strength and shape of the ionizing source of continuum radiation and the geometry of the confined region, in particular the variation of density and other parameters with radius from the

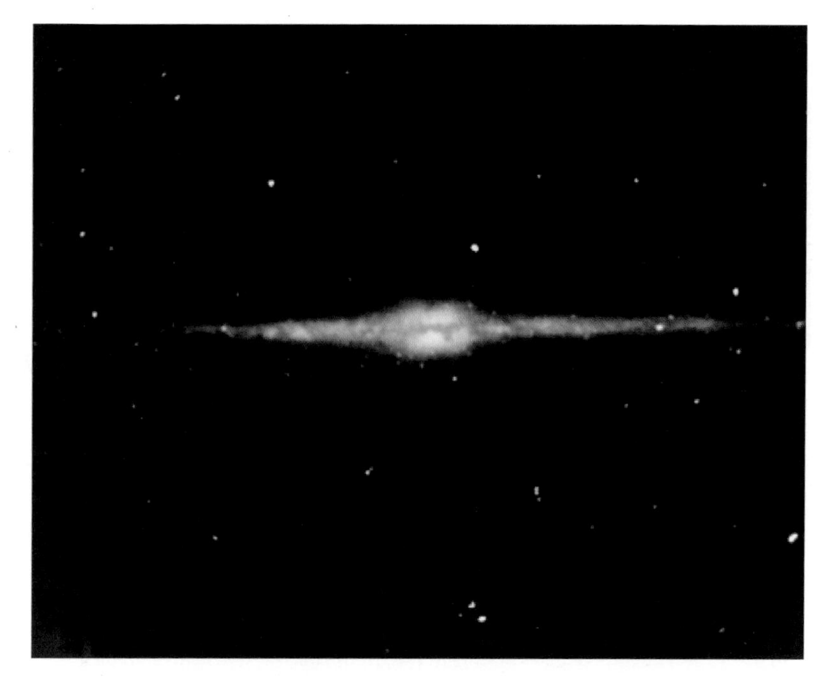

Plate 1. A composite 3-colour infrared image showing the central bulge and the narrow plane of the Milky Way Galaxy. (Courtesy of NASA/ Goddard Space Flight Center "COBE Science Working Group".

Plate 2. The radio and optical emission from NGC5128 (Centaurus A). (Radio data courtesy of NRAO/AUT, optical photograph courtesy of the AAO).

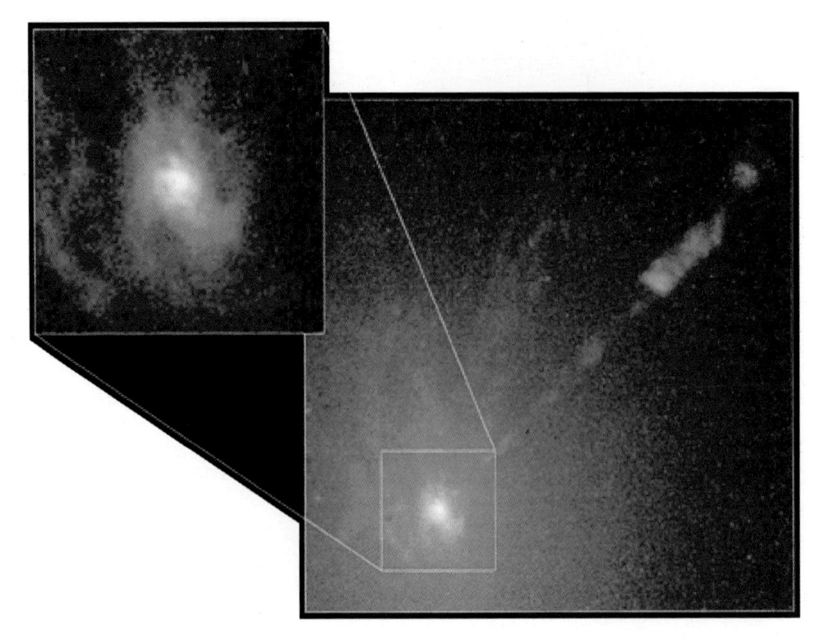

Plate 3. A Hubble Space Telescope (HST) image of the central region of M87 showing the synchrotron jet and the nuclear core. (Figure courtesy of STScI).

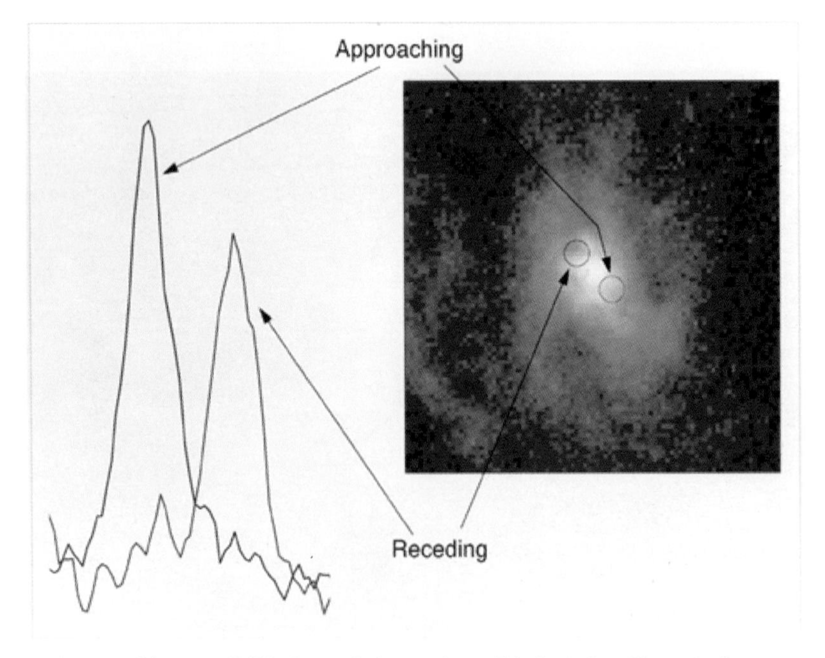

Plate 4. Close up HST view of the nuclear disk in M87. The spiral configuration is clearly visible and the spectroscopic data strongly suggest the presence of a supermassive black hole in the nucleus. One arcsecond corresponds to a distance of about 77pc in M87. (Figure courtesy of STScI).

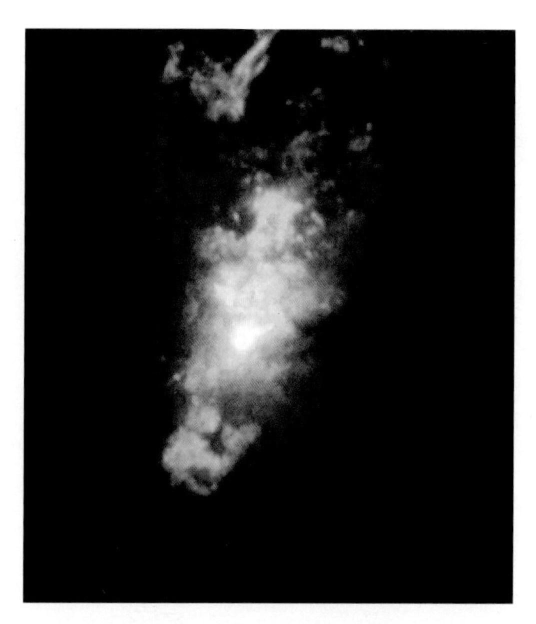

Plate 5. HST composite image of the central regions of the Seyfert 2 galaxy NGC1068 in which the ultraviolet continuum is coloured blue, the emission through an ultraviolet filter is shown as red and the emission from [OIII] is coloured green. The NLR is seen to be extremely complex in structure and the anisotropic emission is readily apparent. (Courtesy of Macchetto *et al.,* Astrophys.J., **435**, L15, 1994).

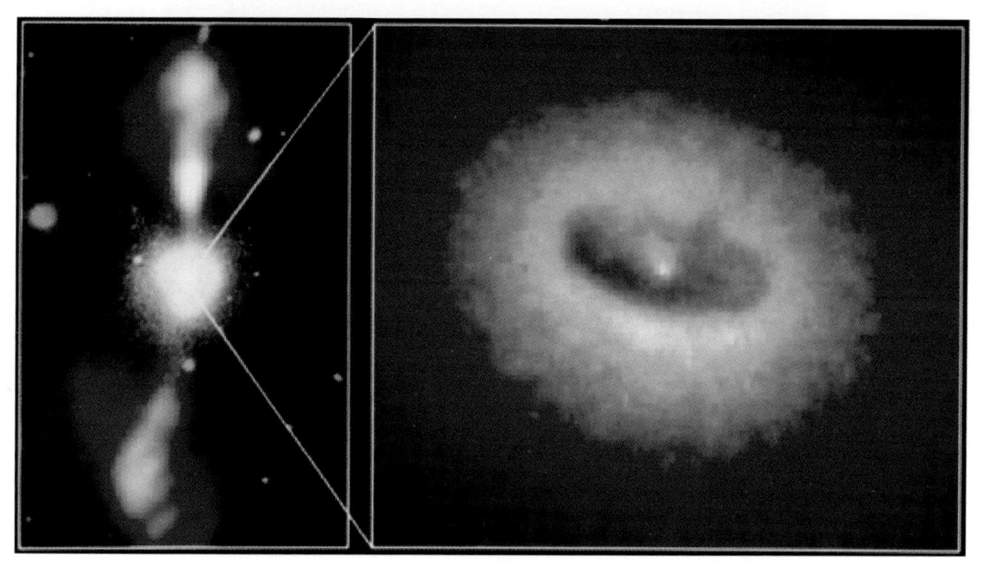

Plate 6. HST image of the nuclear regions of NGC4261 showing a disk of gas and dust lying perpendicular to the axis of the radio emission which is shown in the expanded view on the left by the twin jets and plumes. The optical galaxy is shown located between the radio jets. (Courtesy of STScI).

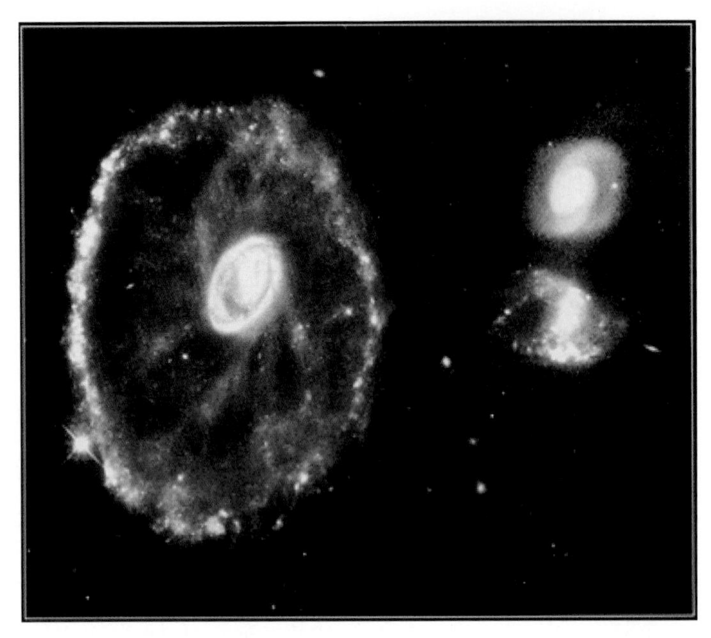

Plate 7. A spectacular HST image of the result of a collision between a spiral (target) galaxy and a smaller (projectile) galaxy which has scored a direct hit, leaving behind a ring of gas which shows extensive star formation. The 'ring' galaxy is known as the Cartwheel and one of the galaxies to the right is probably the impacting object. (Courtesy of STScI).

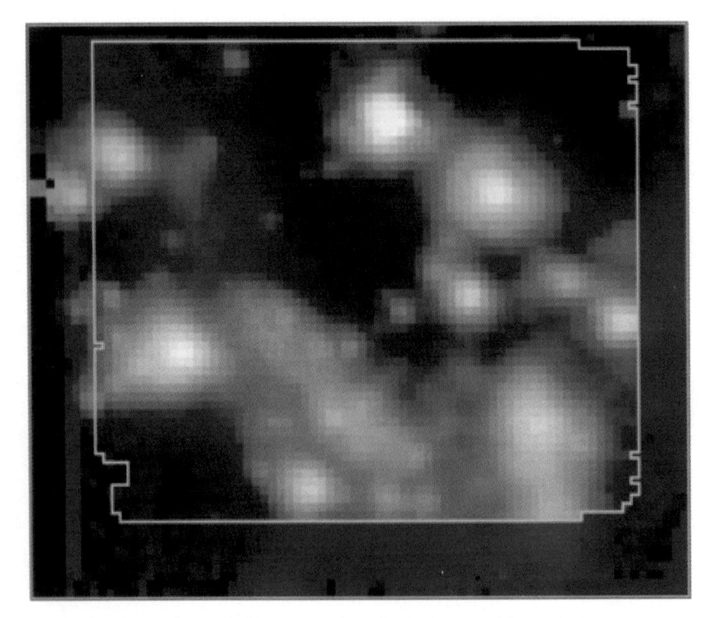

Plate 8. Fabulous three-colour composite (2.2μm-blue, 3.8μm-green and 4.5μm-red) infrared images of the central 0.5 pc of the Galactic Centre region. The blue object at the top is IRS7 and the IRS16 complex is the group of faint blue objects in the centre. (Courtesy of Tom Herbst, from Herbst, Beckwith & Shure, Astrophys.J, **411**, L21, 1993).

central source. These are the key outputs.

But to what observational spectrum should the model spectra be fitted? Given that quasar and Seyfert 1 line spectra nearly all differ in the fine details, is there an 'average' spectrum that could be said to be representative of the BLR? This could be derived by either taking a large and carefully selected sample of objects and averaging their spectra, or, alternatively, selecting a truly 'average' object and using its spectrum. This fundamental starting point for model fitting is by no means trivial, given that the objects range over a large volume of redshift space and one might *a priori* expect that evolution effects could be important, especially in the abundances of the heavier elements. Perhaps surprisingly, this does not seem to be the case. From the highest redshift quasars to the more local ones there seems to be little evidence of chemical evolution (indicating that substantial star formation and mixing occurred at redshifts earlier than $z \approx 4$ in the Universe).

Another difficulty of determining an average emission line spectrum is the degree to which adjacent lines in the spectrum merge due to their very broad profiles. This is known by an exquisite technical term called blending. This can be particularly important for the weak lines as these are often swamped by the broad wings of strong adjacent lines. Another hazard is the blending of many lines from a single species. This was highlighted by the 1985 discovery of the 'iron forest', when it was shown that emission from singly ionized iron (FeII) produces a very dense forest of emission lines from the BLR. When taken together, these contributed nearly as much energy to the cooling budget as found in the Lα flux. This iron flux is observed in a number of wavelength regimes, the main one lying between 200–300 nm, and it contributes to the 'small blue bump' observed in this region of the spectrum as we saw in section 5.2.3.

Another important detail of the model pertains to the wavelength range over which the continuum spectrum is to be fitted. Not surprisingly the optical waveband is the main region. Although we know that the infrared wavelengths carry little energy, they could be extremely important because they are relatively impervious to the effects of dust absorption. By contrast, the UV is a high-energy and vitally important region, revealing the powering energetics of the zone. But perversely, lines in this region are very severely affected by absorption from any dust present. We saw in section 6.2 that comparison of UV to optical line ratios is hazardous to say the least. Another problem with the UV lines is that there is a dearth of telescopes with which one can make observations, mainly IUE and now the HST. This is why observations of ultraviolet lines are somewhat simpler for medium- to high-redshift quasars, where the lines are redshifted into the optical, although absorption by intervening gas (the Lyman forest) then poses further problems of measurement and interpretation. IUE and HST observations of low redshift quasars are at least free of this problem.

To a first approximation the 'average' spectrum of the BLR suggests that the emission mechanisms and conditions are broadly similar in a wide variety of objects. It reveals a gas of moderate levels of ionization but including a wide range of ionization states, with what appears to be a wide range of chemical elements having approximately solar-type abundances. It is tempting to assume that the average spectrum could be interpreted as showing that the densities were about the same value in 'all' AGNs, otherwise different forbidden lines would be observed. However, another possibility is that there is indeed a wide range of density within the region, scaling with radius to some power, and what is *observed* in every BLR is just the average of this range. We shall return to this latter point in chapter 9 when we consider the geometry of the BLR in more detail, as it is more likely that this is what is taking place.

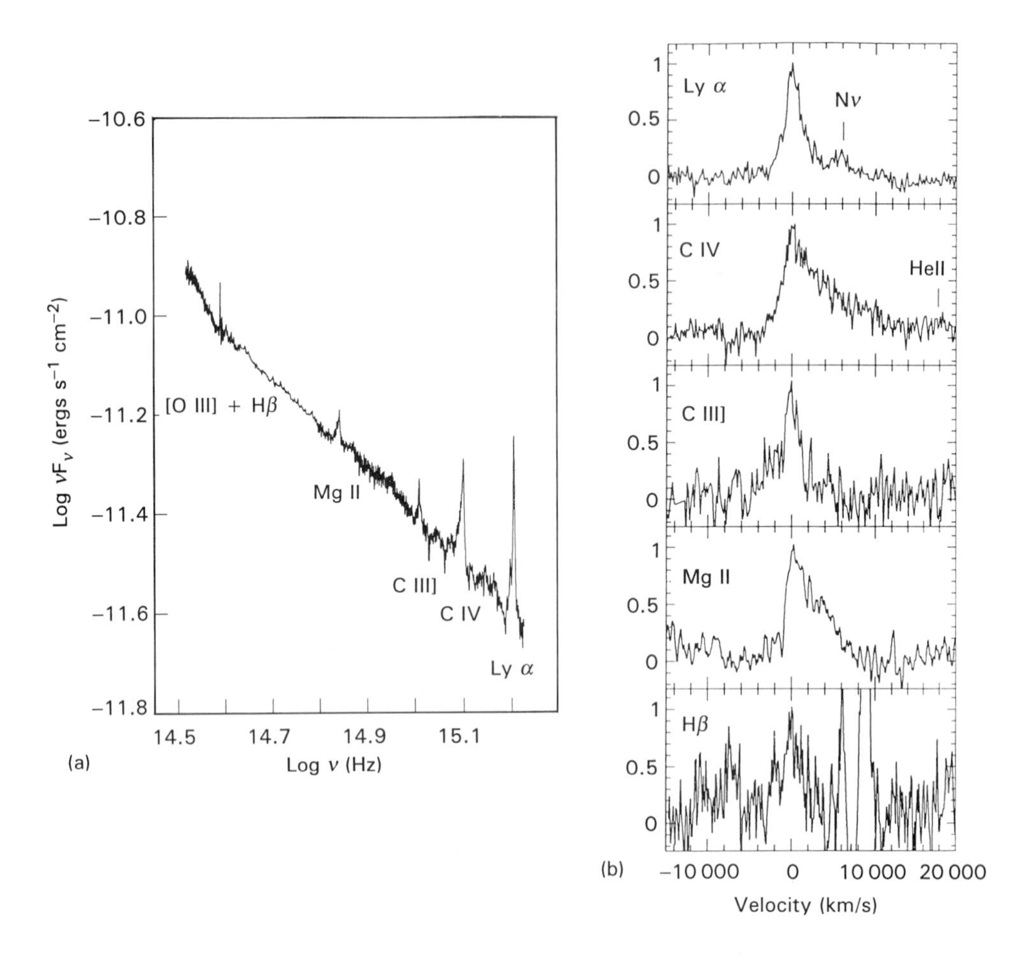

Fig. 6.2. (a) Left, the infrared–UV continuum and emission lines of the OVV quasar 3C279. This plot is in exactly the same units as that of fig. 5.3 to which it can be compared to show the very small zone occupied by the optical part of the spectrum. (b) Right, the optical–UV emission lines of the BLR normalized to the same scale. The profiles of the CIV and MgII lines are seen to be very asymmetric with broad red wings and it is such details that modelling of the BLR must be able to match. (Adapted from Netzer, H. *et al.*, *Astrophys.J.*, **430**, 191, 1994.)

For the moment, let us assume that the BLR is a single zone. Conventional wisdom suggests that the electron density lies somewhere between 10^{14} and 10^{16} m^{-3}. This view is due primarily to the presence of the semi-forbidden CIII] (190.9) line and the absence of the broad component of the [OIII] doublet (with a critical density of 10^{12} m^{-3})

compared to the strength of the narrow component. However, the density must be nearer to the upper end of this range due to the observed relative strengths of CIII] and CIV to Lα. The same results produce a value for the ionization parameter lying between 10^{-2} and 10^{-3}, i.e. the electron density is 100 to 1,000 times the photon density.

Although the range of values for the ionization parameter is generally agreed, there is a great deal of debate about the ionizing continuum itself. This boils down to two aspects, the shape of the continuum and the geometry. The latter point reflects concern that the assumption of spherical symmetry may not be applicable and that different parts of the BLR might see different ionization continuum energies. We will touch on this when we discuss unification scenarios later in this chapter and in chapter 9. What is clear, however, is that using a 'normally observed' continuum spectrum, the Lα/Hβ ratio remains a problem.

The ionizing continuum is usually assumed to have a power-law spectrum ($S_\nu \propto \nu^\alpha$) with spectral index $\alpha = -1$. This index produces equal energy at all frequencies (wavelengths) but the *number* of photons emitted per unit time at higher energies is substantially less than of lower energies. For steeper spectra the drop-off in photon number at the higher energies is increased, whilst for flatter (often called harder) spectra, the fall-off of photon numbers at higher energies is less. As we saw in section 6.2.2, a flatter spectrum (a more energetic ionizing continuum) is able to produce a correct fit to the observed Lα/Hβ ratio. The power-law continuum must extend to photon energies of a few hundred keV, which is in the hard X-ray regime. A very good question to ask is why should a power-law be assumed rather than, say, the spectrum from a very hot blackbody. We shall side-step such an intelligent question for the moment and return later.

A major problem associated with the BLR and the standard model is the confinement of the gas. This takes on two forms. Unless there is some external confinement, the clouds in the BLR are essentially in a vacuum; they expand at the local sound speed and essentially dissipate very quickly (see the discussion on jets in chapter 8). A confining medium would allow the BLR clouds to be long-lived entities. The clouds would then be part of a much larger ensemble of gas in the region, and this gas provides a source of external pressure on the clouds. However, why only components of this gas become photoionized remains a problem. The extended molecular torus (see section 6.5) cannot provide this confinement. Gravitational binding could retain the clouds in the region, but cannot prevent cloud dissipation. As a result, the consensus view is that the gas clouds are transient phenomena, being replenished by material flowing into the central region from the surrounding interstellar medium. This mechanism allows new gas to maintain the population of BLR clouds.

6.3.3 Predictions about the BLR

In principle we can now determine the size and mass of the line-emitting gas. Furthermore, we can show that this gas must be in the form of discrete clouds rather than a uniform medium. How can we do this? The ionization parameter gives an indication of the number of ionizing photons, but the upper limit to the integral of eqn. 6.2 remains something of a guess. Sometimes this limit is taken to be the wavelength at which photons would ionize helium (22.8 nm); however, this is rather arbitrary.

Nevertheless, taking the canonical power-law spectrum of $\alpha = -1$ and the energetics of the region, the ratio of the model-predicted Hα flux to the continuum flux can be determined and compared with observation. The observed ratio is found to be much less than expected if the gas absorbed all the continuum energy. The clear conclusion is that some of the continuum radiation must be escaping directly into space without interacting with the gas. *Therefore the gas in the BLR cannot be uniformly distributed in a spherical shell, but must be severely clumpy or in a non-spherical geometry such as a torus.*

We will now investigate this further and introduce another parameter, called the *covering factor*. This is the fraction of the surface of a sphere that is covered by the absorbing clouds as seen by the continuum source. If all the lines of sight outwards from the centre intercept a cloud, then the covering factor is unity, i.e. there are no gaps through which the line of sight can pass unimpeded. The covering factor is quite uncertain but limits can be set. A value of the covering factor greater than about 20% would lead to observable discontinuities in the continuum at the redshifted wavelength of the Lyman absorption edge. The fact that no such discontinuities have been found from a number of observational results, including observations of the Lyman continuum for high-redshift quasars, gives support for a much lower value of the covering factor. Furthermore, because almost all Seyfert 1 galaxies are detected in soft X-rays without significant intrinsic absorption, the covering factor must again be low. It is now generally agreed to be about 10% for high luminosity quasars and Seyfert 1s but much higher for low-luminosity AGNs.

The small covering factor also implies that the volume filling must be very low, between 10^{-3} to 10^{-6}, although this argument requires other information (such as the column density) to be known or assumed. Hence we have arrived at an important conclusion. *The gas in the BLR is very clumpy*, with either many small clumps of optically thick gas absorbing a small fraction of the continuum radiation, or alternatively, an even larger number of optically thin clouds that appear optically thick when seen together along the line-of-sight. The former view is the standard belief and a schematic of the broad-line region is given in fig. 6.3.

Working through the emission measures of the various lines yields a numerical value for the emitting mass of gas. This is obviously of great importance and so we will follow through the arguments here. The luminosity of a line is given by

$$L = j V \tag{6.3}$$

where L is the luminosity of the line which has emission coefficient, j, and occupies a volume, V. The luminosity can be determined for the hydrogen lines and it is usual to use Hα. However, we do not know the volume precisely and so we resort to use of the number density, N, and the mass of the particle, m. Hence for a hydrogen dominated gas and assuming constant density in the region, we have $V = M/N_p m_p$, where M is the total mass of hydrogen, N_p is the proton number density and m_p is the proton mass. The electron can be ignored as its mass is only about 1/2000 of the mass of the proton. Equation 6.3 then becomes

$$L(H\alpha) = j(H\alpha) \frac{M(H)}{N_p m_p} . \tag{6.4}$$

Numerically calculations suggest that $j(\text{H}\alpha) = 3.6 \times 10^{-38}\ \text{N}_p^{\ 2}$ W m^{-3} in the regions under question. For a hydrogen dominated gas, the density of protons is roughly equivalent to the electron density, and so $\text{N}_p \sim \text{N}_e$, hence we have

$$\text{L}(\text{H}\alpha) \quad = \quad 3.6 \times 10^{-38}\,\text{N}_e\,\frac{\text{M(H)}}{\text{m}_p} \qquad \text{(watts)} \qquad (6.5)$$

which can be reduced to units of solar mass, $\text{M}_\odot$, to give

$$\text{L}(\text{H}\alpha) \quad = \quad 4.3 \times 10^{19}\,\text{N}_e\,\text{M}_\odot \qquad \text{(watts).} \qquad (6.6)$$

Observationally, we know that the Hα luminosity for quasars and Seyfert 1s lies in the range 10^{31} to 10^{39} W. Assuming an upper limit for the electron density, N_e, of $10^{16}\ \text{m}^{-3}$, we can therefore calculate that the most powerful quasars require a mass of ionized hydrogen of only about $10^3\ \text{M}_\odot$ to produce their BLR outputs. This is quite an astonishing result; 1,000 solar masses is not a great deal of mass. This is put into even greater perspective when it is realised that for the least luminous of the quasar/Seyfert 1 BLRs, the ionized gas mass is less than 1 $\text{M}_\odot$. However, we should note that in the above calculation we have implicitly assumed that the gas is of uniform density. If the gas is of a filamentary or clumpy type of structure, then the answer will be larger. We know that the BLR must be clumpy and when this is taken into account the mass of ionized material increases. But by how much? It turns out by less than an order of magnitude. This then gives ionized gas masses of the BLR ranging from 1 to $10^4\ \text{M}_\odot$.

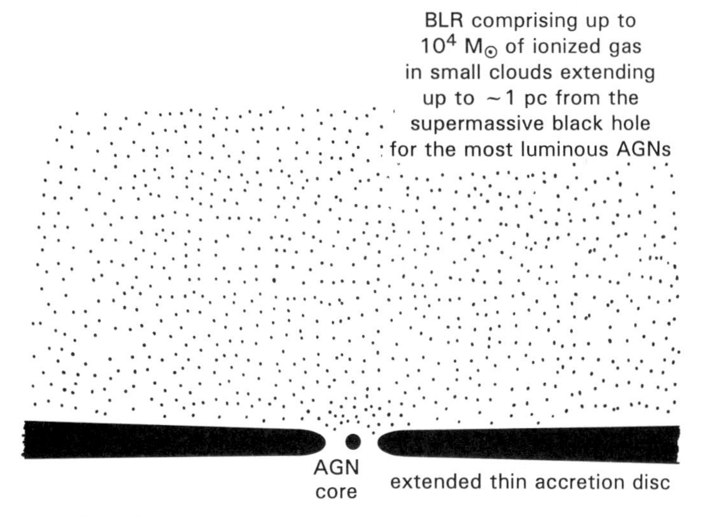

Fig. 6.3. Schematic of the broad-line region.

We can now investigate the volume occupied by this small quantity of gas. Eqns. 6.3–6.5 can be used to obtain the volume directly. Numerically this gives

$$\text{V} \quad = \quad \frac{\text{L}(\text{H}\alpha)}{3.6 \times 10^{-38}\,\text{N}_e^{\ 2}} \qquad (\text{m}^3)\,. \qquad (6.7)$$

For $\text{N}_e < 10^{16}$ and the most powerful quasars, $\text{V} = 2.8 \times 10^{44}\ \text{m}^3$. Assuming homogeneity and isotropy for the region gives a radius of 4×10^{14} m, or $\sim 10^{-2}$ pc. This will obviously be a minimum value because we assumed that the region is homogeneously filled with ionized gas. Once the clumpy nature is taken into account, the radius increases to a value of ≤ 1 pc. Amazingly enough, the minimum radius for the least powerful quasars and Seyfert galaxies and assuming a 100% filling factor turns out to be a mere 3×10^{-5} pc.

Yet another handle can be used to calculate the size of the BLR, the ionization parameter. Assuming all the other factors are known, eqn. 6.1 can be inverted and solved for the radius r. This is not quite as circular as it sounds because for some of the well-studied astrophysical sources, the required parameters (N_e, U, L, etc.) have been determined from observations of many different line strengths. Although only applicable to a small number of objects, the radius of the BLR again comes out to be about 1 pc for the most luminous objects and 0.1 pc for the least luminous. We note that the value of ~ 1 pc is about 100 times larger than the value derived above, assuming strict homogeneity. This in turn predicts a volume filling factor of only about 10^{-6}, which provides another powerful argument supporting the picture of the emitting gas being very clumpy in structure. It is also in agreement with the estimate of a covering factor of around 10%.

Finally, we should also remember the work referred to in section 5.4.4, the reverberation 'mapping' studies of the BLR. These give results that are in keeping with those derived above. While indicating that the BLR is actually smaller, they nevertheless provide strong support for the broad-brush scale-size derived for the BLR. However, we will see in the next section that further complications arise, and these have been driven to some extent by the data from the reverberation mapping studies.

6.3.4 Extensions to the standard model

Notwithstanding the apparent success in describing the overall properties of the BLR, despite heroic attempts on the part of the proponents the standard model has been unable to explain the observed line ratios. Nevertheless, the continued refinement of the model has produced better matches to the observations and has undoubtedly produced a more physical interpretation of the region. The latest improvement is to consider the BLR not as a homogeneous ensemble of gas, but as a stratified system. Although computationally more complex, it is physically more reasonable. On the other hand, we must be careful to point out that these refinements have not solved all the problems of the 'standard model'. Difficulties still remain.

For the purposes of computational modelling the BLR is split into two zones, denoted by HIL for the high ionization lines and LIL for the low ionization lines. The former lie closer to the central engine and reflect a region of electron density $n_e \sim 10^{15}$ m^{-3}. The upper limit to the density comes from the semi-forbidden line of CIII], which becomes collisionally de-excited above densities of a few times 10^{15} m^{-3}. This region is believed to be about 10^{-3} to 10^{-2} pc in size. Surrounding this is a region of much higher electron density, with n_e in excess of 10^{17} m^{-3}, but only partially ionized. These two regions can satisfactorily explain the problems with the differing line ratios. The high ionization lines of Lα, CIII], CIV, HeI, HeII, and NV come from the inner zone, whereas the lower ionization lines such as the Balmer series of hydrogen, MgII, CII and

FeII derive from the outer zone. But what does this look like in a real AGN and how is the very high density obtained in the medium?

Armed with these questions, theoreticians moved on to consider gas flows in the immediate vicinity of a black hole and effects of the accretion disk. The latest series of models have intimate links between the broad-line clouds and the X-ray emission. An important clue establishing this link has come from observations in a number of AGNs that the HIL are always systematically blueshifted with respect to the LIL, although not by the same amount. It appears to depend either on luminosity or redshift, or perhaps even both, but so far it has not been possible to disentangle these effects. For bright, high redshift quasars the blueshift is generally up to ~ 1,000 km s^{-1}, with some objects showing even larger figures.

Taken at face value this is very puzzling. One possible solution is that the LIL are more or less at rest with respect to the dynamics of the emitting system but that the HIL are moving in a radial direction with velocities of the order of 10^3 km s^{-3}. But why should we see only blueshifts rather than redshifts? This can only come about by an orientation effect, whereby the accretion disk makes a substantial angle to the line-of-sight, and thereby blocks out the receding clouds that lie behind the disk. So, from where do the LIL originate? A recent suggestion is that they come from the atmosphere of the accretion disk itself.

Observational evidence to support a stratification model for the BLR comes from the reverberation mapping techniques introduced in section 5.4.4. These studies monitored the response of a number of lines in the BLR to changes in the UV driving continuum. The Seyfert 1 galaxies NGC5548 and NGC3783 have been the target of these studies. The results clearly show that the higher ionization lines have less lag than the lower ionization lines, giving strong support to the stratification picture. In the case of NGC3783, the lines of HeII and NV could not be separated from the continuum (a lag of less than 4 days), Lyα, CIV, Si IV and OIV responded in about 4 days, while Hβ took about twice this time. It was not possible to separate adequately the MgII lines from the underlying FeII lines and the UV continuum in order to determine a unique lag, although a delay of ~ 9 days was found. This is consistent with the Hβ value. A variability time delay of 4 days represents a scale size of 0.003 pc for the middle part of the HIL zone. For NGC5548, the Hβ lag was 18 days behind the continuum. This source has been monitored over four observing seasons and shows a very strong lag peak at 18 days for the Hβ emission, giving a scale size of ~ 0.01 pc for the LIL.

An intriguing aspect of these studies which is certainly providing clues to the make-up of the region is that the line profiles and variability (in response to continuum changes) do not reveal a well-ordered velocity field, but rather the superposition of a random set of velocities. Perhaps even the stratification model is dominated by a clumpy mass distribution in a chaotic or turbulent medium rather than a homogeneous distribution in a smooth velocity field.

6.3.5 New models

The above modification to the standard model may sound somewhat exotic, but has found general acceptance. As such, it is extremely important and forms the basis for even more detailed descriptions of the inner zone of the central engine, the region where the accretion disk and BLR clouds merge.

Indeed, theoretical studies of plasma winds in a centrally ionized zone point to the requirement for a hypersonic flow of gas, the details of which we will not tackle here. However, what is crucial to consider is how this hypersonic wind interacts with the broad-line clouds. This has led to the introduction of a new picture of this region, one which retains the origin of the high ionization lines in small dense clumps of plasma surrounding the black hole and accretion disk, illuminated by the inner zone of the accretion disk (see fig. 6.4). The difference arises from the fact that the hypersonic wind interacts with the broad-line clouds and strong radiative shocks are formed. These shocks radiate isotropically at X-ray energies and back-illuminate the surface of the accretion disk which is heated above its normal thermalized accretion rate temperature.

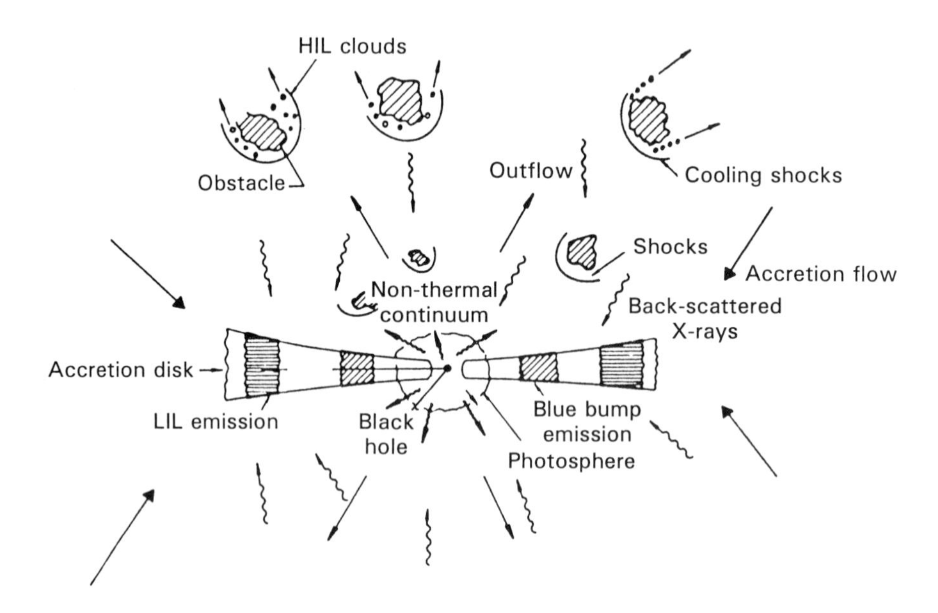

Fig. 6.4. Representation of the complex situation which might be occurring in the BLR. The HIL excitation lines come from cooling gas behind shocks in hypersonic infalling material, whereas the LIL emission lines might originate in the atmosphere of the accretion disk. (adapted from Collin-Souffrin *et al., Mon.Not.R.Astron.Soc.* **232**, 539, 1988).

Remember that the accretion disk is a gaseous structure, so we are looking at heated gas rather than a solid body. The gas in the disk is orbiting the black hole, which neatly explains the velocity spread and the line widths. Furthermore, the very high density and confinement are solved, being the outer part of the accretion disk itself. In this picture, the HIL broad-line clouds still require replenishment from the plasma flow deriving from inward stream of material into the inner zone but they may also be confined by this hypersonic flow. In passing one can note that one particular suggestion for this gas supply assumes a local cluster of hot stars with wind-driven envelopes.

6.4 THE EXTENDED EMISSION ZONES

6.4.1 The narrow-line region (NLR)

Although the NLR is perhaps regarded as the less glamorous of the two line-emitting regions, its importance should not be underestimated. All AGNs show a NLR, providing evidence for activity from a region much farther from the central source than the BLR. Indeed, lying at 10–100 pc from the central engine, the difference in distance is two or three orders of magnitude. We will now give a brief overview of the main features of the NLR, and in later chapters we will see that observation of the gas in NLRs plays a key part in unification scenarios and obscured BLRs.

In the study of the NLR, a key aspect is the separation of those emission line regions that are photoionized by a central, strong, non-stellar continuum source (the true NLR), from those regions that can be powered by local ensembles of OB stars. In the lower-luminosity LINER and Seyfert 2 categories, there is expected to be overlap between central, continuum-powered emission and more local starburst power. This is especially the case when the narrow emission line regions become very extended and in some cases merge with large star formation regions. However, we will retain LINERs as active galaxies. Likewise, we shall choose to come down on the side of those who argue that the major heating mechanism for LINERs is photoionization, rather than shock heating of the gas (which was the rage for a number of years).

What are the main features of the NLR gas? The emission lines show significantly lower levels of photoionization than the HIL region of the BLR gas and they have correspondingly much lower FWHM, with linewidths typically being a few hundred kilometres per second. The NLR is sufficiently extended that it can be resolved spatially in the nearest active galaxies. The key lines of study are usually the prominent [OIII] 500.7, 495.9 doublet. One can undertake the same analysis as for the BLR and hence determine mass ranges and scale sizes. The densities within the NLR are generally believed to lie in the range of 10^{12} m^{-3} to 10^{10} m^{-3} from measurements of [SII], [OIII], etc. However, it should also be noted that this density range can be found within a single active galaxy. This again emphasizes the much more extended structure of the narrow-line regions and the variability in density as a function of radius. The discovery of this range of density solved an earlier puzzle, as we shall now see.

In the 1970s and early 1980s there was perceived to be a problem with the [OIII](500.7, 495.9) to [OIII](436.3) ratio. Observations of forbidden SII lines determined that the density in the NLR was of order $10^8 - 10^9$ m^{-3}. At these densities, the [OIII] ratio depends on temperature alone and the observed ratio requires an electron temperature > 50,000 K. But as we saw earlier, photoionized equilibrium electron temperatures are more in the range of 10,000–20,000 K, hence the problem for photoionization models. This apparent excess electron temperature was taken as evidence of shock heating. However, the problem was solved with observations of further diagnostic lines, which revealed the large range of density present in the objects. A classic study is the 1985 paper by Alex Filippenko where conclusive evidence is presented that high-density regions exist in the NLR, with highest values in excess of 10^{13} m^{-3}. This satisfactorily explains the observed line ratios based on photoionization by a central continuum source. It also demonstrates that the [OIII] emission originates from a different zone in the NLR than the [SII]. This suggests either a radial density gradient, severe clumpiness or again, as in the BLR, some form of stratification.

Given the range of density, upper and lower limits can be determined for the mass of ionized gas. This is often achieved by using the luminosity of the Hα line, which is very strong in many objects. Indeed, the narrow line can have just as much energy as the broad-line component. For the most powerful of the quasars, the narrow-line Hα luminosity is about 10^{39} W and using eqn. 6.6 we see that the mass of ionized gas comes out to be of order 10^9 $M_\odot$. Assuming homogeneity and spherical symmetry for simplicity, the volume exceeds 10^{56} m^3, giving a radius of about 140 pc. But for a more physically reasonable picture of a non-homogeneous medium (either clumpy or stratified) or of a non-spherically symmetric region, this calculation gives only a lower limit to the radius which is therefore expected to be significantly larger. Using ground-based observations the NLR has been spatially resolved and well studied in a number of the nearest active galaxies, and the results give values of radii of a few hundred parsecs upwards (but can be as low as 10 pc for high-density regions).

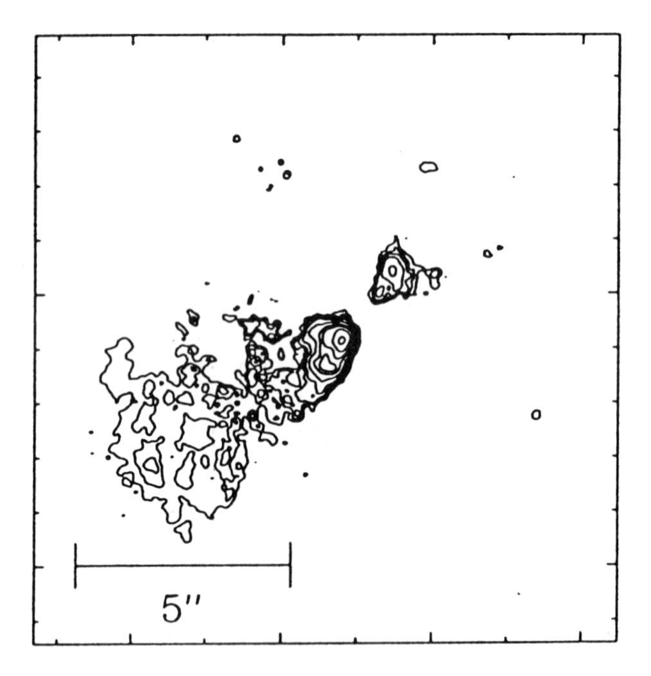

Fig. 6.5. An HST image in [OIII] of the NLR in the Seyfert 2 galaxy NGC5728 showing the dramatic bi-conical ionization structure. (From Wilson *et al.*, *Astrophys.J.*, **419**, L61, 1993.)

The HST with its angular resolution of 0.1″ has made tremendous inroads on this field of study. A good example is the HST observations of the Seyfert 2 galaxy NGC2110, which lies about 50 Mpc distant. The HST resolution at this distance is 23 pc, which is ample to sample the NLR in detail. The innermost pixels reveal red colours which are consistent with reddening due to dust rather than being due to predominantly red stars. Nevertheless, the resulting extinction of only ~5 magnitudes is at odds with the huge extinction (37 magnitudes) deduced from the hard X-ray column densities. This suggests that the HST is still suffering from aperture dilution and that the heavy obscuration must be on a scale much less than 23 pc (see below). The emission gas is found to be anisotropically distributed about the nucleus and furthermore, the structure is not smooth, but shows an S-shaped structure interspersed with knots. This is especially apparent in Hα. The idea of a clumpy and distorted medium rather than a smooth distribution of gas is strengthened. A slight word of caution is appropriate in that NGC2110 is not a typical Seyfert 2 because it also possesses a radio 'jet-like' structure. However, there is excellent evidence based on direct imaging (including the HST work), and supported by the energy balance in the ionized emission and observed continuum, that strong anisotropic emission (predicting ionization cones) must be present.

Some of the most spectacular images to be produced by the corrected HST are of the Seyfert 2 galaxy NGC1068 and one of these is shown in plate 5. This image is produced by combining images taken in ultraviolet continuum, a pseudo U-band filter and the line of [OIII]. The wealth of detail and complexity is readily apparent; the narrow line region delineated by the [OIII] emission breaks up into a clumpy and filamentary structure interlaced with dark lanes, presumably due to obscuration by dust. The NLR ionization cone is also dramatically revealed by the [OIII] emission. This is the northern ionization cone in NGC1068, which although aligned towards us, does not allow us to look directly into it because we do not detect a BLR or a 'naked' nucleus. The southern cone is believed to be aligned away from us and as we observe it through much more dust, it suffers much higher extinction and therefore appears very faint compared to the northern cone. The bright point source in the nucleus is resolved by the HST but it is not the AGN core, which is believed to be deeply obscured. Instead it could be scattered light from the central engine interacting with the surrounding material.

The images in the different filters clearly show that the delineation and apparent orientation of the emission cones are not identical, indicating that their appearance is heavily dependent on the density of gas in the region and any obscuring dust between the gas and the ionizing continuum. We will return to this point when we look at NGC4151 in section 9.2.2.

6.4.2 The extended narrow-line region (ENLR)

As we saw in section 5.3.4, pioneering studies in the mid-1980s using spectrometers with long slits, discovered in a number of the nearer and brighter Seyfert galaxies, narrow-line emission extending over a significant fraction of the galaxy. The extreme example of the admittedly small sample showed line emission extending up to 20 kpc from the nucleus, very much larger distances than attributed to the normal NLR. The term extended narrow-line region (ENLR) has been coined to describe this excited gas, extending a few kpc from the nucleus and believed to be physically distinct from the NLR gas.

Perhaps the most important discovery concerning the ENLR was that it was anisotropic in nature. There was also a strong hint that it was extended along the direction of any nuclear radio emission. We believe that the ENLR is ambient gas in the disk or halo of the galaxy in question, illuminated and photoionized by the central engine. The lack of symmetry of the emission is attributed to the anisotropy of the nuclear radiation as seen by the extended gas in the parent galaxy. This presumably results either from some form of central collimation (like the radio emission) or because of the geometry of the nuclear region. The latter fits in well with the idea that much of the gas in the galaxy is shielded from the ionizing nuclear flux by some unidentified obscuring medium. An obvious candidate for this obscuration is a torus of dust around the central engine.

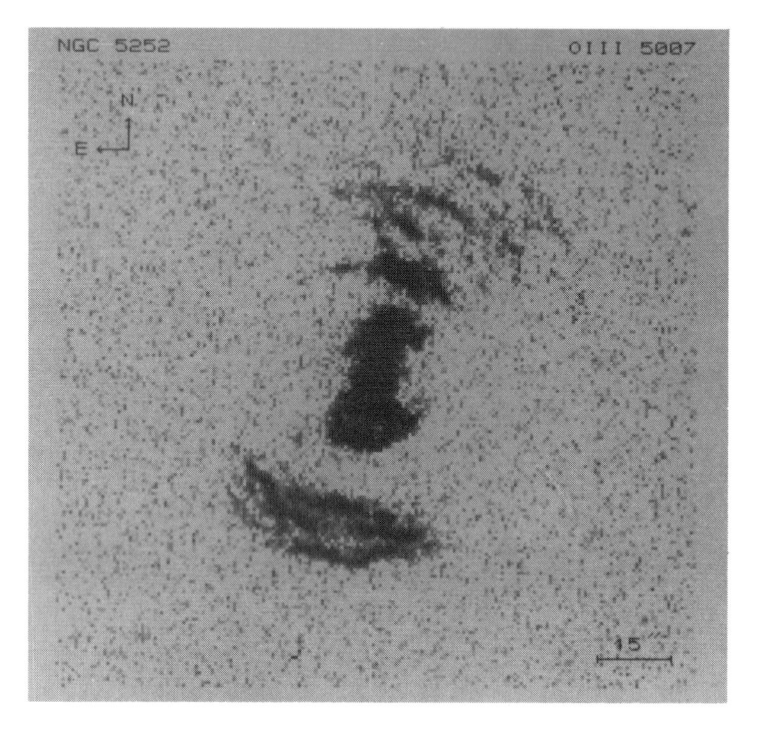

Fig. 6.6. A dramatic image of the anisotropic [OIII] line emission from the Seyfert 2 galaxy NGC5252. The opening angle of the cones is ~70° but the conic axis is misaligned from the major axis of the S0 host galaxy (not visible in this image) by about 30°. Also not apparent in this image is a band of higher obscuration crossing the central region and lying at ~90° to the cone axis. This could be the signature of an obscuring torus. (Adapted from Tadhunter & Tsvetanov, *Nature*, **341**, 422, 1989.)

Further support to this picture came from an extensive study by Mike Penston and collaborators. They showed that in common with a number of other ENLR sources, the ENLR in NGC4151 was illuminated by more photons than could be accounted for by assuming there was an isotropic radiating core that was being directly viewed by the observer. Although this discrepancy could be accounted for by a number of methods, the most appealing was to assume an anisotropic ionizing radiation field. The line-of-sight from the observer to the source of this radiation suffers much more obscuration than the line-of-sight from the ENLR to it. This could either be due to an intrinsically anisotropic source, such as beamed ionizing radiation (as in the jets of chapter 8), or alternatively by radiation that was originally emitted isotropically but whose propagation is modified by intervening material, such as an obscuring torus surrounding the central engine.

6.5 THE MOLECULAR TORUS AND UNIFICATION BY ORIENTATION

We now return to a puzzle introduced in section 3.2, the discovery of a BLR in the classical Seyfert 2 galaxy NGC1068. This was a major breakthrough in the quest for understanding the AGN phenomena and a shot in the arm for the unification pundits. In many ways it can be claimed to be the observation that put AGN central engine unification models on the general astronomical map. The key observation was by 'Ski' Antonucci and Joe Miller in 1985. They found that when the spectrum of NGC1068 was investigated in polarized light (a very difficult experiment to undertake), the resulting spectrum showed a characteristic type 1 appearance rather than a type 2 (fig. 6.7). In other words, they could see a BLR, but only in polarized light.

What mechanisms could exist to polarize light in NGC1068? The simplest is for reflection from a 'mirror' comprising either dust grains or electrons. Indeed the polarized component was very weak compared to the normal type 2 spectrum and fitted in well with predictions of a reflected component. So why do we see a type 2 (NLR) spectrum in direct light but a type 1 (BLR) spectrum in polarized light? The BLR must be hidden from direct view and we only see it by reflection off some form of mirror. The obvious question is what is this mirror. One answer was a thick (compared to the central engine and BLR) disk or torus of obscuring material surrounding the central engine, but inside the NLR. The idea of a central torus of obscuring material had been around for some years, but until this point had never managed to gain widespread acceptance in the AGN field. Since that first observation, the polarized type 1 spectrum has been seen in a number of otherwise 'classical' type 2 AGNs.

We will now investigate the details of this relatively new aspect of AGN study, the molecular torus as it is now called. Dust must be a key ingredient of this torus in order to provide enough extinction to hide the central engine and the BLR. The torus must also have sufficient material to make it opaque to the hard X-rays which are observed in type 1 nuclei but not type 2, as well as the ionizing photons from the hot accretion disk or corona. Values of extinction in excess of 50 magnitudes have been calculated from the observations. Dust and molecular gas are always found together in clouds and there is about $10^9\ M_\odot$ of molecular gas in the well-observed inner kiloparsec of NGC1068. Recent millimetre wave interferometric observations of the molecular gas distribution for the inner kiloparsec of NGC1068 reveal a definite bar-like structure surrounded by a molecular ring associated with the starburst.

At this point we should pause and remember that NGC1068 was one of the prototypes of the starburst–far-infrared galaxies and the presence of this large amount of molecular gas provides a ready supply of material for both the starburst phase and also the molecular torus. Furthermore, the presence of a bar provides a pathway for the molecular material to be channelled to the inner parsec. On the other hand, being molecular in composition, it cannot lie too close to the central engine otherwise the powerful radiation field will immediately dissociate the molecules unless there is extensive dust shielding. Even then, it must still lie at a sufficient distance to prevent evaporation of the grains. This gives an inner radius of order of a parsec in extent. This inner torus must not be confused with much more extended zones of dust and molecular gas emission seen in the far-infrared galaxies on kiloparsec scales. The molecular torus is a very compact phenomenon.

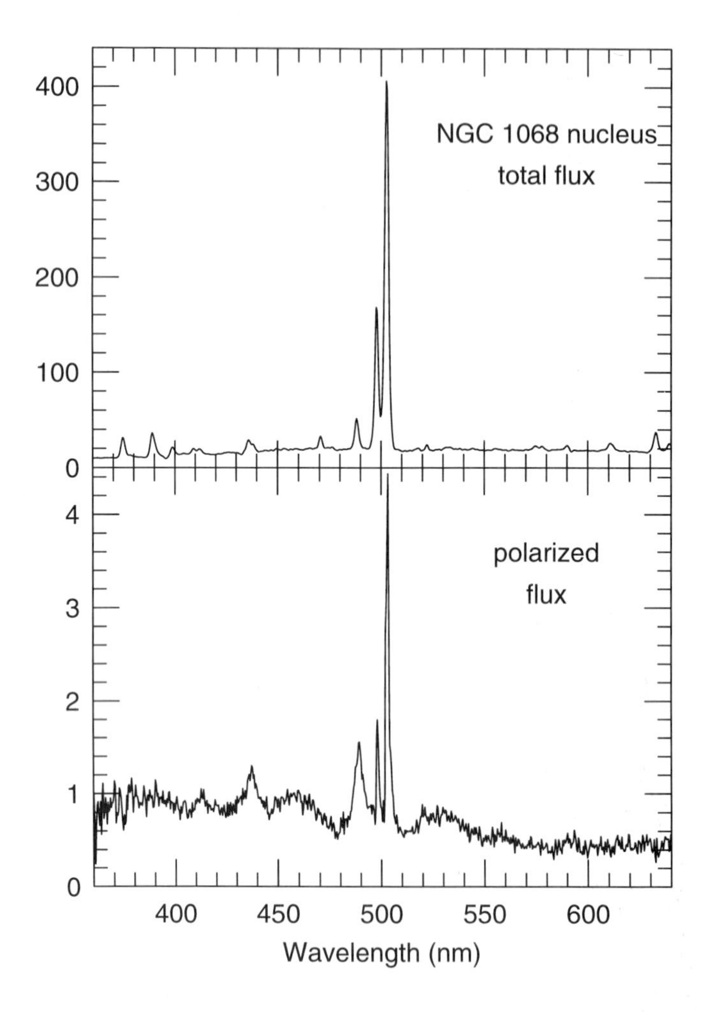

Fig. 6.7. The unpolarized spectrum (top) and the polarized spectrum (bottom) of NGC1068. The top spectrum shows the typical Seyfert 2 characteristics, while the polarized spectrum reveals the characteristic broad lines of Seyfert 1s. The figures on the ordinate show the relative values of the emission. (Adapted from Miller, Goodrich & Mathews, *Astrophys.J.*, **378**, 47, 1991, courtesy of Bob Goodrich.)

The inner face of the molecular torus will be bathed in the intense radiation field of the ionizing continuum from the central engine and it will not only be dissociated but ionized and any dust will probably be evaporated. However, as one moves deeper into the cloud, then, as we saw in chapters 4 and 5, the photon energy density rapidly falls, due to the inverse square law and absorption by dust, and we soon find that the medium is composed of molecular gas and dust. The overall covering factor *in the equatorial plane* must be unity. We can make a guess at the amount of material that is present by assuming that all the X-ray luminosity is absorbed and then re-radiated by the dust.

From observations of the relative luminosities of Seyfert 2 galaxies in the hard X-ray region compared with the infrared ($L_{HX}/L_{IR} \sim 10^{-2}$), the absorbing column density can be calculated to be at least 8×10^{24} cm^{-2}. This is more likely to be in the form of a number of high-density clouds than a uniform medium because self-shielding allows dust to form at much closer distances than otherwise. Furthermore, clouds are required to explain the existence of dust which would otherwise be destroyed by frictional heating due to the general turbulence of the material at the high rotational speeds (>100 km s^{-1}) of the galactic centre regions. So the bulk of the material is hot (~1,000 K) and much more ionized than the normal starburst region.

Because the molecular torus is a long-lived phenomenon, the clouds themselves must also be long lived, or at least replenished in the same form. Taken with the optical depth requirements, extensive calculations have shown that to first order there must be a balance in the cloud size distribution. There cannot be too many small clouds otherwise the shear forces would tear them apart (which would lead to increasing shredding and ultimate dissipation). On the other hand, the larger clouds cannot over-dominate or else they would eventually coalesce and end up as one cloud, which would then be too hot for dust to exist, due to lack of shielding and internal frictional heating.

So far so good, but what about the geometry? The statistics of the ratios of Seyfert 1 and 2 galaxies suggest that the torus must be thick, otherwise we should see more type 1s. We also saw that the NGC1068 observations showed that there must be a relatively dust-free but indirect path out of the central engine, perhaps a single reflection off a mirror. This suggests that scattering from electrons in the poles of the torus might provide the reflecting medium, and this point is still a matter of debate. Although the inner face of a torus provides an obvious mirror, it turns out that the geometry is not ideal for single reflection and this is not believed to be the solution to the mirror.

The next question revolves around why we get a torus rather than a flattened system. In fact producing a thick torus in the conditions expected at a distance of ~1 pc from the centre of an AGN is far from simple. Some pressure source is required to 'inflate' the torus, one of which is from winds from stars and stellar remnants. Supernovae produce fast winds and strong shocks, but they are a very rare phenomenon. Massive stars have large stellar winds and very young stars experience fast wind T-Tauri phases. However, it is very difficult for stars in any form to provide the required degree of stirring of the medium to inflate the torus. The current best suggestion seems to be that the torus is inflated by the frictional heating between the clouds due to the orbital motions. This seems to be just sufficient. A sketch of the torus is shown in fig. 6.8.

Calculations of the dynamics of the rotating clouds in the central parsec or so reveal that intercloud collisions are relatively common, perhaps as much as once per orbit. This process will lead to the loss of energy as the clouds radiatively cool. Therefore there will be a slow drift of material inwards, and this is just as well, otherwise the constant evaporation of the inner surface of the torus would eventually lead to its dissipation. So as long as there is a supply of material to the outer parts of the torus, it will remain a long-lived phenomenon. The molecular gas in the inner kiloparsec of NGC1068 seems an obvious source of such fuel. Just as this book was being completed, commissioning observations in September 1995 of the polarimeter module for the UKIRT infrared camera by the instrument builder, Jim Hough, revealed a spectacular polarized image of the outflow of NGC1068. The shadow of the obscuring torus clearly stands out and this is probably the best imaging evidence to date for a torus. The bulk of the torus is located of order 150 pc from the central engine, just what is expected.

Let us now take a brief look at what the shape of the torus might do to gas flows and winds from the central zone. Due to the radiation pressure within the central 'hole' of the torus, the material evaporated from the inner face provides an immediate source for a poloidal and possible semi-collimated large-angle ionized wind from the AGN nucleus. A further possibility is that equatorial material from the inner zone could end up as fuel for the central engine itself, buried even deeper inside the central hole of the torus.

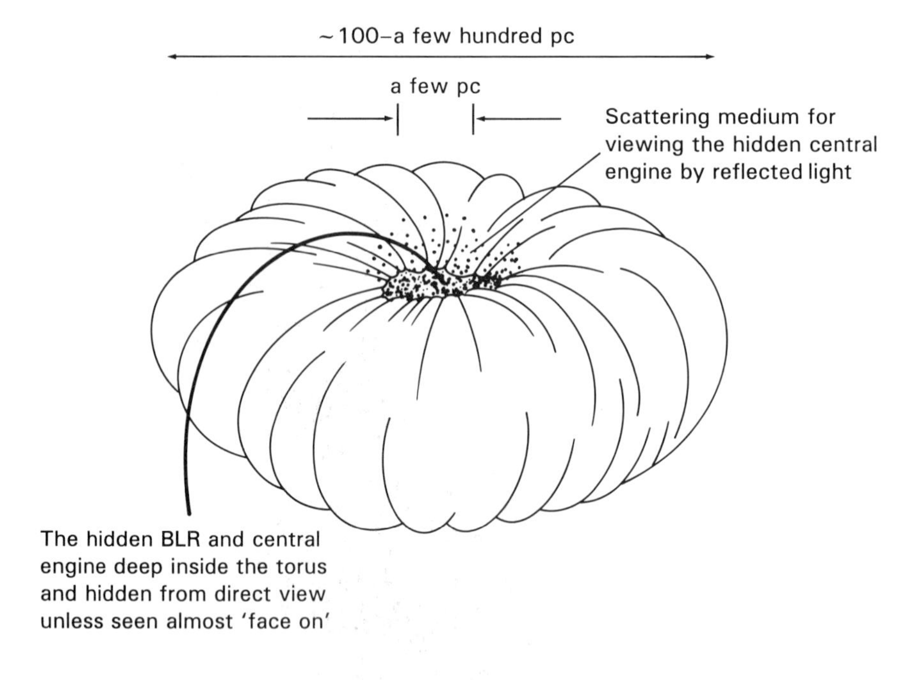

Fig. 6.8. Sketch of the molecular torus surrounding the central engine with the possibility of reflection revealing the otherwise hidden BLR.

The presence of a torus is indicative of circular symmetry, which in itself suggests rotation. This fits in well with our central theme of a supermassive black hole and accretion phenomena. Furthermore, it suggests that this structure must form in a system that is able to relax and convert the angular momentum into a flattened structure. This is clearly the case for spirals, but for ellipticals it is not so clear how flattened parsec and kiloparsec-scale structures form. Very recent evidence has shown that emission from radiating dust and molecular gas is present in the central zones of some of the local elliptical galaxies. So perhaps the core regions of at least some ellipticals have flattened rotational structures.

The concept of a dusty molecular torus immediately suggests powerful orientation selection effects. In the unification picture, if the torus is aligned face-on (i.e. its axis of symmetry is perpendicular to the plane of the sky), then we expect to call it a type 1 object and we will be able to see everything: the narrow-line region, the broad-line region and the unobscured central engine. If, on the other hand, the torus lies in the plane of the sky, then we have major obscuration problems. The presence of the dust is critical to our interpretation of the objects we study. Given sufficient dust in the torus, we would not be able to detect any hard X-ray or UV emission from a central accretion disk. Neither would we be able to see the broad-line region. However, we might expect a much higher possibility of being able to detect gas from the much more extended narrow-line region. We would also expect to see infrared and far-infrared emission from the hot torus. Spectroscopically we would expect to see a type 2 object, showing a NLR only and we would call the object in question a Seyfert 2.

It is apparent that this idea can form the basis for a series of model predictions of what type of an object we would see given various situations of size and dust mass of the molecular torus; power of the central engine; distribution of the NLR and BLR clouds; and crucially, the orientation of the molecular torus with respect to the observer. Notice that we have omitted radio jets, but these must also fit into the picture somehow. We will discuss the orientation effects further in chapters 8 and 9. We have so far failed to consider the orientation of the molecular torus with respect to any rotation axis of the parent galaxy. This is deliberate; it is discussed in chapter 9.

The concept of a dusty torus is extremely attractive to unification pundits. Indeed the need for such a torus was almost required in order to explain the orientation solution to the unification of activity and the observational differences between the two classes of AGN, namely the type 1s with their BLR and NLR and the type 2s with their NLR. But what actual evidence is there for such a torus? In fact, very little. The torus provides an explanation for the 3 μm bump in AGNs as being due to the dust on the point of evaporation in the inner zone but imaging is what is required. Again the problem comes down to angular resolution. Detection of molecular gas comes with insufficient angular resolution to resolve the structures. The mid-infrared emission expected from the torus has proven to be elusive due to contrast and angular resolution difficulties, but very recent infrared observations of the Seyfert 2 galaxy NGC5252 are supportive of the overall picture. The HST has also been hard at work looking for the 'missing' torus and some of the best suggestive evidence is shown in plate 6 for NGC4261.

Data arising from a somewhat unexpected source have finally delivered the best evidence for such a torus, and in passing, have furnished excellent evidence for supermassive black holes (see next section). This evidence comes from observations of maser emission from water vapour molecules in the mildly active galaxy NGC4258. Because the distance of this galaxy is just over 6 Mpc, the maser emission must be powerful to be detectable by VLBI, pointing to some major pumping mechanism at work. We remember from section 3.8 that intense maser emission is associated with some of the far-infrared galaxies and that this is sometimes referred to as megamaser emission due to its enormous power. The masers in NGC4258 are not as powerful as these, but nevertheless are much more powerful than those seen in our Galaxy.

The VLBI observations showed that there were a number of masers coincident with the nucleus of NGC4258 and these masers were distributed along a line, strongly suggestive of a flattened system (figs. 7.4 and 7.5). Furthermore, because the masers radiate a very narrow and wavelength-precise emission line, their respective line-of-sight velocities can be deduced rather accurately from the observed doppler-shifted wavelengths. This result strengthens the disk hypothesis. The overall picture is consistent with a flattened, almost edge-on, rotating disk of molecular material with an inner radius of $\sim$ 0.1 pc and an outer radius of $\sim$ 0.3 pc. We will reserve further discussion of these maser observations to the next section as they give powerful pointers to the mass of the central object. So although we see clear support for a compact molecular disk, the direct evidence for a thick torus is still far from overwhelming.

6.6 THE CENTRAL ENGINE

Now we have much of the 'big picture', the key question is what is happening at the central engine itself. We have seen that there will be a complex cauldron of intense photon fields, energetic charged particles and, potentially, pair production. Numerous high-energy processes are at work, as shown by the accepted views of the reflected component of X-ray Compton scattering from a 'cold or partly ionized—warm' medium. The hard X-ray power-law is suggestive of the primary radiation mechanism and this is what we need to investigate next. Unfortunately, this is an incredibly complex and technical process, the details of which lie beyond the scope of this book. Nevertheless, we shall present a brief overview of the possible mechanisms at work, and continue in the knowledge that none of these have yet found general acceptance in the overall AGN community.

There are basically two models at the current time. One uses electron–positron pairs produced by photon–photon interactions close to the black hole. The resulting pair cascade interactions can than be shown to produce a photon power-law at X-ray energies. The requirement is a very high- photon–particle energy density. The second model is not as exotic in that it makes use of Compton scattering of lower-energy photons by a thermal populations of electrons. These electrons need to have a high temperature to produce the requisite scattering, and so their location must be in a plasma close to the black hole, possibly in the corona or the close to the 'surface' of the accretion disk. Both models are exciting and challenging, and much work is ongoing in this field.

Let us take up this second model as we have already met the physics required to explain it. The complications in all these models lie in the geometries and the details of the fractions of the energy which go into the various interactions. We start off with high-energy electrons in the plasma field of the black hole or the accretion disk corona. These produce the hard X-ray photon power-law by Compton scattering and in the process are cooled. However, some fraction of the hard X-rays will be absorbed by the accretion disk and will emerge from this reprocessing as soft X-ray/UV photons. These photons are then available for Compton scattering by the hot electrons. We can see that there is a feedback mechanism at work and models have now been refined to the point where the process just described can achieve a quasi-stationary situation, with equipartition existing between the high-energy and low-energy photon luminosities. This also has the distinct advantage that the observed spectral slope of the X-ray scattered component of $\alpha \sim -0.9$ can be produced naturally.

We will deliberately postpone further discussions of the central engine itself until the final chapter. By then we will have seen the physical processes taking place close to the supermassive black hole and will have reviewed the jet phenomenon depicted by the spectacular radio galaxies and quasars. In closing we will note that one fundamental question remains. Why do only a small fraction of AGNs show radio jets?

6.7 SUMMARY

The Broad-Line Region (BLR) is a fundamental property of high luminosity AGNs. It is well observed in quasar and Seyfert 1 spectra, and we now know it to be present in at least some Seyfert 2 galaxies. Observations give a very detailed picture of the emission line properties, showing that the lines have very broad wings with velocities ranging from a few thousand to 15,000 km s^{-1}, although the latter represents the very extreme of the range. The ionization levels are high, but not as high as those in the solar corona, having an effective temperature of the order of 10^4 K, but with a much higher ionization due to photoionization than would be expected from collisional effects at this temperature. The very prominent lines of the BLR are a boon for determining the redshift of the object. The ionization parameter tests the strength and shape of the ionizing continuum and although the results are inconclusive, the ionization parameter is deduced to be about 10^{-2} to 10^{-3}. This means that there are 100–1,000 times more electrons than photons in the BLR. Work in this area has stimulated much effort into understanding atomic transitions and cross-sections.

The BLR is composed of gas with approximately solar abundances and is in the form of an ionized region made up of maybe tens of thousands of small, optically thick clumps of gas. The continuum source is assumed to be the only source of heating for the region. Unless the clouds are confined by an external pressure, perhaps due to hot gas, they are transient phenomena, requiring continual replenishment with new material. Assuming a geometry to be spherical and non-stratified allows a crude calculation to reveal that the size of the BLR is restricted to ~1 pc or less for the most powerful quasars and hence cannot be spatially resolved using any current optical facility. The radiating mass of gas ranges from about 10^3 $M_\odot$ for the most powerful BLR emitting quasars to less than 1 $M_\odot$ for the least powerful quasars or Seyfert galaxies. The covering factor for these clouds is of the order of 10% or less.

Attempts to model the BLR emission produced the so-called Standard Model, which assumed a spherically symmetric geometry with a central ionizing radiation source. The model had initial success in explaining the physical processes within the inner few hundred parsecs. With better and more comprehensive line measurements, it subsequently failed in terms of explaining some of the observed line ratios, especially the Balmer decrement and Lα/Hα. The former exceeds the value of 2.87 allowed by strict photoionization theory, and looking to dust absorption to explain this effect has not been successful, while the latter is an order of magnitude down from the predicted value of 13. A number of ways out of this general line ratio puzzle included suggestions of Balmer line self-absorption, a flatter continuum energy spectrum for the ionizing source and a multi-zoned (or smoothly varying) region.

The standard model has been improved and in its latest guise is a multi-zone model, the stratification into a high ionization (HIL) and partially ionized (LIL) zone alleviating the line ratio problems. A consistent picture is now emerging, but on the other hand, a significant increase in electron density, up to 10^{17} m^{-3}, is required in order to explain the lower ionization lines. One of the very latest theories is that these originate at the outer surface, or atmosphere of the accretion disk. The electrons are heated by X-ray illumination from the hot corona surrounding the inner funnel of the accretion disk. The corona can be produced by a number of physical processes. An alternative possibility is that the high ionization broad-line clouds are sources of strong shocks in an outward supersonic wind from the black hole, and these shocks are strongly radiative, producing the X-ray 'back-heating' of the disk.

Surrounding the BLR is the NLR that extends from a few tens of parsecs to over a kiloparsec. The states of ionization are much less than the HIL region of the BLR and it has correspondingly lower densities, of the order of 10^8 to 10^{12} m^{-3}. Due to the large extent of the NLR there are bound to be significant density and ionization variations, and at some point the NLR merges into the extended narrow-line region. The NLR contains up to 10^9 $M_{\odot}$ of ionized gas. For far-infrared emitting AGNs, including radio-quiet quasars, Seyferts, and starbursts, there is an excess of 10^{10} $M_{\odot}$ of gas in this zone just outside the NLR. For the most powerful far-infrared emitters, this can be increased by three orders of magnitude. In the nearer AGNs, the NLR has now been imaged, supporting the above picture.

The idea of a dense and compact molecular torus surrounding the central engine provides a an elegant source of orientation-dependent obscuration and the promise of unification in terms of type 1 and 2 AGNs. The torus in general lies just outside the BLR but inside the NLR and the spectacular data from the HST have shown that the torus is in fact complex. The dust at the inner face is most likely the source of the near infrared emission from AGNs. Calculations of the radiation processes occurring in the torus as it intercepts the ionizing photon field from the central engine are very complex. For a number of sources, such as NGC1068, detailed modelling of the torus is now more than just a theoretical pastime. Recent observations of water maser emission in a local galaxy have strengthened the case for a torus.

We have seen that viable models exist to explain the central engine photon production spectrum, the most popular at the current time is Compton scattering of UV–low-energy X-ray photons by a thermal population of hot electrons and some fraction of the high-energy X-rays are then reprocessed in the accretion disk to provide the photons for the Compton scattering. This process appears to be able to produce the hard X-ray power-law which is observed and points to a complex series of interactions in the accretion disk zone.

6.8 FURTHER READING

General and reviews

Athat,G., *Radiation transport in spectral lines*, Reidel, 1972. (Challenging.)

Weedman,D.W., *Quasar Astronomy,* Cambridge University Press, 1988. (An interesting treatment of the BLR and line radiation.)

Osterbrock,D.E., *Astrophysics of Gaseous Nebulae and Active Galaxies*, University Science Books, 1989. (An excellent text book, a must to read.)

Specialized

The BLR

Goodrich,R.W., 'Dust in the Broad-Line regions of several Seyfert Galaxies' *Astrophys.J.*, **440**, 141-150, 1995.

MacAlpine, G.M., 'II. Gas in active galactic nuclei', *Pub.Astron.Soc.Pacific*, **98**, 134, 1986.

Peterson, B.M., *et al*, 'Steps toward determination of the size and structure of the broad-line region in active galactic nuclei. VII variability of the optical spectrum of NGC5548 over four years', *Astrophys.J.*, **425**, 622, 1994.

Peterson, B.M., 'Emission-line variability in Seyfert Galaxies', *Pub.Astron.Soc.Pacific*, **100**, 18, 1988.

Stirpe,G.M., *et al*, 'Steps toward determination of the size and structure of the broad-line region in active galactic nuclei. VI variability of NGC3783 from ground-based data', *Astrophys.J.*, **425**, 609, 1994.

Wills, B.J., Steidel,C.C., & Sargent,W.L.W., 'Statistics of QSO broad emission line profiles I: The CIV λ1549 line and the λ1400 feature', *Astrophys.J.*, **415**, 563, 1993.

Wills,B.J., Netzer,H., & Wills,D., 'Broad emission features in QSOs and active galactic nuclei II. New observations and theory of FeII and HI emission', *Astrophys.J.*, **288**, 94, 1985.

The standard model and extensions

Collin-Soufrin,S., *et al.*, 'The environment of active galactic nuclei - I. a two-component broad emission model', *Mon.Not.R.Astr.Soc.*, **232**, 539, 1988.

Davidson,K., & Netzer,H., 'The emission lines of quasars and similar objects', *Rev.Mod Phys.*, **51**, 715. 1979. (The Bible of photoionized line emission.)

Filippenko,A.V., 'New evidence for photoionization as the dominant excitation mechanisms in LINERs', *Astrophys.J.*, **289,** 475, 1985.

Kwan,J., & Krolik,J.H., 'The formation of emission lines in quasars and Seyfert galaxy nuclei', *Astrophys.J.*, **250**, 478, 1981.

Krolik,J.H., 'The origin of broad emission line clouds in active galactic nuclei', *Astrophys.J.*, **325**, 148, 1988.

The extended emission zones

Bower,G.A., *et al.*, 'HST Images of the Seyfert Galaxies NGC5929 and MCG8-11-11', *Astronom.J.*, 1994.

Haniff,C.A., Wilson,A.S., & Ward, M., 'High-resolution emission-line imaging of Seyfert Galaxies. I. Observations', *Astrophys.J.*, **334**, 104, 1988.

Macchetto,F. *et al.*, 'HST/FOC Imaging of the narrow-line region of NGC 1068', *Astrophys.J.*, **435**, L15, 1994.

Pedlar,A. *et al.*, 'A neutral hydrogen study of NGC4151', *Mon.Not.R.Astron.Soc.*, **259**, 369, 1992.

Mulchaey, J.S. *et al.*, 'Hubble space telescope imaging of the Seyfert 2 Galaxy NGC2110', *Astrophys.J.*, **433**, 625, 1994.

Penston,M.V., *et al.*, 'The extended narrow line region of NGC4151 I - emission line ratios and their implications', *Astron.Astrophys.*, **236**, 53, 1990.

Perez,E., Tadhunter,C., & Tsvetanov,Z., 'The complex narrow-line region in NGC 4151', *Mon.Not.R.Astron.Soc.*, **241**, 31P, 1989.

Unger,S.W *et al.* 'The extended narrow-line region in radio Seyferts: evidence for a collimated UV field?', *Mon.Not.R.Astron.Soc.*, **228,** 671, 1987.

Wilson,A.S., Ward,M., & Haniff,C.A., 'High-resolution emission-line imaging of Seyfert Galaxies. II. Evidence for anisotropic ionizing radiation', *Astrophys.J.*, **334**, 121, 1988.

The molecular torus and unfication

Antonucci,R., and Miller,J., 'Spectropolarimetry and the nature of NGC1068', *Astrophys.J.*, **297**, 621, 1985.

Antonucci, R., Hurt, T., & Miller, J., 'HST ultraviolet spectropolarimetry of NGC 1068', *Astrophys.J.*, **430**, 210, 1994.

Krolik, J.H., & Begelman, M.C., 'Molecular Tori in Seyfert Galaxies: Feeding the monster and hiding it', *Astrophys.J.*, **329**, 702, 1988.

Mulchaey,J.S. *et al.*, 'Multiwavelength tests of the dusty torus model for Seyfert galaxies', *Astrophys.J.*, **436**, 586, 1994.

Tran,H.D., 'The nature of Seyfert 2 Galaxies with obscured broad-line regions II Individual Objects', *Astrophys.J.*, **440**, 578, 1994.

Tran,H.D., 'The nature of Seyfert 2 Galaxies with obscured broad-line regions III Interpretation', *Astrophys.J.*, **440**, 597, 1995.

Ward, M.J., Ed., *Proceedings of the Oxford Torus Workshop*, 1994.

7

Black holes, accretion disks and the evolution of AGNs

7.1 INTRODUCTION

Throughout this book we have made the assumption that the central engine of an AGN is a supermassive black hole. The gravitational field of this massive body is able to explain the resulting luminosity of AGNs and the evidence in favour of this picture is now overwhelming. In this chapter we shall investigate the black hole phenomenon and how it plays such a crucial role in our understandings of the workings of AGNs. However, in many areas we shall restrict ourselves to hand-waving arguments in order to avoid the detailed physics of general relativity, which is a prerequisite to a full understanding of the concepts at work. But we will not duck the issue of general relativity totally, a brief overview of some of the concepts of general relativity will be presented as we need them.

Einstein produced his amazing *general theory of relativity* in 1915. His earlier, *special theory*, dealt with velocities but not accelerations, whereas the general theory encompassed accelerations and, most importantly for this book, the effects of gravitational forces. The velocity of light lies at the cornerstone of both theories and the implications of both these new theories posed interesting, if not apparently bizarre, consequences. In section 3.9 we were introduced to a key feature of general relativity; the presence of matter distorts space-time. Flat (Euclidean) space is distorted, or curved, by the presence of mass. Light rays trace out the geometry of space-time, and in general relativity light no longer always travels in what we picture as a straight line, it travels in what are called geodesics. These are straight lines in the space-time in which the light is travelling, but only a straight line, as we typically understand it, in flat space. Once space-time becomes curved, perhaps due to the presence of a large mass, then the light paths as seen by us appear to be bent.

Strong supporting evidence for Einstein's major new theory came within four years of its announcement. Direct observations of the predicted effects of space curvature were made during a solar eclipse in 1919 by Sir Arthur Eddington. The path of the Sun against the background stars at the time of the eclipse was carefully predicted and precise measurements were made of the relative positions of those stars (a number of months beforehand). During the eclipse, when the sky was dark, the relative positions of the stars were again measured. They had changed, and by the amount predicted by Einstein's new theory (although the uncertainty in the measurements was reasonably high). These observations showed that the apparent positions of stars changed when the Sun was placed between them and the observer, proving that space-time was curved in the presence of a large gravitating mass (the Sun). Nowadays, precise radio interferometric observations of the changing position of quasars as the Sun passes close to their position on the sky give very precise tests of general relativity. Further examples of the distortion of space-time arose from the 1979 discovery of gravitational lensing with the famous 'double quasar' (see section 3.9). The bending of space-time by a condensed gravitating mass is a cornerstone of black hole physics.

We will commence the chapter with a description of the main aspects of supermassive black holes, including accretion disks. We should remember that the evidence we have for supermassive black holes powering AGNs is secondary in nature, (although highly supportive) and that no accretion disk has ever been detected in such bodies. In fact, the presence of a supermassive black hole is probably on better observational grounds than the presence of an accretion disk in AGNs, but nevertheless, we shall assume both in our discussions, although pointing out where problems still remain.

We will explore the supermassive black hole concept for powering AGNs in detail and we will see how the luminosity of the black hole is linked to parameters such as the mass of the black hole, the fuel supply and accretion rate. This leads us naturally into considerations of the formation and evolution of black holes in an AGN, and this is intimately linked with the switch-on time of an AGN and its subsequent evolution. We will then start to flesh out ideas as to why we might see such a range of AGN luminosity and how it might be linked to parameters of the black hole. Although we will touch on the origin of the spectacular radio jets, much of the discussion on this topic will be reserved for the final two chapters.

Before leaving this chapter we will also review an alternative model to that of a supermassive black hole, a highly luminous cluster of stars. Although nearly all astronomers accept that this idea cannot explain the activity seen in objects such as quasars and Seyfert galaxies, it may find a place in explaining the energy source in some of those galaxies which border on being called active, in particular some of the low ionization galaxies and far-infrared galaxies. This is the starburst model, albeit in a slightly different guise.

7.2 SUPERMASSIVE BLACK HOLES AND ACCRETION DISKS

7.2.1 The physics of black holes

The concept of a black hole is not new. As long ago as the eighteenth century, the great French physicist Laplace calculated that if light behaved as corpuscles (rather than waves) and these were attracted by gravity like ordinary matter, and if a body was sufficiently massive, then the escape velocity could reach the speed of light. Hence light could never escape from the body. It would be hidden from the rest of the Universe; it would be a black hole. So we see that gravity holds the key.

The modern concept of a black hole originated in 1939 and was built on the new physics of the twentieth century. These included the Pauli exclusion principle, quantum mechanics and Einstein's theory of general relativity. Oppenheimer and Snyder argued that once a sufficiently massive star has exhausted its nuclear fuel it suffers catastrophic collapse that does not find an end point in a stable state of matter (a white dwarf or neutron star) but continues to collapse indefinitely. However, at some point towards the end of this collapse there will be a time when the effects of general relativity dominate. The collapsed star distorts space-time to such an extent that it is effectively cut off from the rest of the Universe. Light inside this volume is forever trapped—and this is a black hole as we know it.

Let us venture to investigate this concept in a little more detail. A black hole is a volume of space in which the gravitating body has curved space-time to such an extent that light (and all electromagnetic radiation) can no longer escape the region. In Newtonian terms, the escape velocity from the 'surface' of the object can be thought of as exceeding the speed of light. One way of thinking about this is to trace the imaginary path of the light emitted from a torch pointed vertically upwards from the surface of a mythical object inside a black hole. The light would not follow a straight line perpendicular to the surface as it does on normal bodies like the Earth. Instead it would curve round and follow a trajectory that never allowed it to escape from the body. Seen by an external observer the black hole radiates nothing; all emissions are contained inside an outer horizon and the object is completely black, hence the name. The idea of a 'hole' comes from the fact that material falls into the region due to its gravitational attraction, but is then lost forever to the Universe. Nothing can escape; therefore it is a like a hole in space. The term itself is newer than the concept, being originated by John Wheeler at a conference in 1967.

Technically, the outer horizon of a black hole is called the Schwarzschild radius, referred to as R_S. This is named after Karl Schwarzschild who derived exact solutions to Einstein's field equations in 1916, defining the gravitational radius for a point, non-rotating mass. The Schwarzschild radius is also called the event horizon, being the last horizon seen by an external observer at which events can be witnessed. Material outside this surface can escape the black hole, but anything inside is trapped forever and we have no information of what goes on inside. Indeed, modern physics cannot describe the conditions at the centre of a black hole.

In classical terms it is easy to calculate R_S using the definition that it is the radius at which the escape velocity equals the speed of light. We can equate the potential energy of the gravitating mass to the kinetic energy of an escaping test particle to obtain

$$\frac{1}{2}mv^2 = \frac{GMm}{R}, \tag{7.1}$$

where m is the mass of the test particle of velocity v, M is the mass of the body of radius R, and G is the gravitation constant. For a photon, we can crudely replace v by c to give

$$\frac{1}{2}mc^2 = \frac{GMm}{R}. \tag{7.2}$$

This can be re-arranged to give the Schwarzschild radius R_S

$$R_S = \frac{2GM}{c^2}. \tag{7.3}$$

In fact rigorous analysis produces the same result so we can proceed with confidence.

Using eqn. 7.3 we can show that a $1 M_{\odot}$ black hole has $R_S = 3$ km, while a 10^8 $M_{\odot}$ black hole has $R_S = 3 \times 10^8$ km. This latter value is only ~2 AU, so we see that a supermassive black hole would comfortably fit inside the inner region of the solar system.

In passing we also note that a neutron star has a mass similar to that of the Sun and is believed to have a radius of only 10 km or so. This is not that much bigger than R_S and points to the fact that a tremendously strong gravitational field is expected to exist around such an object. This is sufficient to distort space-time significantly, and precise observations of the slow-down of a binary pulsar further verified Einstein's work and supported the concept of gravitational radiation. This pulsar timing experiment led to the 1993 Nobel Prize for Physics for astronomers Joseph Taylor and Russell Hulse. Just like white dwarfs, neutron stars have an upper limit to their mass, but in this case it is about 2 or 3 $M_{\odot}$. If a collapsing core is more massive a neutron star cannot form. Also, if a neutron star accretes mass until it exceeds this limit, then it will undergo a catastrophic collapse to a black hole. Such events are expected to occur and might be observable as bursts of X-rays or gamma-rays.

Now that we have set the scene for a black hole, the perceptive reader might have noted a problem for our favoured energy source of active galaxies. If everything disappears into a black hole, and we now know that the black hole does not radiate energy into space, then how can it be a source of energy for active galaxies? Not to worry, this is not a show-stopper and the answer is relatively simple. The theories of black holes show that they retain certain fundamental parameters of formation. These are spin, gravitating mass and electric charge. We shall ignore the latter. However, the fact that the black hole retains the gravitational field of the collapsed body is crucially important. The simplest way to imagine this is to picture the black hole as causing a warp in space to such an extent that inside the Schwarzschild radius the space folds up on itself and vanishes. But, just exterior to this the space is severely curved and objects passing by will follow curved trajectories as they are attracted by the gravitational influence of the hole. Depending on the relative velocities and impact parameter, these particles can end up being captured and engulfed by the black hole. Therefore in this process the gravitational field converts potential to kinetic energy but the key question is how to extract this energy in a useful form before it is lost inside the event horizon.

Matter falling directly onto the black hole represents an accretion flow, with a timescale equivalent to the free-fall time from their location to the event horizon. However, it is more than likely that material will intersect the gravitational field of the black hole with glancing trajectories rather than a direct infall and the picture will be much more complex. The details of the infall resulting from ranges of impact parameters and the subsequent orbital modifications, physical interactions and re-radiation of the infalling plasma are too advanced for this book and so we will restrict ourselves to giving an overview of the general ideas. In the next section we shall develop this further and suggest mechanisms by which the central engine radiates and, in some very special cases, how highly collimated jets of relativistic plasma are formed. These are the spectacular radio jets that we will describe in the next chapter.

Before leaving the physics of black holes themselves, we will take a look at the second fundamental parameter of the black hole, the spin. Let us now consider the case of a rotating black hole. This is also highly complex and introduces us to exotic phenomena such as the black hole ergosphere. In 1963, Roy Kerr derived a solution of Einstein's field equations for a rotating system. Six years later, Roger Penrose showed that a rotating black hole possessed another important radius in addition to the Schwarzschild radius. This is called the static radius and lies outside R_S. Within this radius a particle cannot be stationary but must rotate with the black hole. The intriguing point is that while lying inside this radius but outside R_S the particle is still able to escape if it possesses a source of power (like a rocket engine). This region of space is called the ergosphere of the spinning black hole.

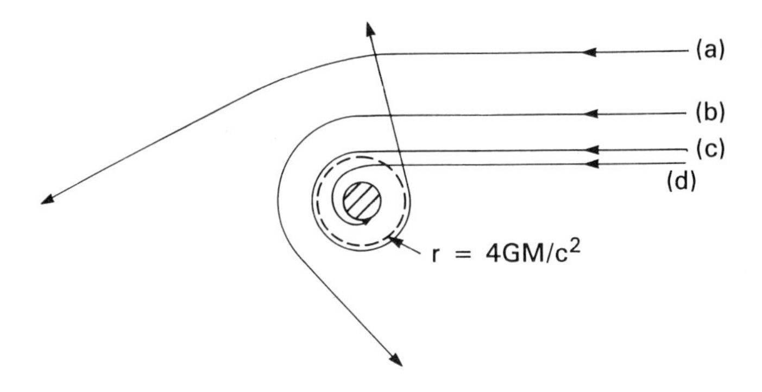

Fig. 7.1. Trajectories of a test particle interacting with a black hole. In the first two cases (a,b) the particle is deflected and escapes to infinity, in case (c) the particle approaches very close to the last stable orbit (the capture radius of 6 GM/c^2) and orbits many times before escaping to infinity, while in (d) the particle is captured and spirals towards the event horizon (2 GM/c^2) shown by the shaded circle.

So far so good. Let us now resort to a thought experiment. Consider what might happen if a particle fell into the ergosphere and subsequently decayed into two other particles before it reached the event horizon. For further simplification we can take the special case where one of the decay products crosses the event horizon in a direction directly opposed to the rotation of the black hole. This particle is lost to the Universe forever, but in this case the other decay particle escapes from the ergosphere. Amazingly, and against all intuition, it turns out that the escaping particle can have more energy than that of the original particle that entered the ergosphere. How is this possible? It seems to violate the hallowed principle of the conservation of energy. Fortunately, this is not the case. The total energy of the system is indeed conserved and the additional energy for the escaping particle comes from the rotation of the black hole. A consequence of this work on rotating black holes (often referred to as Kerr black holes) is a parameter called the irreducible mass. A fundamental theorem of black hole physics is that the irreducible mass must either remain constant or increase; it can never decrease. The irreducible mass is proportional to the square root of the surface area of the event horizon and depends on the product of the mass and angular momentum. So a rotating black hole is a potential power source. Theoretically, up to 29% of the energy of a spinning black hole can be extracted in this way; making it the most efficient energy-generation mechanism yet found.

We shall not venture into the idea of a black hole having 'no hair', nor the Cosmic Censorship Hypothesis, nor the evaporation of black holes, as they have little relevance for AGNs. Instead we provide the interested reader with suitable references in section 7.6. We shall now investigate the infall of particles towards a black hole and the energy generation mechanism. In leaving this section we should stress that although we have seen how black holes can generate energy, in the next section we shall see how yet another process may be responsible for much of the emission we see.

7.2.2 Accretion disks

We now consider accretion in a little more detail because it is from this process that we believe much of the energy powering the AGN derives. The gravitational field of the supermassive black hole will attract gas clouds in the central parsec or so of an AGN. In a general hand-waving manner we can see that the gas will accelerate towards the black hole from a range of directions. Eventually, there will be so many gas clouds in the vicinity of the black hole that as they make their closest approach on their orbital trajectory (see fig. 7.1) they will collide with each other. In this process they will lose kinetic energy which is transformed into frictional heating. As the gas particles continue to lose energy and velocity through this frictional interaction, they will eventually have insufficient energy to escape from the zone. At this point they are captured, perhaps with highly elliptical orbits just like the short-period comets orbiting around the Sun. If the original gas started out in a flattened rotating system, which was still enormously large compared to the size of the black hole, then the gas will tend to retain this rotation. As this collisional process and increasing viscosity continue, the orbits of the gas become more circular and the system will tend to become more flattened. In this process the angular momentum is conserved, with the vertical component of velocity being neutralized by conversion of kinetic energy into potential energy, and ultimately heat.

This rotating disk of material is termed the accretion disk. The individual gas clouds or particles within the disk continue to interact, the result being that the accretion disk becomes even hotter with a temperature gradient from the inner to outer edge. This process of continued particle interaction is called viscous drag. What is its detailed cause? In a Keplerian system the gas in the inner parts of the accretion disk orbits faster, but because the gas particles interact with each other, the particles in the inner zone suffer a drag from the slower orbiting particles of the next outer radial zone. This drag makes the inner particles orbit slower than their required Keplerian orbital velocity for the distance from the black hole. The particles must then either speed up or change orbit. Without an additional source of energy their only option is to change orbit and they do this by using the gravitational field to gain energy, moving inwards and becoming stable again. In a similar way low Earth orbit artificial satellites eventually fall back to Earth soon after encountering the drag of the outer layers of our atmosphere, unless they are boosted to higher orbits by on-board rocket engines. The viscous drag is a continuous process resulting in a slow infall as angular momentum is transferred outwards. We know that the closest stable orbit to a non-rotating black hole is 3 R_S and this distance shrinks to 1.5 R_S for a rotating black hole.

The obvious result of the viscous drag within the accretion disk is to cause it to heat up, especially the innermost part, which becomes an extremely hot source of thermal emission. The energy of a particle at this last stable orbit can be calculated by letting the gain in kinetic energy due to the gravitational infall be manifest as thermal energy (heat). This results in particle temperatures ranging upwards from 10^5 K.

It is interesting to speculate about the fate of the resulting angular momentum loss in the accretion disk. It must go somewhere. In the absence of an external coupling agent the accretion disk spreads out, due to the transfer of the angular momentum to the outer regimes. It is also possible that there is coupling to some external medium, such as magnetic fields. Indeed, this latter aspect is arousing considerable interest and maybe a magnetic coupling produces a torque on the accretion disk similar to what we see in the solar system. Magnetic fields also have the attraction in that they would appear to be a requirement for the production of collimated radio jets. However, we should stress that the picture is far from clear.

The next question to probe is the internal physics of the accretion disk, an entire discipline of current astrophysical study. Accretion disks come in two flavours: thin disks or inflated tori. The simplest case is the thin disk which obtains its pressure support from gas pressure at the outer radius and by radiation pressure towards the inner boundary. The thin disk is basically the low-energy case where the material of the disk is at a relatively low temperature, of the order of 10^4 K, and the sound speed in the gas is much lower than the rotation velocity of the disk. This allows the disk to continuously re-adjust with a slow inward drift of material. The result is a flat pancake-like disk (fig. 7.2(a)) with an aspect ratio (diameter to thickness) dependent on the ratio of the rotation speed to sound speed. The accretion disk can extend inwards to a distance from the black hole of no closer than 3 R_S.

The simplest predicted continuum radiation spectrum from a thin accretion disk is the superposition of the radiation from the blackbodies of a range of temperatures within the disk. This produces a power-law of $S(\nu) \propto \nu^{1/3}$. However, when electron scattering is taken into account for higher accretion flows, the output spectrum is modified at the high-energy end before the onset of the Wien cut-off (section 4.2). However, the spectral shape predictions are generally not well fitted by the observations, and the lack of high-energy UV to low-energy X-ray coverage remains a problem for a statistical treatment. Nevertheless, because of the flat appearance of the optical–UV part of the spectrum from some AGNs (section 5.2.3), we can see the obvious attraction for thin disk accretion models as a potential explanation for the big blue bump.

The second, and much more complicated picture, is the thick accretion disk. The definition of thick is that the aspect ratio approaches unity, requiring the thickness of the disk to be of the same order as its diameter. This is often referred to as an accretion torus, as this paints a much better picture of its geometry. The term radiation torus is sometimes used as radiation provides the source of the large internal pressure that causes the material in the disk to become 'puffed-up'. Another variant on this is the ion-supported torus. At very large distances from the black hole, and depending on the particular circumstances, the disk can either remain thick or it can cool and collapse to revert to a thin disk. In both cases the luminosity of the accretion disk is totally dominated by the extremely hot inner torus of temperature exceeding 10^5 K.

The thick disk model breaks down into two cases depending on the source of its pressure support, which is either radiation or ion pressure. Here we are dealing with a much higher energy situation than for the thin disk model. Now the disk cannot re-adjust itself sufficiently rapidly with outward transfer of angular momentum. It is this that causes it to inflate; indeed this is to such an extent that the models predict it forms a very steep-walled torus surrounding the supermassive black hole (fig. 7.2(b)). In the radiation-supported torus, the radiation pressure is extreme, at or about what is termed the Eddington limit (see below). The theories of thick disks are extremely complex and no overall agreed picture has yet emerged. Nevertheless, they attract much attention because of the potential for the generation of extreme UV and X-ray emission by the inner steep-walled zone and for this funnel-like geometry potentially being able to provide some means of collimation for the very innermost regions of the nuclear jets.

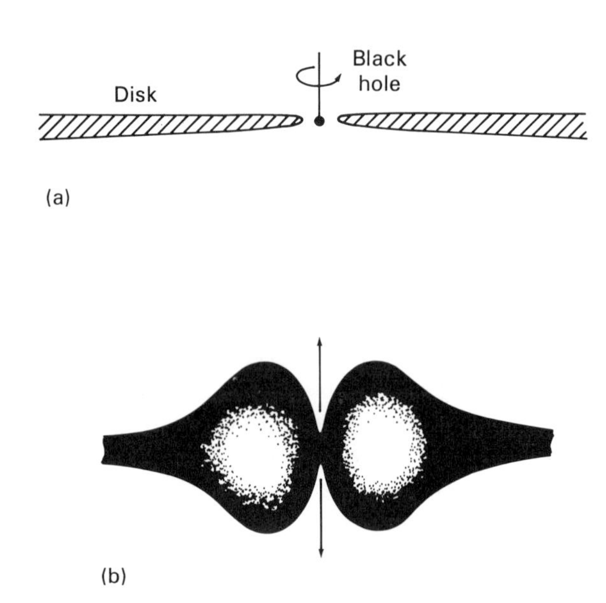

Fig. 7.2. Schematic of (a) a thin accretion disk and (b) a thick accretion torus surrounding a supermassive black hole.

In the radiation supported torus, radiation pressure due to internal opacity provides the entire support for the torus configuration. In the ion supported torus, the pressure support derives from the ions of the gas. This is an unusual situation and comes about due to relativistic and certain other effects that keep the electrons cool *compared to the ions*, which are preferentially heated to the virial temperature. This results in a two-temperature situation. Although the electrons radiate, they can do so only under very specific mechanisms and regimes of temperature and density. In fact their radiation is insufficient to deplete the energy of the torus sufficiently rapidly to cause it to collapse. Instead the torus remains inflated, supported essentially by the gas pressure of the plasma ions. The physical processes are straightforward, but bringing them all together into a single model was only undertaken in a seminal work by most of the giants in this field. The reader is referred to the 1982 *Nature* paper by Rees, Begelman, Blandford and Phinney. Their figure 1 is well worth detailed and lengthy study, for therein lie many of the secrets of accretion disks laid out in terms of physical interaction processes. It is a masterpiece but too complex for presentation here. This paper also outlines many of the reasons why these tori were invented: the feeding of radio lobes of radio galaxies by jets. A breakthrough to the puzzle of the relatively low luminosity of the nuclei of radio galaxies compared to the much higher integrated power in their radio lobes was provided by this solution.

The ion case also leads to interesting effects of electric current generation, dynamo effects and the scope for synchrotron radiation cooling mechanisms. This ushers in the exciting possibility of identifying another mechanism for extracting energy from a spinning black hole, magnetic coupling of the accretion disk to the hole. This of course assumes that both are threaded with the same magnetic field. Two attractive features of magnetic, rotating, ion tori are the production of collimated jets and an easing of the required accretion rate to explain the luminosity.

The deep funnel shape of the very steep inner walls would seem to provide an obvious mechanism for at least some large-angle outflow collimation, leading to an outward radiation field that is no longer isotropic, but cone-shaped of opening angle of some tens of degrees. This might explain the observations of anisotropic ionizing radiation patterns in some AGNs (but is more likely that this is produced by the molecular torus of section 6.5). In terms of collimation the situation is improved even more with a magnetic ion torus, as calculations show that the magnetically induced currents flowing in the central funnel can lead to an even greater degree of self-collimation by a plasma jet. This helpful result for potential jet collimation is further aided by finding that it does not require undue energy to be taken from the accretion disk because the coupling between the jet and the disk remains relatively low. In passing we can note that whatever the geometry of the inner accretion disk, if the main source of the BLR ionizing radiation emanates from a very extended super-hot corona, then we would expect to see isotropic radiation rather than a collimated beam. The debate is still very speculative, but intriguing and we shall return to this in chapter 9.

A second positive feature of magnetic ion tori is in reducing the requisite accretion rate. Here the rotational energy of the black hole is effectively being used as part of the energy source by the mechanisms we saw in section 6.5.1. This in turn means that the primary fuelling energy of the black hole, the accretion of material from the surrounding interstellar medium, can take place at a much reduced rate. This produces a significantly lower accretion rate than would otherwise be required to explain the luminosity, which in turn removes some of the problems of the short lifetimes predicted by luminosity and energy rate considerations (see next section). It has also the added attraction of reducing the well-known difficulty of persuading material to lose its angular momentum and eventually impact on the accretion disk.

In all the above discussions we should not forget that the accretion disk is an incredibly small phenomenon compared to the scale-sizes of parsecs to kiloparsecs of a galactic nucleus. This is all too easy to fail to appreciate. Let us now refresh our knowledge on the sizes of accretion disks. We saw that the inner radius is around one to a few times R_S depending on the type of disk, thin disks giving around 5–10 R_S. Therefore the accretion disk for a 10^8 solar mass black hole has an inner radius of between ~1 to 20 AU. But what about the outer radius? We noted above that as long as certain temperature/density conditions are met the radiation accretion torus remains thick but will eventually collapse to thin disks on a scale size of around 50 to 100 R_S. The ion tori may continue in the inflated state to much larger radii, but eventually, perhaps at thousands of R_S, the disk will revert to the thin case. It is possible that a thin disk can extend to perhaps 10^4 R_S depending on the accretion flow and black hole luminosity. However, although accretion disks can be very extended, the luminosity is dominated by the very hot inner zone, and this just about matches the scale of the solar system. We should stress that even the extreme outer dimension of 10^4 R_S is only of order 0.05 pc, and this can now be contrasted to the size of the BLR. We now see that there is potential for size overlap and therefore the suggestions made in sections 6.3.4 and 6.3.5 that the LIL lines may originate from the surface of the extended accretion disk are shown to be quite satisfactory in terms of the scale sizes and our concept of an accretion disk.

7.2.3 Energy generation

Let us now return to the black hole itself and consider the mechanisms of the energy generation. The supermassive black hole is a singularity in space-time, cloaked in its event horizon (ignoring the ergosphere) and probably surrounded by an accretion disk. We will now take a more detailed look at two closely related aspects in explaining the observed luminosity from the central engine: the energy generation rate and the fuelling timescales. An immediate question is to ensure that the required accretion rate and energy conversion efficiency to produce the observed luminosity are feasible. It is in the pursuit of this topic that much theoretical study is currently devoted. Although we have stated that the fundamental source of energy is gravitational, what does this actually mean? Specifically, how is the gravitational potential energy converted into the photons that we, the astronomers, ultimately observe?

We will tackle this latter point first, although in a global rather than specific sense. In the accretion scenario, the most obvious means of releasing the gravitational potential energy of the hole is, as we saw earlier, by viscous interactions within the accretion disk. The slow inward movement of the mass transfer produces heating of material through this frictional drag, which then radiates as a photon luminosity. We might naively expect that the heating will be larger the stronger the gravitational field. However, there are undoubtedly more complex processes at work involving magnetic effects and possibly photon-pair production events in the inner regimes. Nevertheless, the fundamental parameter is the central gravitational mass and the requirement of a certain amount of mass being attracted towards the black hole. The key number is the energy conversion efficiency of accreted mass to output photons. This then will tell us the mass accretion rate required to account for the observed luminosity. We assume that gas (which includes stars) from the surrounding medium is the fuel. At this stage we will not worry unduly about how the gas loses sufficient angular momentum to sink to the very central regions of the galaxy and ultimately the accretion disk.

So what is the conversion efficiency of gravitational energy to photons? Efficiency calculations are notoriously difficult to undertake with any precision due to the complexities of the medium in question and the number of unknown variables at work. However, to explain the X-ray and optical flares seen in dwarf novae (by way of accretion disks), much progress in this area has been made. Theorists now believe that a conversion efficiency of a few per cent should be attainable, with possibly as much as 10% of the rest mass energy of the particles being converted into photons. We should note that these values are dramatically larger than the energy release due to nuclear fusion, which has an efficiency of only of 0.7% of the rest mass energy. Indeed, the discovery of quasars requiring high energy conversion efficiencies led directly to supermassive black holes being the most attractive explanation for their otherwise unexplainable long-term luminosity.

Let us assume that the conversion efficiency, Q, is 5%. For a quasar of $10^{12}\ L_\odot$, we have an accretion rate, $\dot{m}_{acc}$, given by

$$\dot{m}_{acc} = \frac{10^{12} L_\odot}{Q c^2} \,. \qquad (7.4)$$

Numerically we have

$$\dot{m}_{acc} = \frac{10^{12} \times (3.8 \times 10^{26})}{0.05 \times (3 \times 10^{8})^2} \,, \qquad (7.5)$$

which gives an accretion rate of 8.44×10^{22} kg s^{-1}. In terms of solar masses this is an accretion rate of 1.35 $M_\odot\ y^{-1}$, which at first sight does not seem very much. However, we can put this into perspective by considering the mass of the potential fuel supply. Even if there is $\sim 10^{10}\ M_\odot$ of material in the central zone of a galaxy, this immediately limits the lifetime of the quasar to a few times 10^9 y, if it radiates at a constant luminosity and no new material enters the central zone.

This simple situation is complicated by three factors. As the black hole accretes matter, both its mass and accretion rate increase. However, there is a natural upper limit on the accretion rate of a black hole given by the Eddington limit. This is the point reached when the luminosity from a radiating object is so high that the outward electromagnetic radiation pressure just balances the gravitational attraction of the object. Accreting material is effectively held at bay by the immense outward radiation pressure and the accretion reaches a steady state. We can see that this is a balance between the luminosity and the accretion because if the accretion drops, then the luminosity falls and the accretion resumes again. In this way accretion rates can, in practice, be close to the Eddington limit for spherical accretion, but not above it. The third problem is how to get sufficient gas into the very central regions of the galaxy and subsequently onto the accretion disk. This comes down to the problem of removal of angular momentum.

Let us make a start with the problems by looking at the Eddington limit in more detail, because in principle it should provide a firm foundation to figures for the limiting mass of a black hole for an AGN. At the Eddington limit the radiation pressure on a particle from the hot inner edge of the accretion disk exactly balances the gravitational attraction of the black hole. The radiation pressure depends directly on the luminosity and falls off as the inverse square of the distance from the black hole. The gravitational attraction also falls off as the inverse square of the distance and so the distance terms cancel, leaving the Eddington luminosity depending only on a single variable parameter of the gravitating object, its mass. For a spherical source of mass, M, accreting spherically symmetrically at a steady rate, the Eddington luminosity, L_{Edd} is given by

$$L_{Edd} = \frac{4\pi G M m_p c}{\sigma_T} = 1.3 \times 10^{31} \frac{M}{M_\odot} \quad \text{watts,} \tag{7.6}$$

where m_p is the mass of a proton, c is the velocity of light, σ_T is the Thompson interaction cross-section and G is the gravitational constant. We see that the black hole mass for a highly luminous quasar of $L \sim 10^{12}\ L_\odot$ radiating at the Eddington limit is $2.9 \times 10^7\ M_\odot$. Therefore the black hole must be supermassive. In this context, supermassive means of the order of $10^6\ M_\odot$ or greater. Note that if the black hole is not radiating at the Eddington limit, then the mass will be *even larger*. In principle it seems relatively easy to provide the luminosity of quasars from black holes and accretion disks, but we note that the precise fuelling process is unclear. Given that we see highly luminous quasars at a redshift, $z \sim 4$, we are immediately faced with the puzzle of how $\sim 10^7\ M_\odot$ black holes are produced so early in the life of the Universe. We will return to this fascinating question later when we consider the lifetimes of activity.

7.2.4 Observational evidence for supermassive black holes

The case for supermassive black holes powering powerful AGNs is persuasive. However, can we find direct observational tests for the presence of a supermassive black hole in the nuclei of active galaxies? Because of the size scales concerned we should be clear that it is always going to be difficult to obtain direct imaging data. We again stress that the Schwarzschild radius of a $10^8\ M_\odot$ black hole is only 3×10^8 km (~ 2 AU), and even if the accretion disk were fifty times larger, this would still subtend an angle of less than 2×10^{-4} arcseconds at a distance of 10 Mpc. Even very large accretion disks, of the order of 0.1 pc, are not going to be directly visible with the Hubble Space Telescope. Therefore for the moment, the evidence will have to come from indirect means and we shall now review these in turn.

Perhaps the most powerful clue comes from variability arguments. The observations of rapid temporal variations described in section 5.4 demonstrate that large changes of luminosity are taking place on size scales of light hours and less. Hence we must be dealing with very compact phenomena of an order no larger than the size of the solar system. The only way out of this argument is to reduce the observed luminosity by requiring that the objects be much closer than we suspect or to invoke beaming mechanisms. The former was the 'non-cosmological redshift' argument used against deduced quasar luminosities in the early 1960s. We shall reject this out of hand, and beaming is not a serious contender for the wide range of objects and radiation mechanisms. Therefore, we are left with enormous luminosities emanating from solar system sizes. Only supermassive black holes and gravitational energy conversion have stood the test of time in satisfying these two requirements.

A second indirect method is to measure the effects of the gravitational mass on the surrounding material. This allows the dynamics of the region to be determined and hence the gravitating mass within that zone deduced. When this is coupled with observations of the spatial distribution of the luminosity, astronomers must then decide whether a supermassive black hole is the best, or only solution. This has been a topic of much interest and the recent work has produced great success in the quest for evidence for the existence of supermassive black holes. All the studies have concentrated on determining the gas and/or stellar dynamics in the innermost nuclear regions of relatively nearby galaxies. The observations are restricted to distances of $\leq$20 Mpc due to the limitations of the spatial resolution at greater distances. Although these studies began over twenty-five years ago, the past decade has seen a growing body of evidence that black holes in excess of a million solar masses may be a relatively common phenomenon in the centres of galaxies.

With a spectacular jet emanating directly from its active nucleus and at a distance of only ~15 Mpc, the giant elliptical galaxy M87 (see section 8.5 and plates 3 and 4) has been a favourite target for study. Early evidence suggested that the nucleus of M87 harboured a black hole of mass around 10^9 $M_\odot$ or higher. These results were obtained almost twenty years ago, from observations investigating the distribution of light from the central regions of M87. This revealed a very bright and unresolved central spike (or cusp). Comparison of the light distribution with that of many other elliptical galaxies (from which elegant models had been formulated—mostly due to Andy King) revealed a 'shoulder' of light of a few arcseconds in extent. The profile of the shoulder was such that it fitted the radial density distribution expected from a massive object located at the centre. This work was done in 1978 and was immediately supported by high resolution spectroscopic studies of the dynamics of the stars in the central region of M87. The data showed that the velocity spread of the stars increased steadily towards the centre. Extrapolation of this to the inner arcsecond produced a dynamical situation which could be explained by a central condensation of mass exceeding ~10^9 $M_\odot$. The object was referred to as a supermassive black hole.

This work stimulated much further research on the central regions of galaxies, both observationally and theoretically. As more and better data were obtained, along with a clearer theoretical understanding of the behaviour of material at small galactic radii, the support for a supermassive black hole in M87 see-sawed back and forth. As we are well aware from chapter 2, ground-based optical observations are limited by the 'seeing' of the atmosphere and this minimum scale of around 0.25 arcseconds is too great in spatial resolution terms to provide conclusive proof of the existence of a supermassive black hole.

Solving the question of supermassive black holes in the cores of local galaxies, especially M87, was one of the prime directives of the Hubble Space Telescope. Even with its blurred vision, the HST data showed that within an inner radius of 10 pc, the photon luminosity exceeds 10^3 $L_\odot$ pc^{-3}, and this is superimposed on an unresolved

(<1 pc) central optical spike. The spike is much bluer than the surroundings and at least as blue as the inner jet that is known to be due to synchrotron radiation. Furthermore, the luminosity of the spike is in excellent agreement with an extrapolation of the radio luminosity and the observed radio-optical spectral indexes. This provided strong evidence for the non-thermal nature of the emission from the central spike, and so it is almost certainly not the radiation emanating directly from the hot accretion disk. However, it could be the throat of the inner jet (see sections 8.5 and 8.6).

We saw in section 5.3.3 that in an attempt to clinch the debate once and for all, the refurbished HST was pointed towards M87 as one of its very first targets of study. Observations in February 1994, within two months of the refurbishment, gave the results that the astronomical community had been waiting for. Sufficient evidence was produced to support the supposition that there was a supermassive black hole in M87. Furthermore, this provided the first conclusive evidence for the existence of supermassive black holes. Wide Field Planetary Camera 2 images of the inner core of M87 are shown in plates 3 and 4. In the images, which are taken through a special passband filter to isolate the Hα and [NII] lines, we can clearly see that the gas in the centre of M87 is disk-like in structure, and there is the distinct suggestion of inner 'spiral arms' emanating from this elongated disk. The disk is ~1 arcsecond along the major axis, which corresponds to a linear size of about 75 pc at the distance of M87. The spatial resolution of the image is of order 3 pc. The mass of gas in the disk is determined from the observed flux and the known electron density and gives a value of $3.9 \pm 1.3 \times 10^3\ M_\odot$.

The nuclear region was also observed by the Faint Object Spectrograph. Spectra were obtained at the position of the nucleus and two diametric locations 0.25″ away. These showed that the velocities of the gas at these two positions are ±500 km s^{-1} with respect to the nucleus. Supporting emission line data at two other locations show that the gas is rotating in a Keplerian state about a mass of $2.4 \pm 0.7 \times 10^9\ M_\odot$ within the inner 0.25″ (18 pc) of the core of M87. The conclusion that this is a supermassive black hole was immediately accepted by most of the astronomical community but we see that these scale sizes are still enormous compared to those expected from a putative accretion disk.

Following from our earlier discussion we can use the deduced mass of gas in the disk to calculate a lifetime for the AGN core activity in M87. The prominent emission from the core of M87 is radio synchrotron emission, and using a conversion figure of 10% for the gravitational potential energy into radio photons, gives a lifetime for the radio source of around 5×10^6 y. This is very short and turns out to pose a problem for the activity we see in M87, as we discover in section 8.5.

Before the dramatic HST observations of February 1994, the difficulties of unravelling the secrets of M87 had turned attentions to other objects. Evidence supporting the existence of supermassive black holes in the cores of galaxies was provided from rather surprising quarters: work on the 'boring and non-active' local galaxies M31 and M32. This might seem extremely perverse, but let the story continue. These galaxies were studied not because they were active (they are not), but because they are very close and because the starlight is easier to see as there is no contaminating non-stellar continuum radiation from an AGN core. At distances of less than 1 Mpc, they are open to very detailed spatial studies. The mid- 1980s showed that the stars in the central regions of both galaxies were orbiting the nucleus at high velocities and with a high velocity spread. The conclusions seemed to require a dark mass of ~10^7 $M_\odot$ of material in the central 100 pc^3 of both M31 and M32. The corresponding density is in excess of 10^5 $M_\odot$ of material per cubic parsec, which although does not prove the existence of a black hole, requires some extreme forms of modelling and exotica to provide alternatives.

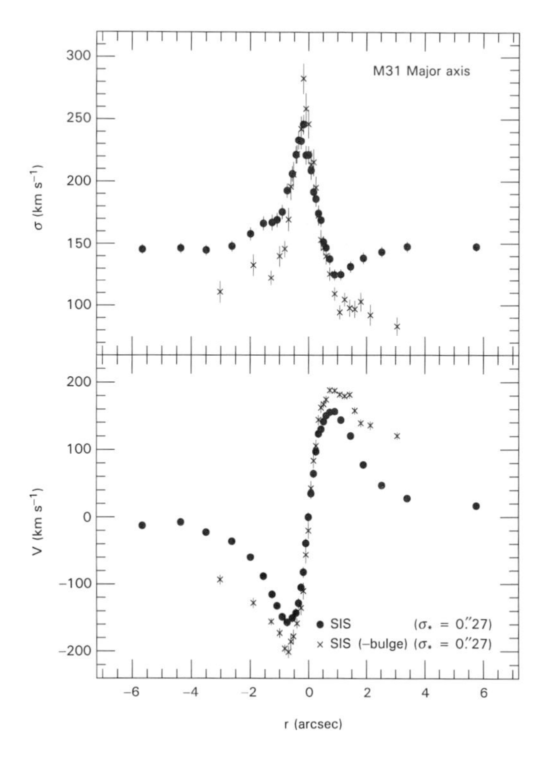

Fig. 7.3. Compelling evidence for the existence of a supermassive black hole of mass > 10^7 $M_\odot$ in the core of M31 obtained from the radial velocity (bottom) and velocity dispersion (top) of stars. The filled circles are for the kinematics of the nucleus plus bulge, the crosses show the kinematics after the bulge component has been removed. (Courtesy of John Kormendy.)

The latest wave of support for M31 housing a supermassive black hole candidate comes from optical observations using the excellent seeing on Mauna Kea and the new CFHT spectrometer with 'tip-tilt' guiding. In 1988, astronomers were able to determine that the stars in the very nuclear region (radii < 4 pc) rotate in a disk structure that is dynamically divorced from the surrounding central stellar bulge. The mass-to-light ratio of the central parsec is 30 ± 5, indicating the presence of a massive and dark object in the core. [The mass-to-light ratio is a useful quantity and references the total mass and luminosity of an object to those of the Sun, which has a mass-to-light ratio of unity. Very high mass-to-light values are indicative of non-luminous material for the mass.] The results of the latest velocity profiling for M31 are shown in fig. 7.3 and agree well with the earlier mass estimate for a central black hole of mass 3×10^7 $M_\odot$. Interestingly enough, the studies on M31 showed that the 'obvious' nucleus does not appear to be at the precise dynamical centre of the galaxy and M31 now appears to have a double nucleus. This complicates matters somewhat, but nevertheless, astronomers are confident that M31 houses a supermassive black hole.

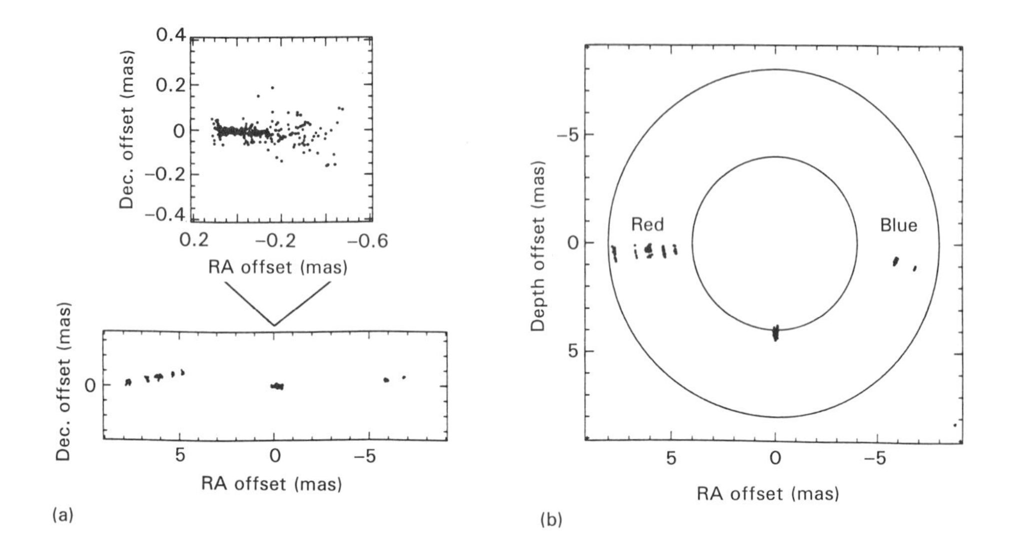

Fig. 7.4. (a) The distribution of the water masers in NGC4258 in units of milli-arcseconds offset from the central region which is also shown in an expanded view above, (b) the calculated positions of the masers along the line-of-sight direction. The circles mark the inner and outer edges of the molecular disk. (From Miyoshi *et al.*, *Nature*, **373**, 128, 1995.)

As we noted in section 6.5, water vapour maser emission has now been imaged by VLBI from a nearby galaxy NGC4258 (fig. 7.4). As well as being very suggestive of a molecular torus (albeit very flattened), the velocity information gives further strong support for a supermassive back hole. The velocities of the maser emissions fit a perfect Keplerian rotation picture, allowing the enclosed gravitating mass to be calculated. This is 3.6×10^7 $M_\odot$ from a region of radius < 0.13 pc, giving a mass density $> 4 \times 10^9$ $M_\odot$ pc^{-3}. Star clusters cannot provide this density, as globular clusters have stellar densities four orders of magnitude less than the required value. An enlarged cluster of single stars of mass ~1 $M_\odot$ would have stellar separations of the order of only 100 AU, and with a collision time $< 10^8$ y it would dissipate rapidly. So it seems that this is very strong evidence indeed for a supermassive black hole. Indeed, these data give the highest central mass density for any object so far observed at the time of writing and as such it is one of the best pieces of evidence for supermassive black holes. Furthermore, the masers show little infall motion and it is interesting to speculate that this might support the view that the accretion rate at this epoch is low. Perhaps this is because the inner regions have been swept clean and NGC4258 is now a semi-dormant, or only mildly active AGN, a mere shadow of what has been in the past.

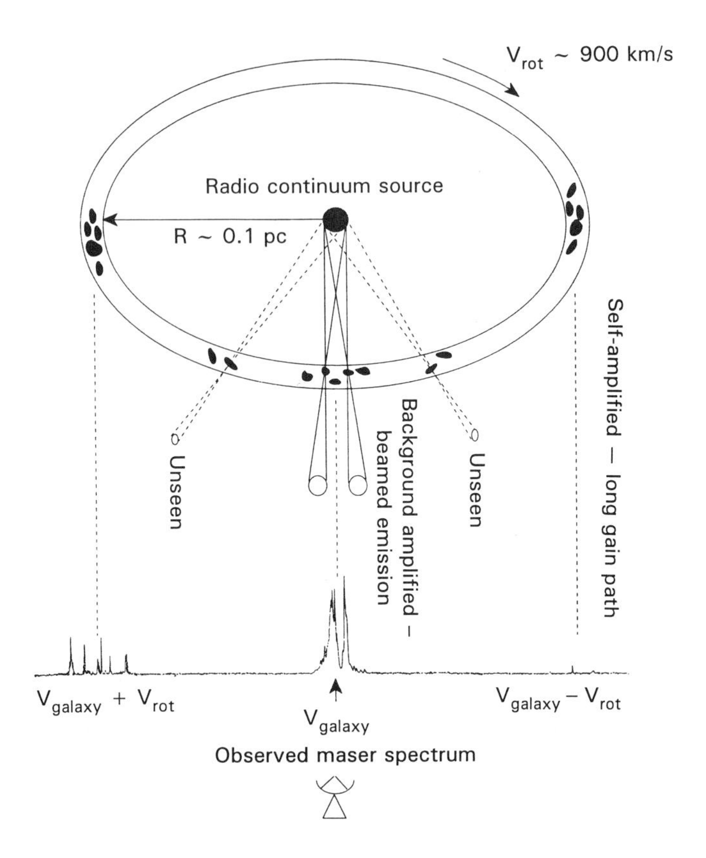

Fig. 7.5. Possible model of the maser geometry emission in NGC4258. (Updated from Greenhill,L.J. *et al. Astrophys.J.*, **440**, 619, 1995 and courtesy of Lincoln Greenhill.)

Another very recent and important observation pertains to the discovery of a gravitationally redshifted X-ray iron Kα fluorescent line in the Seyfert 1, MCG-6-30-15. As we saw in sections 2.6.2 and 5.2.5, the fluorescent iron line has now been seen in a number of AGNs (fig. 5.9) and is usually interpreted as reflection from gas close to and surrounding the central engine (see chapter 9). A very deep (four day) observation by the ASCA satellite showed that the iron line was very broad, with a velocity of around 100,000 km s^{-1} and that it was very asymmetric, being extended to the red (fig. 7.6). The only explanation seems to be that the line originates from gas which is in a very strong gravitational field. A supermassive black hole nicely fits the picture, and calculations suggest that the gas lies between 3 and 10 R_S. If this is the case, and most observers appear to accept the results, this is giving us information from the very inner parts of the accretion disk.

Within the very near future, we might expect to see further observations from the

HST and VLBI maser work revealing more evidence for supermassive black holes in the cores of local galaxies. It is important to stress, that the 'proof' or at least the acceptance of the evidence supporting the existence of supermassive black holes in M87 and NGC4258 goes a long way to showing that the current picture of active galaxies is correct. The gravitationally redshifted iron line observed by the ASCA satellite has given probably the strongest evidence for supermassive black holes in AGNs, and very recently, a similar line discovered in NGC4151 by Bob Warwick has provided even further support.

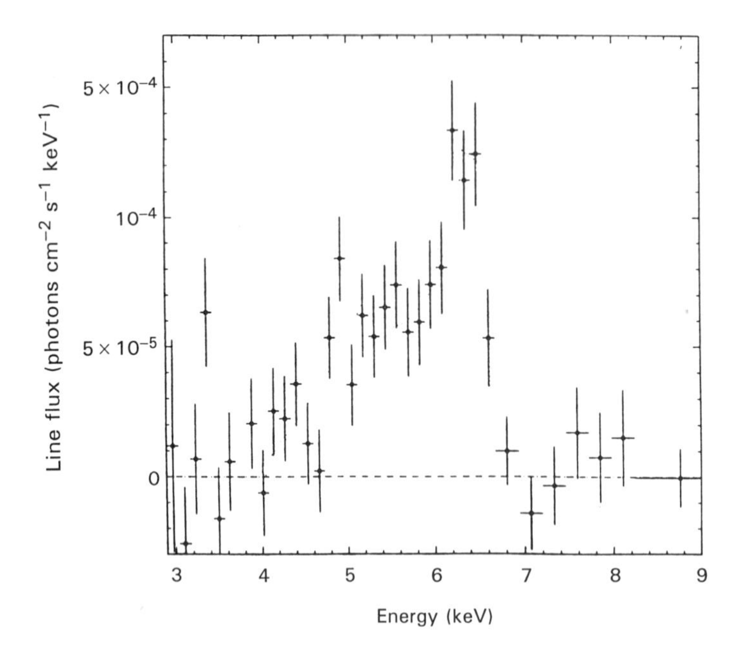

Fig. 7.6. The X-ray spectrum of MCG-6-30-15 after removal of a power-law and reflection components. The fluorescent Kα iron line is clearly visible at an energy of ~6 keV and is very asymmetric, being extended to the red. The velocity width is ~100,000 km s^{-1} and fits with a model of the gas suffering a gravitational redshift in the field of a supermassive black hole. (From Tanaka *et al*, *Nature*, **375**, 659, 1995.)

7.3 FORMATION AND EVOLUTION OF AGNS

We now consider the formation and evolution of an AGN. We will tackle this is three stages: the formation, followed by the subsequent evolution and luminous lifetime, until we reach the dormant or end phase. So far this book has described the events of the second phase, in which the galaxy in question houses a highly luminous AGN, but we have not yet tackled the lifetime of the process. The prospect of an end-phase introduces the possibility of dormant or extinct AGNs masquerading as otherwise normal-looking galaxies at the present epoch. These would harbour a non-radiating supermassive black hole, the subdued remains of a past glory (as we saw for M31 and M32 in the previous section). In the descriptions that follow we will use the term quasar as a general illustration of a luminous active galaxy powered by a supermassive black hole. Before concluding this section we shall take a detailed look at the centre of our Milky Way Galaxy, long thought by many to harbour a supermassive black hole. Was our Galaxy once a quasar or a Seyfert?

7.3.1 The formation of quasars

We have already established the possibility of the formation of quasars at early epochs of the Universe through galaxy mergers. Because of our belief that the fundamental power source is a supermassive black hole, the question of AGN formation is intimately linked to the formation and development of supermassive black holes. Indeed, an obvious first question is to ask how supermassive black holes can form at such early epochs in the Universe.

We observe quasars at redshifts of almost 5 and this poses strong constraints on timescales for formation of supermassive black holes. A fundamental question to ask is whether they can be formed in an isolated galaxy from rapid growth of seed black holes, or are interactions required in order to trigger central collapse and hence form a supermassive black hole. As early as 1978, Martin Rees proposed a gravitational inescapability diagram (fig. 7.7) outlining scenarios for the formation of a supermassive black hole. This diagram basically survives intact today and the main problem is in actually finding ways that are not too contrived to enable black holes of mass $\sim 10^8\ M_\odot$ to form. This is far from simple.

This brings us to one of the current hot topics of research today, galaxy formation in the early Universe. We cannot venture far into this field as it involves a higher level of the treatment of cosmology than we have given. Instead, we refer the reader to *The Hidden Universe* by Professor Roger Tayler in this series, as well as to a selection of reference papers, and we further remind the reader of the brief overview presented in section 1.6.

Observationally the COBE satellite has revealed the incredible smallness of the small- scale anisotropy of the cosmic background radiation. This shows that the radiation field was incredibly smooth at a redshift of $z \approx 1{,}000$, which in turn restricts the size of any inhomogeneities or initial mass fluctuations from which galaxies could subsequently form. We have also seen that because there is insufficient baryonic (normal) matter in the current Universe, very substantial amounts of unobservable cold dark matter (CDM) are required in order to satisfy the requirements of inflation in the very early Universe. CDM brings attractions for galaxy formation, but with it come other problems that we will not venture into. Nevertheless, it is important to note that whatever the cosmology (a critical density Universe with $q_0 = 0.5$ shown in fig. 1.6 by curve A, or a low density open Universe with $q_0 = 0.1$ shown by curve B), there remains precious little time between the earliest epochs at which we realistically believe that galaxies could form ($z < 20$) and the appearance of highly luminous quasars at $z \sim 5$.

Let us investigate this in more detail because herein lies the crux of the problem. We saw in section 1.6 that using an Einstein–de Sitter model the look-back time could be related to the age of the Universe by eqn. 1.16, i.e.:

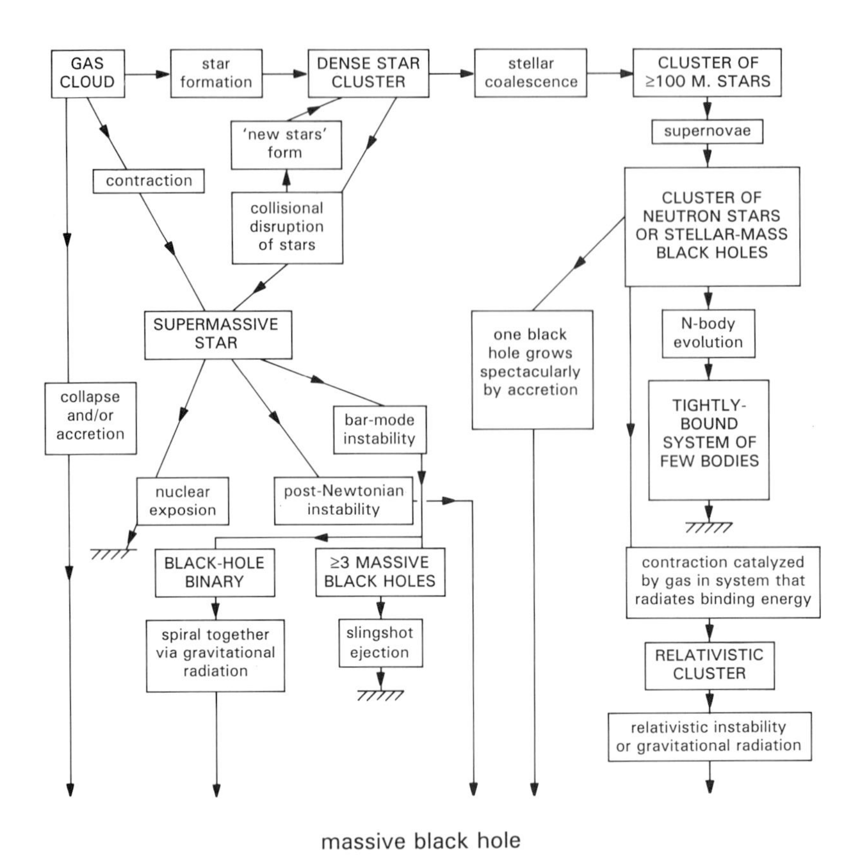

Fig. 7.7. Diagram for the possible formation routes of a black hole. (From Rees, *Ann.Rev. Astron.Astrophys.*, **22**, 471, 1984.)

$$(1+z) = \frac{R_0}{R(t)} = \left(\frac{t_0}{t}\right)^{2/3} . \tag{7.7}$$

A quasar at a redshift of z = 5 corresponds to a look-back time of 93% of the age of the Universe. We immediately see the staggering importance of these timescales, a fact frequently not fully appreciated. It is worth re-emphasizing this point because although a redshift of z ~ 5 might seem relatively small, in fact it is much closer in time to the age of decoupling at z ≈ 1,000 than it is to the current epoch. When we look at redshifts ~5, we really are sampling the early Universe in terms of phenomena such as galaxies and large-scale baryonic structures. For a Hubble constant of $H_0 = 50$ km s^{-1} Mpc^{-1}, a redshift 5 quasar has an age of only 9.1×10^8 y and the cosmic background epoch of z ≈ 1,000 represents an epoch of only ~ 3×10^5 y since the big bang.

In all of this we need to remember that timescales are intimately linked with the particular cosmological model being used. There are really only two extreme models used today: the so-called low density (open) model and the critical density model of inflation noted above. We must also bring in the Hubble Constant and, as we saw in section 1.6, where the outcome of the calculation depends significantly on the precise value of the Hubble constant, the variable h is used. The critical density model with a value of the Hubble constant of 100 km s^{-1} Mpc^{-1} gives the minimum time for processes to take place prior to a redshift of ~5, whereas a small value of q_0 and $H_0 = 50$ km s^{-1} Mpc^{-1} maximizes the time available for galaxy formation in the early Universe. (See *The Hidden Universe* for a full discussion of these topics.)

Whichever is correct, the shortness of the timescales in question is readily apparent, especially for supermassive black hole formation. If we assume that supermassive black holes have grown to their current size (say by a redshift $z \sim 5$), then an obvious question is by what rate could this growth have been accomplished and from what size must it have started. Black holes grow by accreting matter but there are limits to this process.

We can define an Eddington timescale, t_E, as the time for the mass of a black hole to increase by a factor of 1/e while it is accreting at the Eddington limit. This can be written as

$$t_E = \frac{\sigma_T \, c}{4 \pi \, G \, m_p} = 4.4 \times 10^8 \, y \,, \qquad (7.8)$$

The Eddington timescale is just about the same as the dynamical timescale for a typical galaxy, and also the age of the Universe at that epoch. Is this a coincidence?

This short timescale poses severe problems, it does not allow sufficient time for supermassive black holes to have been built up from small-size black holes by mass accretion limited by radiation pressure (the so-called Eddington accretion limit). Therefore we require either super-Eddington accretion rates or a large seed for a putative massive black hole. How could we obtain super-Eddington rates? One way is to assume that the accretion is very anisotropic, such as produced via an accretion disk, but even here, the outgoing radiation must be well separated from the infalling material. Furthermore, this radiation–matter separation cannot derive from the interaction of the accreting material and the radiating photons, or we revert back to the Eddington luminosity. Determining the precise solution for a super-Eddington accretion rate appears elusive, on the other hand, the requirement for radiating at the Eddington luminosity is attractive in that it minimises the mass of the black hole required to explain the luminosity.

A black hole radiating at the Eddington limit, L_E, has a luminosity directly proportional to its mass, M (eqn. 7.6), and we can identify a characteristic accretion time, t_E, by

$$t_E = \frac{M \, c^2}{L_E} \,. \qquad (7.9)$$

We now need to introduce the radiative efficiency to get a better handle on the timescales. We can define a radiative efficiency term, ε, which is linked to the luminosity and accretion rate, $\dot{m}_{acc}$ by

$$L = \varepsilon \, c^2 \, \dot{m}_{acc} \qquad (7.10)$$

where L is the radiating luminosity. Taking the specific case of $L = L_E$ we obtain the relationship between the accretion rate and the mass of the black hole given by

$$\dot{m}_{acc} = \frac{M}{\varepsilon t_E} \quad . \tag{7.11}$$

We can then determine the accretion timescales, t, from

$$\frac{\dot{m}_{acc}}{M} = \frac{1}{\varepsilon t_E} \quad , \tag{7.12}$$

which gives

$$t = \varepsilon t_E \, ln\,(M/M_o) \quad . \tag{7.13}$$

This can be rewritten as

$$ln\left(\frac{M}{M_0}\right) = \frac{t}{\varepsilon t_E} \quad , \tag{7.14}$$

or

$$M = M_0 \exp[t/\varepsilon t_E], \tag{7.15}$$

with M_o being the initial mass at time $t = 0$. Bringing this all together we can see that the radiating luminosity

$$L \propto M \propto \dot{m}_{acc} \propto \exp(t/\varepsilon t_E)\,, \tag{7.16}$$

which holds the key to quasar fuelling and lifetimes.

For a particular luminosity, we can now calculate the mass of the black hole required to power the source (assuming $L = L_E$). For example, the luminosities of the redshift 4–5 quasars are of order 10^{13} $L_\odot$, which leads to black hole mass requirement of 10^8 to 10^9 $M_\odot$. The necessary short timescale for this process to have built up can only be obtained by having an extremely large accretion rate, or by having a large initial mass, M_0. However, the former is only feasible as long as the radiative efficiency, ε, is small. So we are still left with the problem of forming at least 10^8 $M_\odot$ on timescales of the order of a few times 10^8 years.

This is far from simple and requires ingenuity on the part of theoreticians. The only free parameters in this process are the initial mass, M_0, and the radiative efficiency, ε, but even here there are problems. Making ε large, say close to its maximum theoretical permitted value of 0.42, means that black hole masses will grow too slowly in the time available unless they start off with a very high mass M_0, which just moves the problem back by asking where these proto-black holes came from. Another variant on this is to suppose that the quasars formed at a very high redshift, perhaps of order 20, which seems most improbable given the observed decline in quasar number density beyond $z \sim 2$–3 (see fig. 7.8), and in any case still leaves the problem of how to form the initial high mass on an even shorter timescale. This requires entirely new models and is not favoured.

The key to fast black hole growth is to have ε very small. This leads to a very high accretion rate and it allows sufficient time (just) for the black holes to accrete the required mass and radiate at the observed level. However, in the process it rapidly depletes the fuel supply. This scenario then demands that there must be a copious and easily accessible fuel supply in the host galaxy at the time of the black hole formation and growth. Perhaps mergers might help in this part of the story, but even here there are good theoretical reasons to suggest that the radiative efficiency is dominated by effects very close to the accretion disk rather than by the available supply of fuel. Lack of fuel would be a severe problem, but excess fuel would not change the efficiency by itself. Indeed, most workers in the field seem to have settled on a value for ε of around 0.1. (A good description of this process along with an excellent reference list is given in the paper by Turner listed in section 7.6.)

However, there might be a way out of the dilemma of timescales and fuelling by allowing ε to vary with time. The trick is to enable the black hole to become very massive very quickly, requiring an initial small value of ε. The resulting fuelling problem is reduced by assuming that a plentiful fuel supply is available at the earliest times in order to sustain the required high accretion rates. This massive accretion phase takes place before the quasar has fully switched on, and for our example this occurs at a redshift greater than 5. Following this black hole gorging phase, the luminous 'quasar' switches on (in our case at a redshift of $z \sim 5$), and now there comes the time of black hole famine. The radiative efficiency now increases and is accompanied by a dramatic decline in the accretion rate. The luminous quasar now enters an extended radiating lifetime phase. However, although this scenario produces a possible explanation, it is still very hard to see how the black hole manages to get to the required mass in the available time. More work is undoubtedly required from the theoreticians.

This brings us to the question of how we might test the many theories of quasar (and galaxy) formation. In principle, one of the key discriminators is the observed luminosity function of the object in question. Let us remind ourselves of the luminosity function. It is the number density expressed in terms of the luminosity range of the objects in question, i.e. it is the number of objects per co-moving volume of space per luminosity interval. Unfortunately, we immediately hit a big problem. The luminosity function for quasars is uncertain at low redshifts (there are no powerful nearby quasars), and very uncertain at redshifts beyond about $z \sim 3$. The situation for Seyferts is worse. Hence without a good knowledge of the luminosity function, especially at high redshifts corresponding to the epoch of galaxy and/or quasar formation, the theories are largely unconstrained. Luminosity functions frequently fail to paint the true picture for the student, and the dramatic differences are masked by the use of the logarithmic time-scale. To help make this point the co-moving density of quasars is plotted in fig. 7.8 on a linear time-scale as well as the usual redshift form.

The current general consensus is that quasars display both luminosity and density evolution and that there is a peak in the density function at a redshift of between 2 and 3. At lower redshifts the quasar population declines notably. Although the situation is uncertain for higher redshifts, the available evidence strongly suggests another decline although not as rapid as the decrease to the current epoch.

We again note that recent evidence demonstrates the existence of powerful quasars at redshifts of almost 5, and furthermore, some have prodigious amounts of dust. The presence of dust has two effects: one is to cause difficulty in identifying quasars (see below) and the second has important consequences on the rate of stellar evolution in the early Universe. We have already seen above that the existence of quasars at a redshift of 5 allows precious little time for galaxy and black hole formation on cosmic timescales, but in addition we now need to explain the production of copious amounts of heavy elements and dust. The only simple means of satisfying this is through rapid and extensive formation of massive stars.

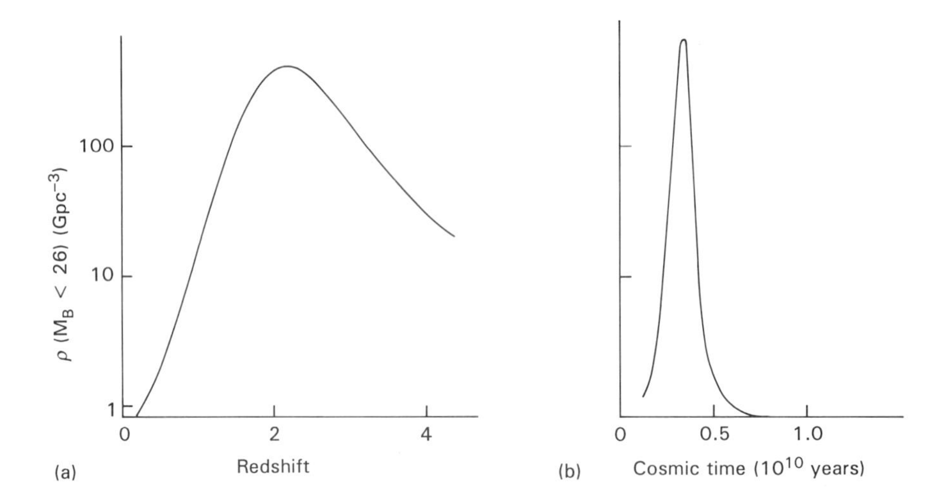

Fig. 7.8. Sketch of the quasar density plotted (a) as a function of redshift (the normal notation) *from* the current epoch and (b') as time *to* the current epoch from the big bang. The right hand plot dramatically shows the very early epoch of quasar formation and short era of quasar dominance.

Let us now resume our speculation of quasar formation. Given the evidence that points to modest redshift quasars occurring preferentially in merging or merged systems, we can paint a scenario of quasar formation triggered by collisions in the early Universe. It is true to say that the following picture is undoubtedly one that is viewed through rose-coloured spectacles; the real situation is bound to be more complex. Nevertheless, we will note these potential complexities in passing and we shall also put aside the problem of the growth of a supermassive black hole. One of the keys to quasar and galaxy formation is that both can be expected to be accompanied by extensive and rapid star formation and plate 7 shows that even direct collisions produce extensive star formation. However, this again brings up the question of lifetimes, timescales and observability. If this phase of massive star formation is extremely short (as we might reasonably expect), then it will be very difficult to observe because the number of objects in the Universe undergoing this phase at any one time will be small. On the other hand, we can expect rapid star formation to be triggered by mergers, *after* galaxies have formed. This will be a later phase and so offers a potentially better chance of observations, but again, the respective lifetime of the star formation phase is crucial.

Assuming that galaxies have formed, with or without a central massive black hole, we can now test how the merger hypothesis fares. In our scenario, we begin from a starting point that firmly stacks the deck in our favour. Take the case of two spiral galaxies in the early Universe; both are gas rich and probably residing in a rich environment of close neighbouring galaxies. Collisions would be common. What happens when two gas-rich spirals collide? Much work has been done using numerical simulations and computational modelling to provide spectacular movies of the resulting interactions. In the collision process, the best end-point result from our perspective is to have a merger rather than a glancing interaction. This is a more efficient star-forming mode as it ensures that the disks of the two galaxies will overlap and increases the molecular cloud collisions leading to a higher level of star formation. Spiral galaxies maximize the burst of star formation due to the large gas content and the fact that the gas is centrally condensed. From what we know about molecular clouds and star formation in the local Universe, a timescale for this process can be calculated. Most workers in this field agree that estimates of the order of 10^7 to 10^8 years seem reasonable. This scenario predicts that the merger will result in a luminous infrared galaxy, rich in star formation and producing copious amounts of heavy elements and dust.

Simulations of the merger process can, in certain circumstances depending on orientation and orbital impact parameters, lead to the production of a centrally condensed super starburst before the merger has been completed. The overall timescale for the super starburst process to take place is also around 10^7 to 10^8 years, at the end of which the merger is complete. This predicts that super starbursts should precede the merging process and that the early Universe should be rich in infrared luminous objects. IRAS showed that these objects do indeed exist, but the survey failed to probe deep enough to prove the hypothesis. IRAS 10214+4724, at $z = 2.3$, shows the promise for future study. Ground-based studies in the submillimetre with sensitive detectors like SCUBA on the James Clerk Maxwell Telescope are already making inroads into this exciting field and the ISO satellite should be a major player in these studies. My prediction is that most quasars will be dust enshrouded.

Note that in testing the above scenario we need to take into account not only the expected temperature distribution of the dust in the starburst, but also the redshift of the galaxy in question. As the redshift increases the peak of the dust emission will be moved to longer wavelengths by a factor $(1 + z)$ due to the expansion of the Universe. At a redshift of 4, dust emission peaking at 100 μm in the rest frame of the object will be moved to 500 μm in our rest frame, providing a perfect target for a submillimetre telescopes like the JCMT. Two landmark papers by Dave Sanders are cited in 7.6. These and the references in them give an excellent summary of progress in this field in terms of the observations and the linking theories.

But what about the AGN in our galaxy merger scenario? If either of the galaxies possessed a central black hole, then this can act as a seed for further growth. However, even if neither had a black hole before the galaxy interaction, calculations show that the merging and starburst phases are long enough for a massive black hole to form and subsequently grow (bearing in mind the caveats discussed above). A key parameter in this process is the dissipation of angular momentum, allowing the material to sink to the galaxy centre and thereby aiding the formation of a black hole. It turns out that the formation of a central molecular gas disk is an efficient mechanism for this to occur and evidence for such a process is seen in Arp 220. Indeed, similarly to the general case for star formation and the critical density for gas collapse (the Jeans mass), a critical density can be calculated. This depends on the kinetic energy of the gas compared with the gravitational field in which it resides and it is found that the gas in the inner disk of Arp 220 is well in excess of this value. Therefore in this case there appears to be ample time for a massive black hole to form and for the surrounding material to sink towards it, providing a source of fuel. During this time, the central AGN begins to contribute to the overall luminosity of the galaxy and as time progresses, the super starburst rapidly fades, to be replaced by the luminosity from the quasar. Testing this hypothesis will require extensive observations of statistically large samples of ultraluminous galaxies and radio-quiet quasars at a range of redshifts to determine if the overall continuum spectra show changes in keeping with this picture.

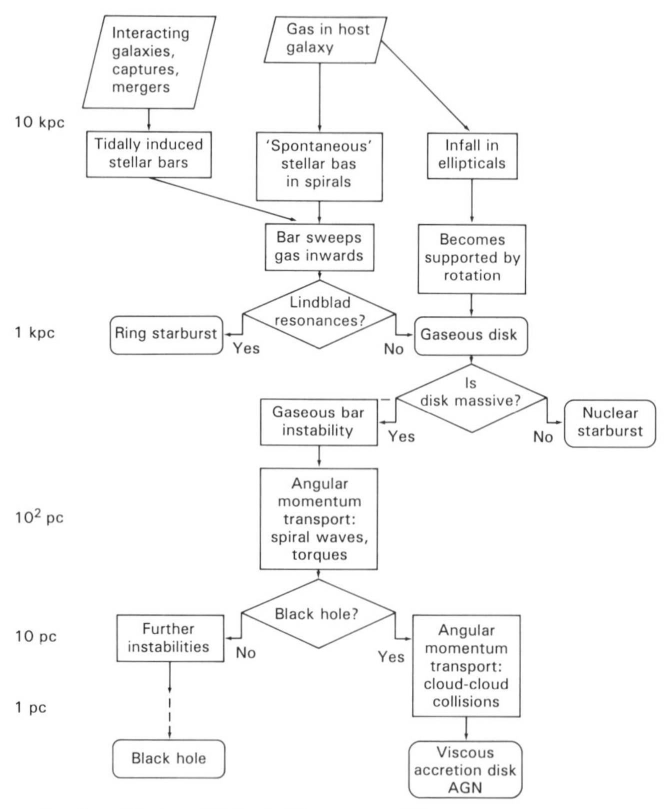

Fig. 7.9. Schematic of a unified model for fuelling AGN: different pathways by which host galaxies can acquire a massive black hole and become an AGN or a starburst galaxy. (From Schlosman, Begelman and Frank, *Nature*, **345**, 679, 1990.)

Although this has been a generally superficial argument, there is much in its favour. Observationally we know that quasars at medium redshifts tend to lie in clusters. We have seen that the deep optical and infrared imaging of quasars and radio galaxies in the redshift range $z \sim 0$ - 2 reveal that many are distorted in appearance, often with multiple nuclei in the galaxy, and show 'companions' close-by on the sky. Because we do not have redshift information for these companions we cannot yet say whether these emission components are true companions, or a chance line-of-sight effect. However, statistically speaking the companion argument offers the best interpretation. Merging is now a well-observed phenomenon and many of the weird shapes observed in the Universe can faithfully be replicated by the numerical simulations. To improve the contrast in the images and to show the distribution of the ionized gas, imaging of quasars is undertaken using filters to isolate a single line (such as [OIII]). These images also point to very disturbed gas distributions and in one or two cases provide evidence for tidal tails. In the future, the tidal tails for the high redshift quasar population might be revealed by deep imaging with the new generation of 8 and 10 m ground-based telescopes as well as the HST.

Indeed, one of the main targets for the high imaging quality of the HST is to investigate the host galaxies of quasars. However, it turns out that the HST is not ideally suited to detailed studies of the profiles of host galaxies, mainly because of its relatively small aperture for collecting energy, and because of the need to subtract the effects of the telescope diffraction pattern itself, something which is extremely tricky for the HST. However, what has emerged is that the HST is excellent and far superior to ground-based telescopes at picking out bright contrasting features close to the central spike of a quasar.

A number of studies on galaxy evolution and population types have now been published and two main points stand out; normal looking spirals and ellipticals exist at a redshift of ~0.2 and greater (an important finding in itself), and on the whole quasars hosts seem to be found in disturbed systems. These often show multiple nuclei and other features of merged systems. These studies tend to agree with the much more complete infrared and optical ground-based imaging but the greater spatial resolution shows that, while the infrared data tend to show a smooth galaxy profile, the HST data reveal that the galaxies are often much more clumpy, especially when viewed with filters which isolate emission lines such as Hα or [OIII]. This provides further support for the idea that quasars exist predominantly in highly disturbed systems and that mergers may play a significant role in quasar formation and evolution.

On the other hand, as discussed in section 5.3.2, a rather unexpected finding has come out of one of the HST studies in that although taken from only a very small sample, the host galaxies of some of the nearby and bright quasars were much less luminous than expected from the ground-based studies. However, some of these quasars also showed significant companion galaxies and evidence of interactions and one quasar is caught in the act of merging with its companion (fig. 7.10).

Returning to our argument abcut quasars being the products of mergers, we can summarise that as well as the irregular morphology found for many objects, the calculated timescales for merging and starbursts also appear to fit in reasonably well with the overall picture. But what about the problems? In our assumption above, in taking two gas-rich spirals we clearly selected the most beneficial case. The situation for massive star formation is less conducive in the case of the collision between two elliptical galaxies because ellipticals contain very much less gas than spirals at the current epoch (but we are not so sure of the gas content in the early Universe). However, even for two ellipticals colliding, as long as the interaction leads to a merger, then a starburst phase should be seen. In both cases there is always the statistics of the impact parameters to consider and the subsequent chance of an interaction leading to a merger or just gas stripping. Although the latter is interesting, it is precisely what we do not want for our scenario of forming quasars. We need gas dumping, not gas stripping.

Fig. 7.10. HST picture of the quasar PKS2349 showing merging between the quasar and a galaxy (the remains of which are the wisps directly surrounding the bright quasar). At least three other companion galaxies are also evident in the photograph. STScI-PR95-04.

Finally, let us ask whether all quasars are formed from mergers and go through the ultraluminous infrared galaxy phase. Statistics are very poor at this stage, but the relative number density of quasars compared to ultraluminous FIRGs with luminosities in excess of 10^{12} $L_{\odot}$ in the relatively local Universe is indeed sufficient to satisfy this idea. Although locally there are significantly more ultraluminous infrared galaxies than quasars, as we go to higher redshifts, their number density plummets. This is probably attributable to an observational selection effect because the infrared surveys have yet to probe to the faint flux levels where these galaxies would be readily detected. It is also possible that the traditional blue optical sky surveys are not selecting them because they are probably very red objects and devoid of broad lines due to enshrouded dust.

A final comment for this section is that the results from one of the key projects of the Hubble Space Telescope, the Medium Deep Survey, show that at the faintest limits probed by the HST, small, blue, irregular galaxies outnumber spiral and giant ellipticals. These surveys probe to redshifts of $z \approx 1$. Because we see nothing like this number of blue dwarf galaxies at the current epoch, the question remains as to what has happened to them. The fact that they are very blue suggests intensive star formation, yet without significant dust. Perhaps they are in the very first phase of massive star formation out of which the enriched elements are recycled through subsequent generations of stars formation to form dust. Are they the building blocks for 'classical' galaxies? The fact that they are irregular also imposes constraints on their formation. However, we shall ignore these exciting consequences and move on to consider the lifetimes of quasars.

7.3.2 Quasar lifetimes and fuelling

Let us now consider the lifetime for a quasar phase of a galaxy harbouring a central supermassive black hole. In terms of the quasar lifetime, then as long as a supply of fuel is readily available, the quasar can continue to radiate at, or close to the Eddington limit. The timescale then becomes a function of the black hole mass and its accretion rate. There is a relatively simple relationship for the quasar lifetime, T_Q, assuming it is radiating with Eddington luminosity, L_{Edd}, given by:

$$T_Q \propto \frac{M_F\, \varepsilon}{L_{Edd}} \qquad (7.17)$$

where M_F is the mass of the fuel. Taking an efficiency of $\varepsilon = 0.1$ and assuming that a significant fraction of the mass of a galaxy (say $10^9\ M_\odot$) can be converted into fuel, then the lifetime for a $10^{12}\ L_\odot$ quasar turns out to be only of the order of 6×10^8 y. To extend this lifetime requires (a) the availability of more fuel, (b) a black hole radiating below the Eddington luminosity (but this immediately requires a larger mass black hole to explain the high luminosities observed), or (c) a variable radiating efficiency producing a decline in the luminosity along with a slow growth in the black hole mass.

Another way of looking at this question of the fuel supply which probably gives a better feel for the problem is to ask what accretion rate is required to drive the observed luminosity. If we take our $10^{12}\ L_\odot$ quasar as an example, then for $\varepsilon = 0.1$, an accretion rate of around $2\ M_\odot\ y^{-1}$ is required. This is a modest number and with an available fuel supply of $10^9\ M_\odot$ we again obtain a lifetime of around 5×10^8 y. This is very much less than the age of the Sun, and indeed, very much less than the age of the Universe. Therefore, we expect quasars to decline from some preferred epoch because they simply run out of fuel. This is an important conclusion.

Let us return to the luminosity function for quasars shown in fig. 7.8. This clearly shows the dramatic epoch of quasar prominence. However, what is far from clear is whether the life of a quasar consists of radiating at a given luminosity followed by a once-and-for-all fading, or whether there are repeated periods of high luminosity followed by periods of lower level output. Because we can never hope to see global changes on these cosmic timescales, we have to revert to statistical arguments about how many quasars are observed and in what luminosity state they are in, i.e., the luminosity function. Because this seems to be a very smooth function, there appear to be no strong arguments for quasars having an oscillating existence. It is more likely that they radiate at a fairly steady luminosity during their youthful stage of life.

At their peak epoch, which we know lies somewhere between redshifts of 2 and 3, the quasar density is well determined. The population of bright galaxies is also known at this epoch and a simple comparison reveals that quasars form about 1–2% of the total population. This means that at least 1% of bright galaxies possessed a supermassive black hole at this epoch. Let us now move to the current epoch and look at Seyferts, which we firmly believe are merely lower-luminosity quasars. We know that Seyferts comprise around 10% of bright galaxies and we can speculate that these same Seyferts must be present at the quasar epoch, but perhaps unobservable because of their lower luminosity. This leads us to ask whether Seyferts, or low luminosity quasars, were formed at the same epoch as quasars, or whether they took longer to form. A different line of attack is to speculate that perhaps some Seyferts were the same type of objects as are seen at redshift 2, but they have now faded. In this model, Seyferts are examples of dying quasars. We can put some flesh on the latter speculative suggestion. Unless quasars have even shorter lifetimes than estimated above, then we observe too many Seyferts for them all to have been quasars in the past (from a comparison of the number densities). Nevertheless, there is no reason to believe that some fraction of the Seyferts might not represent quasars in their declining phase.

We can summarise by saying that because the radiating lifetimes of the lower luminosity AGNs will be significantly longer than for quasars, then even if Seyferts (for example) were formed at very distant epochs, it is no great surprise that they are still plentiful today. Therefore we are left with the certainty that there must be a population of objects at the current epoch that were quasars in the past. Can we identify what they are? Perhaps the M87's of this world might be candidates, but even if they were once quasars, there are still insufficient to explain the old quasar population. The next sub section provides some of the answers to this question.

7.3.3 Dormant quasars

Section 7.2.4 provided accumulating evidence that some of the nearby and apparently normal galaxies possess a supermassive black hole in their cores. We will now investigate this further. Probably the best evidence for the presence of a supermassive black hole in the core of a dormant galaxy is for M31. The importance of determining this fact with absolute certainty cannot be understated. To all but the most penetrating of observations, M31 is the typical, normal, dormant and rather boring (to AGN pundits) neighbourhood spiral galaxy. Even with high resolution VLBI and optical observations, along with UV and X-ray monitoring, M31 still seems remarkably docile. The luminosity from M31 is basically the starlight of the galaxy; the lack of any evidence for even a mini-quasar puts limits to its central luminosity of less than 10^{-5} of an average quasar. There are no broad emission lines, no central bright compact radio source and no unexplained X-ray emission. (Although we observe X-ray emission from M31, this is readily understood as radiation from individual supernova remnants and X-ray binary stars, and does not require an unidentified central source that we have come to know as a marker for AGN activity.) Therefore, as we saw in section 7.2.4, only the observations of the mass distribution point to the black hole explanation being favoured (and even then, not by all astronomers working in the field).

What might a dormant quasar look like? Note that we have used the word dormant, in the same sense as for a volcano. The world's greatest complex of ground-based telescopes reside atop a dormant volcano, Mauna Kea in Hawaii. However, it is strongly believed from knowledge of the motion of the underground hot spot in the Earth's crust that Mauna Kea will not erupt again. However, the geological definition of an extinct volcano requires it to have been dormant for a period that happens to be longer than the time since the last eruption of Mauna Kea. Therefore technically it is dormant, but all of us who have telescopes on Mauna Kea are happy to believe that it is extinct. The same is true for dormant quasars such as M31.

But if M31 houses a supermassive black hole of mass $\sim 10^7$ $M_\odot$, we can legitimately ask why we do not see radiation of some description indicating its presence. The answer is presumably because of insufficient fuel, or rather that the fuel interacts with the black hole in such a way that the emissions are either of low luminosity, or very rare. Let us expand on this. There are two obvious sources of fuel, gas and stars. If gas is present in the central zone, then we might expect that it should ultimately spiral into the accretion disk and in so doing radiate. Because it clearly does not, we infer that either the gas supply is very small, and/or the accretion disk has dispersed. In this case, the gas would then spiral directly into the hole and although it would radiate in the process, much of the radiative heating effect of the viscosity of the disk (providing the luminosity) would be lost. Either way, lack of emission suggests a depleted gas supply in the very central region.

Perhaps the quasar has swept this inner zone clear over time. The gas might then lie at a distance such that the orbits are stable and Keplerian unless perturbed by external influences, such as pressure waves from supernovae, or gravitational interaction with a nearby galaxy (such as its neighbour M32). Calculations suggest that neither is very likely to provide sufficient perturbation in the case of M31. Therefore the 'sweeping up of available material' stands as a possibility for quasar quenching and ultimately emission suppression. If this is the case, then a great number of normal galaxies could house supermassive black holes.

But what about stars? This question has generated a whole industry of analytic and numerical calculations of the interaction between a star and a supermassive black hole. The interaction depends on many factors: the mass of the black hole, the impact parameter, the size of the star, the velocity of the orbit and so on. In 1983, when I casually proposed that a possible explanation for an observed large optical–infrared–millimetre flare in the quasar 3C273 might be due to a supermassive black hole swallowing a star, little did I think how complex the situation would actually turn out to be.

The key parameter in most interactions is the impact parameter, and for a black hole we can extend this to take into account the tidal radius (the radius inside which an object is torn apart by the differential gravitational field on its opposite sides being greater than the object's self-gravity). The tidal radius of a black hole is proportional to the radius, r^{-3}, whereas the mass is directly proportional to r. Hence the tidal forces at the event horizon are much smaller for a massive black hole than for a small black hole. Bodies are therefore torn apart before they pass into a small black hole but can pass into a massive black hole unscathed because the tidal radius is inside the event horizon. For an infalling star like the Sun, the dividing line between being either torn apart or being swallowed whole is for a black hole mass a few times 10^8 $M_\odot$. So for supermassive black holes, we might not expect to observe anything extraordinary arising from the eating of solar mass stars in the centres of galaxies.

We also note that the black hole will grow by the mass accretion process, and as it does the tidal radius will decrease with respect to the event horizon. At some point the mass of the supermassive black hole will exceed the tidal limit for stellar disruption and subsequently, quasar luminosity fuelled by stellar accretion will also decline. This may play a part in explaining the turn-down of quasar luminosity and the subsequent decline to the dormant phase.

For smaller supermassive black holes, stars will be torn apart before entering the event horizon. Therefore the gas provides a fuel source and a luminosity. However, even here the picture is far from simple. In the case of M31, the central density of stars is reasonably well known and calculations point to the conclusion that the chance of a sufficiently close interaction of a star with the tidal radius of the putative black hole is very small. Statistically speaking, such an event might happen on a timescale of tens of thousands of years. Therefore we would not expect to see much activity from the core of M31 due to star fuelling. Further calculations on the stellar disruption processes, the subsequent fate of released material, and the lack of observed steady emission from the core of M31 suggest that if disruption does take place, it will more likely result in a transient event, such as a high-energy (X-ray) flare lasting only of order a year, rather than providing a source of long-term luminosity. However, on the positive side, the release of energy is high, and luminosities of order 10^{12} $L_{\odot}$ are anticipated. If this is released at optical wavelengths it would certainly make a big splash in the nucleus of M31, the core of which would brighten to a spectacular level, shining nearly as brightly as Sirius. One suspects that this would do wonders for astronomy funding and so there is definite merit in being on the lookout for star-eating flares in galaxies. But, as we have seen, these events are rare, and just as in supernova searches, to make acceptable progress will require telescopes dedicated to such searches. However, as long as enough energy is emitted in the optical rather than the X-ray region, then large flares in the nearby galaxies could be detected by the small army of dedicated amateur astronomers with their telescopes.

In closing we can resort to further speculations of what could happen to supermassive black holes, which, albeit requiring very special circumstances, could have a bearing on the fate of some quasars. Martin Rees pointed out the highly surprising fact that special situations could exist in which supermassive black holes were ejected from a galaxy. He noted that this could happen as a result of the recoil from collisions between two black holes of unequal mass. Although sounding far-fetched, there is the possibility that if multiple black holes are common (which is far from being an accepted concept) and these slingshot collisions are also common, then this is a further potential mechanism for killing off quasars. The expelled black holes will then hide in the Universe as completely unobservable objects apart from their gravitational interaction with surrounding matter or radiation.

7.3.4 A black hole at the centre of our Galaxy?

We know that our Galactic Centre is neither a quasar, nor a Seyfert nor any other kind of active galactic nucleus that we have considered in this book due to the simple fact that we do not observe the required emission components or luminosity. However if M31 might harbour a supermassive black hole at its centre, it is natural to ask if our Galaxy might do the same. Might our Galaxy have been an AGN in the past?

The Galactic Centre is only 8 kpc distant, and therefore in principle we might think that it would be relatively simple to determine whether or not a supermassive black hole is lurking at the core. Unfortunately, this is far from the case, and it brings us back to our discussions in section 1.2.2, where we saw that the extensive dust in the plane of the Milky Way produces ~ 30 magnitudes of visual extinction to the Galactic Centre, and restricts our view in this direction to less than 2 kpc. Therefore it is only at those wavelengths that can penetrate the dust and reach the Solar System that we can study the secrets that are hidden to optical telescopes. These are primarily the wavelengths of radio and the infrared, supported by high-energy X-ray and a single very important gamma-ray observation.

At a distance of 8 kpc, 1 arcsecond subtends a linear scale of 0.039 pc at the Galactic Centre, and so we have the benefit in principle of resolving structures on very small linear scales. Do we know precisely where the Galactic Centre lies? This answer would seem to be obvious, but in fact is not so simple. Usually the centre of a galaxy is defined by the kinematic centroid of the motion of the stars of the galaxy. However, because of the large extinction, it is only possible to see the stars of the Galactic Centre region from near infrared observations. These studies have shown that the light from stars (which are probably K and M type giants) peaks towards a central region of radius less than 1 pc, at the centre of which lies an extended source at what is assumed to be the dynamical centroid of the stellar distribution. This source is called IRS16. For a spherical distribution of stars in the inner 0.1 pc, the density exceeds $10^6\ M_\odot$.

The central 10 pc or so is termed Sgr A, a region containing molecular gas, ionized gas and a source of extremely compact radio emission. Concentrating only on the very inner part of the Sgr A region, we find that the structure is dominated by a ring of molecular gas and dust. The inner edge of this ring (or extended disk as it more properly should be termed as it extends out to radial distances of around 8 pc) lies at a radial distance of 2 pc and is well defined. The dust and gas in this disk are far from uniform, showing clumpy structure, and a number of HII regions are located on the inner edge, suggesting that the inner surface is generally photoionized. From measurements of the mass of dust and the radiating temperature, the energy of the central illuminating source providing the heating UV radiation can be calculated. This turns out to be of the order of 10^6 to $10^7\ L_\odot$, the precise value depending on the geometry of the ring. There is ~ $10^4\ M_\odot$ of gas in the ring, at temperatures well above that of the normal interstellar medium. This shows that the gas has been heated by some process, but the precise mechanism is not clear and could range from either collisional process or photoionization from a central source. Observations of the dynamics of the molecular gas in the dust ring suggest that it is rotating about the Galactic Centre in the same sense as the general rotation of the Galaxy and with a velocity of just over 100 km s^{-1}, but with strong local variations in velocity structure presumably revealing a turbulent disk structure. This can be used to give the mass of the material interior to a radius of 1 pc, which again turns out to be a few times $10^6\ M_\odot$. The disk is not flat, but warped, which poses problems in understanding how it could be stable for long timescales. Also, the sharpness of the inner edge of the disk poses further questions in that there appears to be insufficient outward external pressure generated by the inner 1 pc to maintain this sharp surface and it should smear out on timescales of order 10^5 y. Hence we are either talking about short-lived phenomena or the requirement for an additional pressure support, perhaps from magnetic fields.

The region interior to the circum-nuclear ring has almost been swept clear of both gas and dust; the densities are down by factors of 10 and 100 respectively. However, what dust remains in the inner zone is much hotter and the remaining gas is ionized. Although there are excellent observations of these components, the precise dynamics and geometry are incredibly complex to disentangle. There are streamers and arcs flying out of the plane of the Galaxy and the inner core appears to show a bar-type of structure, which some have likened to an inner mini-spiral of gas. The best probes of this region are the infrared lines from ionized atomic species such as H, He, S, Ne and Ar. The singly ionized neon lines are particularly useful in providing excellent spatial resolution and velocity information. Ionization theory and the observed line ratios suggest that any central ionizing source must have a temperature of order 30,000 K to 35,000 K, but not hotter, and requires a luminosity in Lyman continuum photons of 1.3×10^6 $L_\odot$. The corresponding bolometric luminosity is a few times 10^7 $L_\odot$, fitting in well with that deduced from the dust emission.

So far so good, but no evidence requires an exotic object to be lurking at the centre. However, the emission lines have further shown that the dynamics are very strange and the bar-like feature seems to show either gas infall or outflow from its ends. Evidence for outflow has been claimed from observation of very broad hydrogen and helium lines, with velocity widths of ~ 1500 km s^{-1}. These originate in the inner 0.1 pc and are centred on IRS16. Similar line widths to these are only found in our Galaxy in high mass-loss supergiant Wolf–Rayet stars. The required mass loss rate is around 10^{-3} $M_\odot$ y^{-1}, which is high by any standards as Wolf–Rayet stars have mass-loss rates of order 10^{-5} $M_\odot$ y^{-1}. However, the outflow interpretation has difficulties in explaining other observations, and a number of workers favour an inflow onto a massive object to give the required velocity width, a sort of mini-BLR. The outflow camp uses the observation of the ionized gas streamers perpendicular to the Galactic Plane as evidence for an expulsion mechanism at the centre, but these are far from conclusive. To summarize, we can say that the picture for the geometry and dynamics of the inner parsec is still confused.

Now let us turn our attention to the candidates for the Galactic Centre role, the sources IRS7, IRS16 and Sgr A*. The easy source to start off with is IRS7 because this is strongly believed to be a single M2I supergiant lying close to the Galactic Centre. Such a star is expected to show mass-loss rates of order 10^{-5} $M_\odot$ y^{-1} and cannot be much older than a million years or so.

The infrared study of IRS16 has been plagued by the limitations on angular resolution imposed by the Earth's atmosphere. Tantalisingly, the solution always seems just that extra sub-pixel away from the researchers' grasp. Although initially identified as a single extended object, IRS16 is now known to be a cluster of many objects, and a number of these have been examined both photometrically and spectroscopically. IRS16 was at first thought to be a group of star clusters, but now the brightest components have been measured to be less than 0.02″ in diameter (see plate 8 and fig. 7.11) corresponding to a size less than 160 AU. The luminosity of the two brightest components of IRS16 exceeds 10^6 $L_\odot$, which is quite acceptable for a group or small cluster of O or B stars. As we noted above, stars, or a single massive star, might be the origin of the observed high gas velocity through mass-loss stellar winds. The best suggestion at this time is that the central parsec is occupied by a cluster of very luminous and helium-rich stars, which are either supergiants or Wolf–Rayets. However, although these can provide the correct order of radiation pressure driven velocities of around 700 km s^{-1}, in order to account for the amount of gas observed in the inner 0.1 pc requires rather a large number of these extremely rare stars, or for a smaller number to have extremely high masses. From our figures presented above, something in the range of 40 to 100 stars are required, or a factor of two fewer if masses up to 100 $M_{\odot}$ are included. Is this too extreme? If the massive star cluster model is correct, then the epoch of the star formation burst which produced them must have occurred less than a few million years ago.

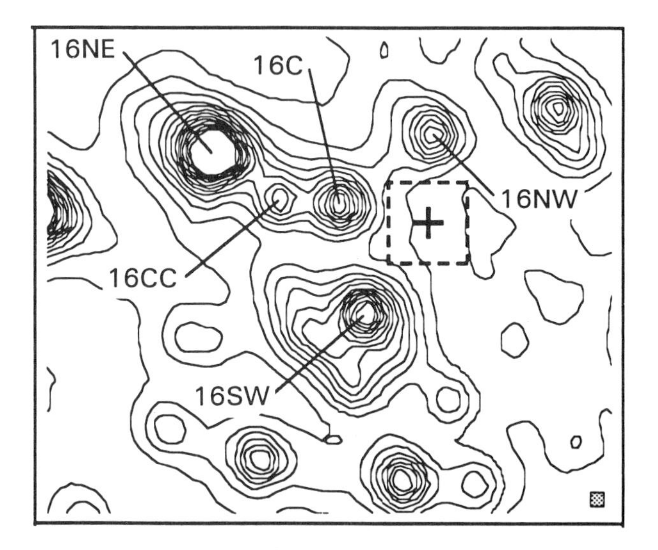

Fig. 7.11. A contour plot of the 2.2 μm image of the inner 0.5 pc of the Galactic Centre Region with the brighter members of the IRS16 cluster identified. The cross is the position of Sgr A* with the length of the arm representing the uncertainty in position. The sides of the surrounding square (shown dashed) are 1 arcsecond in length. See also plate 8. (From Herbst *et al.* *Astrophys.J.*, **411**, L21, 1993.)

These hot supergiants would also provide sufficient power to explain the ionization of the inner zone. Indeed, support for strong winds from the IRS16 complex comes from very recent VLA studies which, after maximum entropy deconvolution, reveal a comet-like tail emanating from IRS7. One explanation of this shape is that it is a bow shock produced when the envelope of the stellar wind from IRS7 interacts with the wind from another source. The apex of the comet-like appearance does indeed point towards IRS16, suggestive that the IRS16 cluster is the source of a strong wind. Calculations suggest that a wind of about 10^{-3} $M_{\odot}$ y^{-1} might provide a possible fit to the putative shock, which again fits in well with the stellar wind model. Nevertheless, this interpretation is not yet accepted by all workers in the field.

Located close to, but not apparently coincident with any of the more obvious sub-components of IRS16, is an intriguing radio source termed Sgr A^{*}. The radio luminosity of Sgr A* is ~ 10^{28} W, which is some ten orders of magnitude less than the Eddington luminosity of a 10^6 $M_\odot$ black hole. Its size has been determined from 1.35 cm VLBI observations to be less than 16 AU. At the longer wavelength of 3.6 cm, it is somewhat larger and shows an elongated structure. This extremely compact radio source has been found to vary on timescales of around a day. As one light day corresponds to a distance of 170 AU, this fits in well with the radio source size and also indicates that the variability takes place with velocities a significant fraction of the speed of light. This brings us back to the realm of relativistic particle acceleration. Very careful astrometry of Sgr A* with respect to background quasars has shown that it has a velocity across the line-of-sight less than 38 km s^{-1}, suggesting that it must be close to the centre of the Galaxy, and/or it is a massive object.

Perhaps one of the most exciting observations concerning the Galactic Centre region concerns the detection of the 0.511 MeV gamma-ray line emission from the annihilation of electrons and positrons. The luminosity in this line is in excess of 10^4 $L_\odot$ but the very narrow linewidth means that the electrons and positrons annihilate in a thermalised environment rather than a high-energy medium. In the Galactic Centre region, this translates to a region of gas with velocities less than the 700 km s^{-1} noted earlier, posing a potential problem. Evidence of variability of the annihilation line on timescales of six months imposes further restrictions on the source size. This in turn gives a lower limit to the density of the medium in which the line is produced.

When this line was first discovered, the angular resolution of the gamma-ray detectors was inadequate to verify that the Galactic Centre was the precise source. Later experiments, employing much higher spatial resolution (using coded mask detectors), showed that much, if not all the narrow component of this line comes from a bright high-energy X-ray source known as 1E 1740.7-2942. The continuum spectrum of this source fits that of Compton emission from a hot region of an accretion disk. Furthermore, during a period of flaring, the signature of the annihilation line was clearly observed. Very deep VLA radio maps have revealed the presence of parsec-scale, double-sided radio jets on both sides of the X-ray source. However, 1E 1740.7-2942 is *not* located at the Galactic Centre, but is almost 100 pc away (in the plane of the sky). Therefore the problems associated with the narrow-line emission are solved, and although it has nothing to do with the Galactic Centre, nevertheless it may represent a stellar-sized black hole accreting material. A model which seems to explain the observations is for electron–positron jets (radiating by synchrotron emission) losing energy as they interact with the surrounding molecular cloud and in so doing, form the bound electron-positron (positronium) state, that subsequently annihilates in the form of a very narrow line.

So what do we make of all the data pertaining to the centre of our Galaxy? Does it help us in our study of AGNs? Depressingly the answer is that even with all the available evidence we still cannot decide between a 'dormant' central engine and a star cluster model. Workers in this field have split into two distinct camps: those for and against a black hole. However, a number of questions remain to be answered by both camps. The star cluster group needs to explain why IRS16 appears to be a unique object. There is no evidence for any star formation occurring in the region, so IRS16 must be a very young phenomenon. Also, the extremely compact radio source SgrA* remains to be explained. Again, SgrA* is unique to our Galaxy but closely resembles radio sources found in the nuclei of two local spiral galaxies, M81 and M104. For the black hole pundits we can ask why the emission is so weak. Where is the evidence that SgrA* is a supermassive black hole allowing our Galaxy to have a dormant central engine? It is much easier to accept the fact that it is a small black hole than a massive one, and on no grounds can it have a mass exceeding about 10^6 $M_\odot$. The spectroscopic studies on the IRS16 cluster members produce an enclosed mass figure of around 3×10^6 $M_\odot$ residing within the inner 0.14 pc of the dynamical centre of the Galaxy. Therefore one thing is certain: our Galaxy was not a quasar in the past. Indeed, if it was active, then with only a million solar mass black hole at the most, it must have been a rather low luminosity AGN.

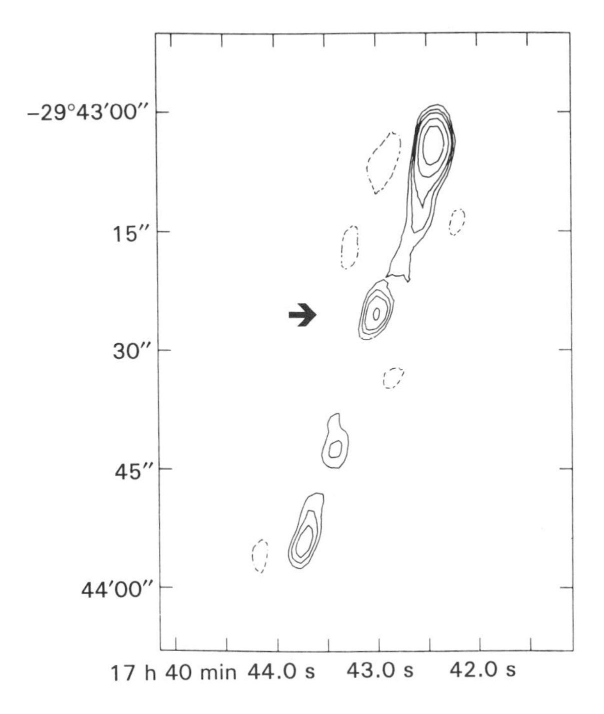

Fig. 7.12. 6 cm VLA map of the radio emission from the hard X-ray source 1E1740.7-2942 showing the edge-brightened double jet structure. This source is believed to be a stellar-sized black hole candidate. The radio core is indicated by an arrow. (From Mirabel *et al., Nature,* **358**, 215, 1991.)

Is there a massive black hole or not at the centre of our Galaxy? There is no doubt that the jury remains firmly out on this question. We will have to await better studies, perhaps from the Hubble Space Telescope equipped with the infrared camera. However, the degree of complexity of the gas motions in the inner parsec, with streamers, bridges, arcs, bars and mini-spirals (all descriptions from the literature), demonstrates vividly that our simplistic models of the inner workings of AGN are just that: global simplifications and horrendously lacking in details. Nevertheless, if we have got the big picture correct, and we believe we have, then in spite of the limitations this is a tremendous triumph of science. As surely as day follows night, the details will ensue.

7.4 AN ALTERNATIVE MODEL

The standard model of a supermassive black hole and accompanying accretion disk have been accepted by most astrophysicists as the underlying power-source for AGNs, and as we have seen above, the evidence from M87, NGC4258 and MCG-6-30-15 is now overwhelming. Furthermore, alternative models explaining the power sources of the most luminous of the AGNs and the jet-sources such as blazars and radio galaxies have not survived the test of time. On the other hand, models for powering the lower luminosity AGNs, such as the Seyfert and LINER galaxies, without recourse to black holes have been proposed and defended vigorously for a number of years. The far-infrared galaxies could also come under this banner. The models rely on the source of energy as a massive star cluster at the nucleus of the galaxy. Why are these models so attractive to their authors? It is because they do not require a supermassive black hole and can make do with massive star formation, which is a relatively well known and observed phenomenon.

The one key element for these models is that they can, under some circumstances, explain the power source for the BLR in the lower luminosity AGNs. They do this by using a starburst nuclear region and the resulting Type II supernovae (the explosion of very massive stars at the end of their evolution) provide the power source. In some senses this model also uses gravitational energy. Gravitational attraction produced the very high star density in the nuclear core, and is responsible for the Type II supernovae which ultimately provide the power for the broad-line region in terms. To explain the BLR an intense ionization flux must be produced and supernovae are the sources for this. Supporting evidence for this idea comes from two areas of observation. When a supernova explodes in a dense or ionized medium such an HII region, very broad and highly ionized emission lines can be produced. These resemble the lines seen in the spectra of Seyfert 1.5 galaxies. Furthermore, some of the flaring behaviour of Seyfert galaxies look similar to the light curve of a Type II supernova.

The starburst model pundits have taken this idea and modelled the shocks produced by a supernova and its expanding envelopes. It has been found that when a supernova envelope expands into a dense molecular cloud associated with a region of high star formation, very high radiative luminosities can be produced. At the outset, the expanding remnant has velocities ranging from 10^3 to $\sim 2 \times 10^4$ km s^{-1} and sweeps up material in the immediate vicinity of the explosion. After a period of about a year, when the remnant has evolved to a size of a few light weeks across, a change occurs due to the development of a very strong shock. The material now radiates strongly and most of the kinetic energy of the shock front is converted into radiation as the remnant undergoes a complex series of shocks. This radiation is calculated to be in the UV and X-ray region and therefore serves as a potential power source for a putative BLR. The timescales are such that the maximum luminosity occurs some 3 years after the explosion but is rather short-lived, with most of the stored kinetic energy in the expanding shell being radiated away after a further five years or so. Starting with a Type

II supernova energy of 10^{44} J, the peak luminosity corresponds to a few times 10^9 $L_\odot$, with much of it being in the form of X-ray emission.

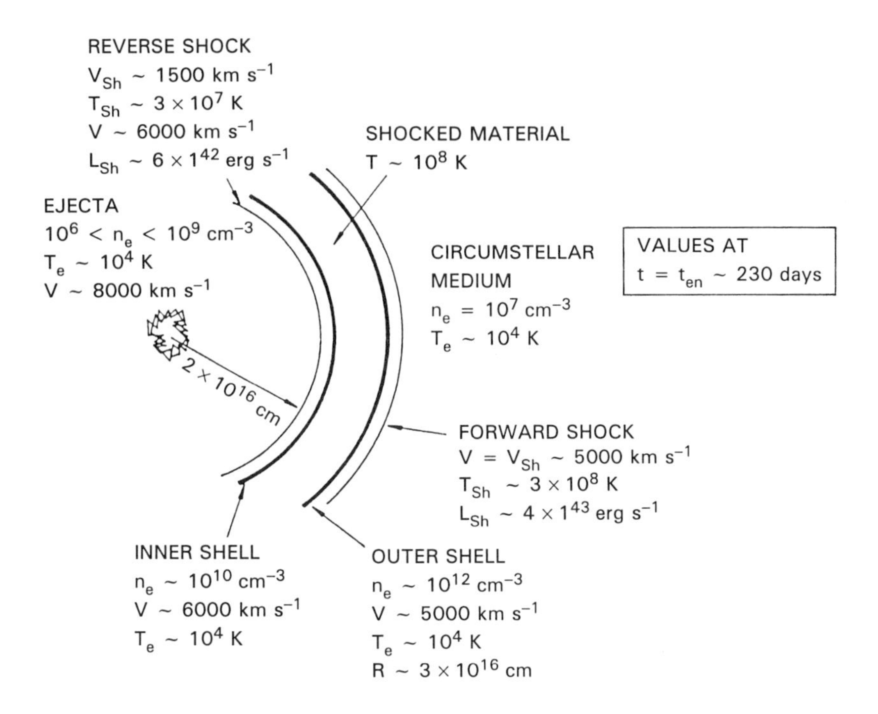

Fig. 7.13. Schematic representation of the evolution of the supernova remnant in the starburst model for AGNs. (From Terlevich *et al.* *Mon.Not.R.Astron.Soc.*, **255**, 713, 1992.)

Observationally speaking, very good agreement can be obtained for the observed line ratios. As we have seen, the difficulty of simultaneously explaining the iron lines and certain hydrogen line ratios compared with carbon ratios has led to an extension of the Standard Model to incorporate a stratified, two-zone regime. In the supernova case, these two zones can easily be identified. The highly ionized region is associated with the inner shell of the remnant (fig. 7.13), for which calculations give $U \sim 4 \times 10^{-3}$ and electron densities of 10^{16} m^{-3}, whereas the low ionization region comes from the outer shell giving values of $U \sim 10^{-4}$ and $n_e \sim 10^{18}$ m^{-3}. The output spectrum for an AGN is the sum of a range of supernovae of different ages. However, this is where the free parameters come into play in terms of selecting a desired supernova rate. It turns out that there is considerable margin in being able to explain the observed line ratios. On the other hand, another success for the model is a neat explanation of the lack of broad forbidden lines. These are expected to occur when the gas densities fall below $\sim 10^{10}$ m^{-3}. In the supernova case it turns out that when these densities are found in the shells, the shock velocity has already fallen to below 500 km s^{-1}, which means that the forbidden lines will be narrow rather than broad.

The key question is to establish whether there are sufficient supernovae to provide the average (constant about some mean) luminosity that is apparently being produced by Seyferts over a reasonable lifetime. (Because we see so many Seyfert galaxies in a reasonably typical state, this must be a relatively long-lived phenomenon.) There are undoubtedly problems with the rate of constant regeneration of supernovae and some match of parameters must be made to give an overall lightcurve with a desired luminosity and an acceptable supernova/star formation rate.

The pundits argue that there are fundamental differences between the steady and flaring AGN states and the starburst model pertains more to the former. While it is true that Seyferts are not flaring all the time, indeed they seem to be rather quiescent in the optical to radio (although not in the UV and X-ray), there is some evidence that some members of the population are variable in their optical lightcurves on timescales of months. Nevertheless, even in their non-flaring states, Seyferts have high luminosities and possess a BLR. Both of these observations must be explainable by any model. We will conclude this section on a positive note by saying that the model successfully reproduces many of the observed features of low-luminosity AGNs. Despite this, it has not found acceptance in the astronomical community at large, which remains firmly wedded to a supermassive black hole power source for all AGNs, the beauty of unification outweighing the problems still to be solved. In the starburst picture, a different phenomenon is required to explain radio emission and it has little explanation for the presence of hard X-rays or the X-ray variability. The requirement for two different physical processes goes against the current trend of observations and philosophy, both of which appear to favour unification rather than a bifurcation.

Before ending this section we will just recap some of the aspects of starburst galaxies. Even before the IRAS satellite survey, galaxies such as NGC1068, NGC253 and M82 were known to have very high rates of star formation, and as we saw in section 3.8 became known as starburst or far-infrared galaxies. However, these galaxies tend to have star formation taking place over a significant volume of the central regions of the galaxy rather than a point-like nucleus. The IRAS satellite showed that there were also super-starburst galaxies, although the suspicion remains that many of these are indeed AGNs in the sense of the word that we have used throughout this book. The nuclear regions are heavily dust enshrouded and it is impossible to determine the scale of the nuclear zone in the optical, or to see the hidden BLR. In chapter 9 we will discuss the suggestion that at least some of the super-starbursts are buried quasars.

7.5 SUMMARY

We have now met the bizarre consequences of general relativity and the main properties of black holes. We did not venture into detail on the more exotic aspects of black holes, such as evaporation, etc., but restricted ourselves to the main aspects pertinent to AGNs. Black holes are exotic phenomena but are a very efficient method of converting gravitational energy into radiant luminosity required to explain the luminosities observed from AGNs. However, supermassive black holes are required in this case, and supermassive refers to masses exceeding $\sim 10^6\ M_\odot$. One of the most crucial properties of a black hole is the event horizon, or Schwarzschild radius. Inside this radius no information can be obtained about the black hole. The Schwarzschild radius is of order

2 AU for a supermassive black hole of mass $\sim 10^8\ M_\odot$, and this gives us some idea of the scale size we are dealing with and the difficulty of directly observing a supermassive black hole.

There are strong theoretical grounds for expecting an accretion disk to surround a black hole and for galactic objects such a phenomenon has now been observed, albeit by indirect means. The accretion disk can be either relatively cool ($\sim 10^4$ K) and thin, or extremely hot (upwards of 10^5 K) and inflated. The latter produces a very steep-walled inner torus that might be responsible for the large-angle collimation of the outward radiation and particle flow. The surroundings of this inner zone lie in the realm of high-energy interaction processes and X-ray and gamma-ray emissions are not only entirely possible, but very probable. There is strong potential for this radiation to back-illuminate the outer atmosphere of the accretion disk, heating it even further.

Obtaining direct evidence for black holes remains elusive, but with scale sizes of the order of the solar system and essentially being invisible (either cloaked by an accretion disk or an opaque hot corona, or just black) this is hardly surprising. Even the accretion disk is too small for imaging. However, there is a growing body of strong supporting evidence for a massive black hole in a small number of local galaxies, which otherwise look remarkably normal, boringly so in fact. The best evidence comes from a study of the dynamics of the material in the gravitational field of the black hole, along with the light distribution in the same region and there is strong suggestive evidence that the core of M31 houses a supermassive black hole that is, for some reason, dormant at the current epoch. A breakthrough came with the refurbished HST observations of a rotating disk of ionized gas in the core of M87. This was soon followed by VLBI observations of maser emission in NGC4258 which showed perfect Keplerian motion about a very high central mass density. The ASCA gravitationally redshifted X-ray emission line of fluorescent iron has probably provided the best evidence to date for a supermassive black hole. All these observations have convinced the vast majority of the astronomical world of the existence of supermassive black holes. Virtually all AGN pundits are wed to the concept of a supermassive black hole and an accretion disk, indeed they have now moved on to study the precise details of the composition, structure and physical processes of these accretion structures.

Quasar were most plentiful at an epoch between redshift 2 and 3. Both before and after this epoch the quasar population declined. We know that quasars exist at a redshift of almost 5, corresponding to a time when the Universe was less than 10% of its current age. This provides very stringent limits on the time available for galaxies and supermassive black holes to form. There is a natural limit to the accretion rate for a radiating body. This is called the Eddington limit and comes about because the outward radiation pressure is sufficiently high to halt the infall. Accreting at the Eddington limit does not seem to allow sufficient time for these black holes to form and so super-Eddington rates are appealing. These are possible only if non-spherical accretion takes place, and it is still not obvious that they can operate over adequate timescales. Reducing the radiative efficiency is a requirement that goes hand-in-hand with super-Eddington rates and an attractive possibility is to resort to a time-variable radiative efficiency. This predicts a massive accretion rate and black hole growth phase, which might just allow enough time for supermassive black holes to form to power quasars by redshift of ~5. This phase is followed by a steady quasar radiation lifetime, leading to eventual decline through lack of fuel. However, many problems remain to be solved in this entire area.

For highly luminous quasars, black hole masses exceeding 10^7 $M_\odot$ are predicted, requiring accretion rates of the order of a few solar masses per year. The *potential* fuel supply available in a galaxy allows lifetimes in excess of 10^8 y. However, lengthy quasar lifetimes pose an additional problem regarding the dynamics of the fuel supply in the vicinity of a supermassive black hole. After a certain length of time, the black hole will have swept clean the inner zone of the galaxy (here we are talking about scales of order parsecs). With no further infall the AGN is effectively quenched. Unless more material can be fed into this zone, the quasar lifetime will be automatically self-limiting. In order to extend AGN lifetimes (or even start the process in the first place), we require some form of perturbation to the otherwise smooth central axisymmetric gravitational potential of the central regions of a galaxy to maintain continual accretion.

What mechanisms of perturbation can we think of? Two immediately spring to mind: the presence of a bar structure in the central regions of the galaxy, or gravitational perturbations due to the close passing of nearby galaxies. The idea of AGNs having barred central regions, or mini-spirals, has gained some favour over recent years as a result of high angular resolution optical imaging (e.g., see plate 4). Also, the idea of perturbations in the form of galaxy interactions (plate 7) forces us to speculate on whether there was a preferred epoch for quasars in the distant Universe, when galaxy interactions were more prevalent. As we noted in chapter 1, perhaps many powerful quasars have now died out after exhausting their central fuel supply due to the sweeping mechanism of the supermassive black hole and lack of further galaxy perturbations to refuel the engine. If this is the case, then the obvious conclusion is that many galaxies might harbour supermassive black holes, but, being unfed, they will be dormant and no longer classed as AGNs. There is no doubt that some galaxies have been quasars in the past, perhaps many of the Seyferts that we see today, or the M87s and such like. It is clear that M31 and M32 were not quasars, and neither was our Galaxy.

The investigations of our Galactic Centre are severely hampered by the dust in the plane of the Milky Way. Nevertheless, major strides have been made. When we investigate the Centre of our Galaxy we find a wealth of phenomena, including an exotic X-ray source that is probably a stellar mass sized black hole with double radio jets and producing (somehow) electron–positron annihilation in its neighbourhood. However, this source is not at the Galactic Centre. The mass contained within the inner 0.14 pc of the centre of the Galaxy is calculated to be no more than $\sim 3 \times 10^6$ $M_\odot$ and within this region lies the unique source referred to as IRS16. This is highly luminous and is probably a cluster of very massive and luminous stars. Believed to be even closer to the Galactic Centre, if not at it, is the tantalizing radio source Sgr A*. This has a size of less than 16 AU and is variable on timescales of a day. Its origin is unknown and although it is also unique in our Galaxy, it has counterparts in the nuclei of the nearby spiral galaxies M81 and M104. Is Sgr A* the signature of the central black hole? We now know that the largest black hole that could be hiding in the centre is less than 10^6 $M_\odot$, which, although showing that our Galaxy has not been a quasar in the past, does not exclude the possibility that it may have been a low-luminosity AGN.

It is salutary to refer to a comment in Martin Rees' 1984 *Annual Reviews* article where he quotes from Zeldovich and Novikov's 1964 paper in which they discussed the radiating mechanism of quasars based on the first data from 3C273. 'They conjectured that for a gravitational source of energy: (a) radiation pressure perhaps balances gravity, so the central mass is approximately 10^8 solar masses, (b) for a likely efficiency of 10%, the accretion rate would be 3 solar masses a year, (c) the radiation would come from an effective "photosphere" at a radius of about 2×10^{10} km, outside of which line opacity would cause radiation to derive a wind, (d) the accretion may be self-regulatory, with a characteristic time-scale of about 3 years. It is amazing how close those far-sighted ideas remain to the study of the power sources of AGN today'. We must echo these thoughts even more so today, some eleven years after Rees's article.

The alternative to black hole powered AGNs is the so-called starburst scenario. However at the outset we should note that these are generally thought to apply (if at all) to radio-quiet and lower-luminosity AGNs. This derives from the well-observed starburst phenomenon and uses the high supernova rate to provide the power source for the BLR. Observations of supernovae showing Seyfert-like spectra have given support to the idea. The expanding supernova remnant ploughs into the dense surrounding medium of a molecular cloud and in the process strong radiative shocks provide the UV–X–ray luminosity to power the BLR, albeit for a time of only a few years. Explaining the long-term luminosity and lifetimes of even the low luminosity AGNs requires this continued supernova rate in dense molecular clouds. Although the supernova rate is not excessive, the cloud lifetime begins to look fragile.

Nevertheless, the model is attractive in that it can reproduce the observed BLR line ratios as well as the two-zone standard model, and it is a relatively well-understood physical process. Current wisdom is not in favour of this model, mainly as it does not explain any of the radio properties, the presence of hard X-ray photons, or with the X-ray variability and it has great difficulty explaining quasar luminosities. It does not provide the beauty of a unification scenario using a single fundamental power source, but suggests that there should be a division at some point between black hole powered and supernova-powered AGNs. It is not clear where this division might lie, because the more recent observations have tended to show more of a unifying theme rather than a bifurcation. We will end with the viewpoint that the 'black holes explain all' version of AGN activity is widely accepted. Nevertheless, we should remember that it still has problems in many areas of detail, as shown by the continued refinements of the standard model which some would say have now become contrived.

7.6 FURTHER READING

General and reviews

Brown,R.L., & Liszt,H.S., 'Sagittarius A and its environment.' *Ann.Rev.Astron. Astrophys.*, **22**, 223, 1984.

Frank,J, King,A., & Raine,D., *Accretion power in astrophysics,* Cambridge University Press, 1992. (An excellent treatise on accretion disks, AGNs and the BLR.)

Genzel,R., & Townes,C.H., 'Physical conditions, dynamics and mass distribution in the Centre of the Galaxy', *Ann.Rev.Astron.Astrophys.*, **25**, 377, 1987.

Kafatos., ed, *Supermassive black holes,* Cambridge University Press, 1988. (A series of papers on many aspects of black holes and AGNs - a must.)

Kormendy,J., & Richstone,D., 'Inward bound: the search for supermassive black holes in galaxy nuclei', *Ann.Rev.Astron.Astrophys.*, **34**, in press, 1995. (Another must.)

Longair,M.S. *High energy astrophysics, vol. 2,* Cambridge University Press, (An excellent treatment of accretion disks—along with other physics.)

Rees,M.J., 'Black hole models for active galactic nuclei', *Ann.Rev.Astron.Astrophys.*, **22**, 471, 1984. (The gospel according to the master—wonderful reading.)

Thorne,K. *Black holes and warped space-time*, (Interesting and stimulating.)

Specialized

Supermassive black holes and accretion disks

Gaskell,C.M., 'Direct evidence for gravitational domination of the motion of gas within one light-week of the central object in NGC 4151 & the determination of the mass of the probable black hole', *Astrophys.J.*, **325**, 114, 1988.

Greenhill,L.J., *et al.*, 'A subparsec diameter disk in the nucleus of NGC 4258', *Astrophys.J.*, **440**, 619, 1995.

Ford,H.A., *et al.*, 'Narrow-band HST images of M87:evidence for a disk of ionized gas around a massive black hole', *Astrophys.J.*, **435**, L27, 1994.

Harms,R.J., *et al.*, 'HST FOS spectroscopy of M87: evidence for a disk of ionized gas around a massive black hole', *Astrophys.J.*, **435**, L35, 1994.

Halpern,J.P., & Filippenko,A.V., 'A test of the massive binary black hole hypothesis: Arp 102B', *Nature*, **331**, 46, 1988.

Jaffe,W., *et al.*, 'A large nuclear accretion disk in the active galaxy NGC4261', *Nature*, **364**, 213, 1993.

Kormendy,J., 'Evidence for a supermasive black hole in the nucleus of M31', *Astrophys.J.*, **325**, 128, 1988.

Lauer,T.R., *et al.*, 'Planetary camera observations of the M87 stellar cusp', *Astron.J.*, **103**, 703, 1992.

Miyoshi,M., *et al.*, 'Evidence for a black hole from high rotation velocities in a sub-parsec region of NGC4258', *Nature*, **373**, 127, 1995.

Rees,M.J., Begelman,M.C., Blandford,R.D., & Phinney,E.S., 'Ion-supported tori and the origin of radio jets', *Nature*, **295**, 17, 1982.

Rees,M.J., 'Dead Quasars in nearby galaxies ?' *Science*, **16**, 817, 1990.

Saslaw,W.C., 'VIII. Gravitational processes in galactic nuclei', *Pub.Astron.Soc.Pacific*, **98**, 171, 1986.

Stella,L., 'Measuring black hole mass through variable line profiles from accretion disks', *Nature*, **344**, 747, 1990.

Tanaka,Y., *et al*, 'Gravitationally redshifted emission implying an accretion disk and massive black hole in the active galaxy MCG-6-30-15', *Nature*, **375**, 659, 1995.

Sun,W.H., & Malkan,M.A., 'Fitting improved accreion disk models to the multiwavelength continua of qausars and active galactic nuclei', *Astrophys.J.*, **346**, 68, 1989.

Formation and evolution of AGNs

Haehnelt,M.G., & Rees, M.J., 'The formation of nuclei in newly formed galaxies and the evolution of the quasar population', *Mon.Not.R.Astron.Soc.*, **263**, 168, 1993.

MacKenty,J.W. *et al.*, 'Markarian 315: A test case for the active galactic nucleus-merger hypothesis', *Astrophys.J.*, **435**, 71, 1994.

Prestwich,A.H., Joseph,R.D. & Wright,G.S., 'Starburst models of merging galaxies', *Astrophys.J.*, **422**, 73, 1994.

Shlosman, I., Begelman, M.C., & Frank, J., 'The fuelling of active galactic nuclei', *Nature*, **345**, 679, 1990.

Shields,G.A., 'I. Theory of active galactic nuclei', *Pub.Astron.Soc.Pacific*, **98**, 130, 1986.

Small,T.A & Blandford,R.D., 'Quasar evolution and the growth of black holes', *Mon.Not.R.Astron.Soc.*, **259**, 725, 1992.

Turner,E.L., 'Quasars and galaxy formation. I. The z > 4 objects', *Astron.J.*, **101**, 5, 1991.

Sanders,D.B., *et al.*, 'Ultraluminous infrared galaxies and the origin of quasars', *Astrophys.J.*, **325**, 74, 1988.

Sanders,D.B. *et al.*, 'Warm ultralumious galaxies in the IRAS survey: The transition from galaxy to quasar ?' *Astrophys.J.*, **328**, L35, 1988.

The Galactic Centre

Herbst,T.M., Beckwith,S.V.W., & Shure,M., 'Infrared imaging of the Galactic Center and the search for Sagittarius A*', *Astrophys.J.*, **411**, L21, 1993.

Krabbe,A., *et al.*, 'The nuclear cluster of the milky way: Star formation and velocity dispersion in the central 0.5 parsec', *Astrophys.J.*, **447**, L95, 1995.

Mirabel, I.F., *et al.*, 'A double-sided radio jet from the compact Galactic Centre annihilator 1E1740.7 - 2942', *Nature*, **358**, 215, 1992.

Rogers,A.E.E., *et al.*, 'Small scale structure and position of Sgr A* from VLBI at 3mm wavelength', *Astrophys.J.*, **434**, L59, 1994.

Tamblyn,P., & Rieke,G.H., 'IRS 16: The galaxy's central wimp ?' *Astrophys.J.*, **414**, 573, 1993.

The starburst model

Terlevich,R., *et al.*, 'The starburst model for active galactic nuclei : the broad line region as supernova remnants evolving high density medium', *Mon.Not.R.Astron.Soc.*, **255**, 713, 1992.

Terlevich, R., *et al.*, 'The starburst model for active galactiv nuclei - II. The nature of the lag', *Mon.Not.R.Astron.Soc.*, **272**, 198, 1995.

8

Radio galaxies, jets and superluminal motion

8.1 INTRODUCTION

Radio emission from galaxies was discovered during the first series of observations with the first generation of astronomically dedicated radio telescopes. As we saw in section 3.6, the study of the so-called radio galaxies became a major topic of study in the 1950s and onwards. The very extended, twin-lobed radio emission extends dramatically beyond the normal confines of a galaxy as defined by its starlight. In one or two sources this emission extends to distances of a few megaparsecs. In this chapter, we shall investigate three aspects of radio emission: the twin-jet emission leading to the classical radio doubles; orientation effects of the jet with respect to the line-of-sight to the observer giving us *core-dominated sources*; extending this concept to the powerful relativistic jets which give rise to the beaming properties described observationally by the blazar phenomenon and superluminal motion.

In all of this, we will *not* dwell on a description of the lobes, or their interactions with the surrounding medium (see figs. 3.12–15), but provide references in section 8.8 for those who wish to study these aspects further. For a small number of sources, jets are observed at other wavelengths, and in one case, that of M87, we shall compare the optical, UV and radio images of the jet. In so doing we will find that the picture looks remarkably similar, so similar that it presents problems for the hitherto usual explanation.

Nonetheless, although fascinating, we shall avoid becoming immersed in the intricate details of the jets themselves and again refer the reader to other works. We will use the jets as a signpost, highlighting the ultimate source of the power, the activity at the nucleus of the parent galaxy. We shall furthermore take for granted the assumption that the extended radio lobes are fed by the radio jets emanating from the nucleus. In terms of the AGN phenomenon our interest lies in the jets as vehicles and observational clues to the activity in the nucleus, and in the mechanisms that allow such highly collimated and long-lasting entities to form and remain stable. We will briefly dwell on why only a small fraction of active galaxies possess powerful jets. However, we will find that the study of jets, from the classical double sources through single-sided jets to the extremes of the range represented by the superluminal sources, provides us with an excellent introduction to unification scenarios.

A point to remember throughout this chapter is that single-dish radio telescopes or extended telescope arrays such as MERLIN and the VLA give us information on the kiloparsec scale, but Very Long Baseline Interferometry (VLBI) gives us information on the parsec scale. It is these latter observations that are crucial to probing the innermost parts of the jets.

Double, or twin-lobed radio galaxies were introduced in section 3.6.1, with a number of examples being shown in figs. 3.10–3.13. The lobes can extend up to megaparsecs in scale, although in the vast majority of cases they are only hundreds of kiloparsecs long. The radio power in the lobes extends up to 10^{35} W. The radio galaxies have been classified into two groups, referred to as Fanaroff and Riley type I and II. The FRIs have lower radio luminosity and the lobes fade out at the ends, whereas the FRIIs have higher radio luminosity, the lobes are end or edge-brightened, and there is usually a compact radio core lying between the lobes coincident with the nucleus of the parent galaxy. The latter are often referred to as classical doubles in the radio astronomy jargon, although there are notable variations due to a wealth of morphological sub-classifications that we shall avoid. When observed with sufficient sensitivity however, FRIs also show core emission from the parent galaxy.

In terms of the jets feeding the lobes, a large fraction of FRIs show jets which are usually (but not always) twin on the kiloparsec scale and single-sided jets on parsec scales. In FRII radio galaxies jets are more difficult to detect while FRII quasars tend to show a prominent single-sided kiloparsec jet feeding one of the two radio lobes visible. It turns out that the transition between FRI and FRII sources marks an approximate change between intrinsically double- and single-sided jets, although as always, there are a small number of exceptions to the rule and some sources are classed as FRI/II. Quasars are all FRII sources and have single-sided jets.

8.2 THE PROPERTIES OF JETS

Jets are discussed in a wide range of astrophysical phenomena, ranging from solar prominences, outburst on comets as they approach the Sun, star formation regions (bipolar outflow and molecular jets), a spectacular binary star (SS433) and active galaxies. How do we define a jet? A jet is a well-collimated outflow of material or radiation, and is often evidenced in astronomy by a clear structure, separate from the surrounding medium. The jet is not always immediately obvious and sometimes there are 'hidden' jets where the jet is only manifest by identifying blobs of excited emission that trace out the underlying geometry and extent of the jet. This is often the case for star formation. Clearly a jet that is defined by direct radiation from the jet material is much easier to see, although the wavelength must be chosen carefully. (Try looking for jets from radio galaxies in the optical—it is usually a very fruitless exercise.) An important point to note is that we have said nothing about being able to observe the actual outflow of material, we have just said we can observe the radiation from a jet of material. We will return to this vital point later.

The term 'jet' has unfortunately tended to become somewhat misused, with many broad outflows being classed as jets alongside the true jets of the long, thin, pencil-like phenomena. We shall assume that for a structure to be referred to as a jet, its length is many times greater (say > 4) than its width. Although this is rather arbitrary, it seems to work in practice and for the radio jets from AGNs this criterion is fine. Nature is usually very symmetric, and when we see jets it is a clear sign that either an anisotropic process is at work or there is some very extreme form of bipolar collimation. Twin jets seem to fit in well with the simple idea of a rotational phenomenon with the two poles of the rotation axis providing a natural symmetry for back-to-back jets. But what about the observations that show only a single jet? Do we need to introduce new physics, such as a unipolar model, or what? We will opt for simplicity and explain our way out of this potential embarrassment by seizing on the relative orientation of the object to the observer. This is yet another form of the unification hypothesis. Basically the idea is that for all these radio galaxies there is only one fundamental property at work; what it looks like and what we call the object in question depends essentially on how we view it.

The basic theory describing the powering of the extragalactic double-lobed radio sources came from Peter Scheuer, and separately, Roger Blandford and Martin Rees. In 1974 they postulated that the lobes were fed by the transport of energy from the nucleus of the optical galaxy by narrow jets. The jets are produced by the central engine of the galaxy and this was identified as a supermassive black hole. The jets were very narrow and because these back-to-back jets emanating from the central engine resembled the exhaust trail seen in jet aircraft it was dubbed the twin-exhaust model. When the jets plough into the surrounding extragalactic medium a shock front is set up, energy is dissipated at a working surface and the lobes are formed by material bouncing backwards and outwards. Although the details of the original work have been found to be seriously deficient in explaining the collimation process, nevertheless the overall concepts have been spectacularly successful in explaining the general picture of the extended radio emission in galaxies. It should be stressed that the idea was produced *before* parsec-scale jets were observed with radio interferometers. As such it is an excellent example of theoretical prediction being borne out by observation. This basic picture of jet production is now taken for granted and much of the theoretical study has moved on to the finer details of the jet physics and the energy transport processes.

One of the most spectacular demonstrations of a jet fuelling a twin-lobed extragalactic radio source is NGC6251, which we mentioned in section 3.6.1. This FRI source is shown in fig. 8.1 and is sometimes referred to as the cosmic blowtorch for obvious reasons. The galaxy is about 100 Mpc distant and shows jet emission on size scales ranging from 300 kpc to less than 1 pc (assuming the source lies in the plane of the sky). This also demonstrates the dramatic resolution of ~1 milli-arcsecond achieved by VLBI studies (bottom panel). We can clearly see that there is highly collimated emission extending from the innermost parsec of the galaxy to regions way beyond its optical extent. This is a key result. Furthermore, the jet is seen to be a continuous process from the parsec-scales of the core of an AGN to lengths of hundreds of parsecs. This is conclusive evidence that radio lobes are fuelled by core-powered jets in AGNs.

We also note how smooth the right-hand jet appears compared to the left-hand jet. It is both shorter and weaker and is often referred to as the counter-jet. Is this a clue to what is going on and can you think of possible explanations? We know that FRII sources frequently have only one visible jet, so is NGC6251 a lower luminosity version of the same phenomenon? Indeed it is sometimes referred to as a FRI/II intermediate class. Let us explore this single-sidedness in more detail.

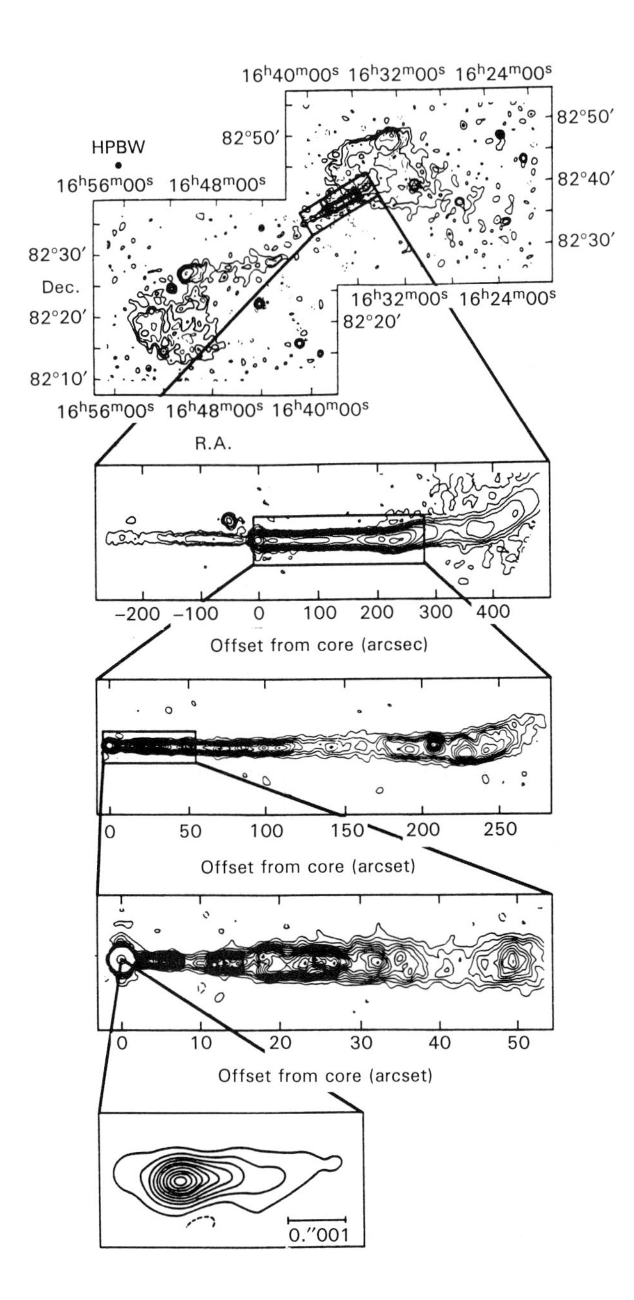

Fig. 8.1. The two jets in the galaxy NGC6251. The right-hand jet is shown at a variety of spatial resolutions taken with different radio interferometers at various wavelengths. The entire region is shown on the top map taken at 49 cm using the Westerbork Synthesis Radio Telescope, three VLA maps at 18 cm, 21 cm and 18 cm are then presented and finally a 2.8 cm VLBI map shows the elongation of the milli-arcsecond core. (From Bridle & Perley, *Ann.Rev. Astron.Astrophys*, **22**, p322, 1984.)

8.2.1 Are all jets double?

There are a number of theories explaining why we only see a single jet on the parsec to kiloparsec scale. One possibility is that the jets are intrinsically asymmetrical in power. In terms of fig. 8.1, the explanation is that more fuel is being pumped into the right-hand jet than the left-hand one and therefore the former is more easily visible. Alternatively, there could be something about one of the jets that makes it radiate more efficiently, giving a single-sided appearance even though equal energy may be flowing through both jets. A variation of this idea is that the receding jet is more suppressed by some mechanism than the forward jet. Free–free absorption is one possible mechanisms to explain this (because of a greater path length in the suppressing medium). Finally, and it must be said the most favoured explanation, is that the material in the jet is moving relativistically and the jet is oriented away from the plane of the sky. This produces a significant relativistic boosting factor, brightening the jet approaching the observer and dimming the receding jet. If this relativistic effect is sufficiently large it could easily explain the resulting single-sided appearance. Yet another possibility that cannot be ignored is that the jet is intrinsically single-sided and repeatedly flips between radiating in diametrically opposing directions. This is the flip-flop model which just about manages to explain most of the observations, although what physical process should produce a single-sided and flip-flopping mechanism is quite unexplained. However, it has severe limitations in that the change of jet direction must happen frequently because we observe so many classical double radio galaxies with symmetric lobes of radio emission. Observations of hotspots evident in both lobes also severely limits the amount of time the jet can be 'off' for that particular lobe, otherwise the hotspot would have decayed away.

Most astronomers feel uncomfortable with this concept and we will not lean towards this explanation. Although there are possibilities of incorporating an unintentional bias against a counter-jet in the highly complex process in which the radio interferometric maps are put together and analysed, this is not believed to be a fundamental reason for the apparent lack of counter jets. Furthermore, there are no known cases of parsec and kiloparsec jets pointing in opposed direction and so no source has ever been caught in the act of 'flipping'. Finally, there are regions with synchrotron lifetimes much less than the light travel time across the source observed on both sides of the nucleus (hot spots in M87—see 8.5) which supports intrinsic double-sidedness of the jets. The relativistic explanation is most favoured and it is this that we shall embrace.

One of the key tests radio astronomers are currently attempting to undertake is to obtain extremely high dynamic range radio maps of the one-sided sources with the confident expectation that they will eventually detect the counter-jet as they probe to fainter flux levels. The arrival of the VLBA (see section 2.4.4) has improved dynamic ranges dramatically and values of many thousands are now routinely possible. These are still early days but clear detections have been found for NGC1275 and 3C338 (see 8.4). It seems to be the case that at a level close to and below the FRI-II division, the number of sources with twin parsec-scale jets increases dramatically but the majority of FRIs are still one-sided. Perhaps free–free absorption by gas, (maybe in the torus) is adding to the Doppler dimming and further decreasing the brightness of the receding jets.

Even if these studies are still inconclusive, there is other very strong evidence for the jets being oriented at an angle to the plane of the sky. This evidence comes from polarization studies and the argument goes as follows. The brighter beam is oriented towards us and therefore the lobe fed by it is closer. The radiation from the jet and lobe will have travelled through less interstellar material in the parent galaxy than its receding jet counterpart (which is often not seen). Therefore, the degree of Faraday rotation of the approaching lobe by the interstellar gas should be less than for the receding lobe. In other words, the brighter beam and lobe should be more strongly polarized. This is known as the Laing–Garrington effect after the discoverers and is now observationally well documented, providing strong evidence for orientation to be a prime explanation for single-sided jets. At longer wavelengths depolarization occurs giving the same result. We shall now ignore further ideas about intrinsically single-sided jets.

8.2.2 Lifetimes and energy requirements

Figure 8.1 also gives us information about the lifetime for the jet activity. The outer lobes are at a distance of 300 kpc from the centre, which means that even if the material in the jet is moving at the velocity of light, the material filling the outermost parts of the lobes must have left the nucleus at least a million years ago. Furthermore, because the right-hand jet is continuous, we can see that this process must also have been continuous over this timescale assuming the velocities remain the same. For jets which are tilted significantly to the line-of-sight, then the actual (de-projected) length of the jets is longer and so the timescale is correspondingly increased.

A reasonable assumption is that the mechanical energy required to inflate the lobes and fill them with hot plasma has derived from the central engine and has been fed through the jets. The total energy in the lobes and timescale of the transportation process can then be used to calculate the minimum mechanical (kinetic) energy generated by the central engine. The uncertainties in this calculation derive from the lack of knowledge of the jet inclination to the plane of the sky and the conversion efficiency of the mechanical energy into radiation. (Calculations based on synchrotron radiation lifetimes and using assumptions about equipartition of energy in the lobes can also give minimum energies generated by the central engine.)

On the other hand, it might be argued that the jets do not *transport* material from the nucleus of the galaxy to the lobes, but that re-acceleration takes place *in the jet*. Nevertheless, even here we have to contend with the fact that information cannot be transmitted faster than the speed of light and so the physical extent of the jet means the timescale argument still holds. For those sources in which the jet is invisible, some estimate of the lifetime of the process can be obtained from the measurement of the total radio power in the lobes. This power must be derived from energy fed by a jet and making assumptions about energy transport produces another lifetime for the process. Again, times in excess of a million years seem to be a reasonable estimate. Even though this large-scale, stable structure is powerful evidence for long-term activity in the nuclei of galaxies, we must remember that timescales of the order one to ten million years are extremely small compared to the age of the Universe of around 10^{10} y.

8.2.3 The physics of jets

Perhaps the most important observation relating to jets comes from VLBI studies which show that the core emission of these jet sources is very compact (< 1 milli-arcsecond). Knowing the distance of the galaxy through the observed redshift, the brightness temperature of the core can be calculated. In many cases this dramatically exceeds the 10^{12} K limit of synchrotron radiation introduced in section 4.4.3. Therefore we can readily predict that there should be observable X-ray emission due to the synchrotron self-Compton process. This is not generally observed, and as we have noted before, the only way out of this dilemma is to assume that the emission is not anisotropic but is highly beamed. This leads directly to the superluminal motion prediction which is now well observed (see section 8.3).

The emission we see from the jets is synchrotron radiation, and this immediately tells us that we must have energetic electrons in the presence of a magnetic field. But this poses a problem. We know from section 4.3.1 that a synchrotron radiating ensemble of electrons has a lifetime proportional to the inverse of the electron energy and the magnetic field strength (eqn. 4.42). The resulting electron lifetime calculations are orders of magnitude less than the time required for the electrons to propagate to the outermost parts of the jet. Therefore, to explain the lifetimes we require either a complex fluid transport of jet material which suppresses synchrotron radiation by some mechanism such as magnetic field isolation, or electron re-acceleration in the jet. We shall return to this problem later. Strong polarization is found in the jets (a further argument for synchrotron radiation) and it turns out that for most of the twin-jet sources the magnetic field is aligned perpendicular to the length of the jet, at least along the central ridge line of the jet.

One of the earliest puzzles concerning jets was to explain why they remained so very thin, like torch beams, for such enormous distances. What process could provide such a high degree of collimation? We must remember that a jet which is not confined by some external pressure (termed a free jet) will expand sideways as it moves forward. It will expand and fan out similarly to the time evolution of vapour-trails of jet aircraft (and even here there is the pressure of the surrounding air). Therefore an extremely long pencil-beam jet must be collimated by some mechanism, either by internal confinement or by an external pressure. We saw in the previous section that there was hope that the initial collimation might be produced by the physical processes linked with the accretion disk or its super-hot corona. At first sight the deep funnel of the thick torus provides some large-scale angular collimation over a small physical distance (solar system scale) and back-to-back jets could derive naturally from this process. But whether this first stage collimation is possible (see later) and how the resulting beam is 'focused' into the very tight collimation of a narrow pencil-beam jet with a diameter of roughly a parsec, remains a unsolved problems. Most workers anticipate that magnetic fields probably come into the picture somewhere, providing some form of refocusing. However, the details remain to be determined and this is a key area that continues to cause perplexity. We will return to this topic in section 8.6.

What do we know about the general properties of jets? We know that the flow in lower velocity 'jets', like the smoke from a high chimney stack, is soon disrupted due to turbulence. The jet rapidly dissipates. A simple explanation for the prevention of dissipation is that the jets are supersonic; the jet material is moving at a velocity exceeding the sound speed in the medium. Supersonic jets will stay collimated for much longer timescales, indeed as long as they remain supersonic. It is also the case that a supersonic jet is less prone to entrainment of gas and a subsequent dissipation of kinetic energy as heat. Over the past decade much progress has been made in studying the physics of jet fluid flows (particularly for FRII jets) by numerical simulations on supercomputers. These can now reproduce jets that exhibit many of the detailed features seen on VLA maps (see fig. 8.14) and the spectacular plumes such as those seen in fig. 8.12. An example of a computer-generated jet is shown in fig. 8.2. Although modelling with supercomputers has been very successful, it should be remembered that the jets we observe span a length scale over 10^8 and this is still far beyond what can be achieved using current modelling techniques.

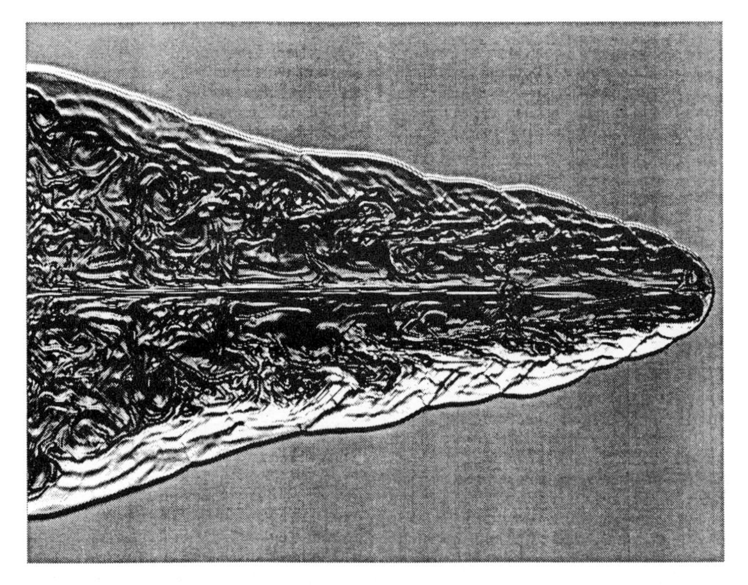

Fig. 8.2. Computer simulation of a hypersonic jet. (From Loken, *et al. Astrophys.J.*, **392**, 54, 1992.)

Two key parameters seem to determine the overall structure of the jets and lobes: (i) how supersonic is the jet, and (ii) the ratio of the density of the gas within the jet to that of the surrounding medium through which the jet is moving. The supersonic level of the jet is described by the Mach number, which is how fast the fluid flow compares to the local speed of sound in the material. Recent work on computer simulations shows that a supersonic jet produces a bow shock, and intriguingly, a cocoon-like return flow consisting of a sheath of subsonic and turbulent plasma is then found on the outside of the jet (fig. 8.3), both of which were predicted in the early work of Scheuer and Blandford and Rees. A number of astrophysical jets have now been well resolved in the direction across the jet and most display a centre-brightened rather than an edge-brightened structure. This supports the view that the emission comes from the material being transported in the jet itself, rather than the surrounding sheath of material being excited by the passage of the jet through the medium.

We now believe that for the FRII sources the plasma in the jets is light (i.e., it is less dense than the surrounding medium) and the jets are highly supersonic (hypersonic), with Mach numbers exceeding ~10. Although FRI jets also have a low density, they possess much lower Mach numbers than for FRIIs while still being supersonic. We will see that this is important for jet structure, and subsequently in considering the blazar class of synchrotron source and separating the BL Lacs from the OVV quasars. In some cases, such as 3C31, there is a rapid widening of the jet close to the nucleus; but it occurs without disruption. This is probably due to the entrainment of the jet at a small Mach number. In other cases, the flaring is probably caused by the transition to a subsonic flow which produces turbulence and rapid decollimation. This is seen for many FRI jets when they have propagated to distances of the order of kiloparsecs. This may also coincide with the outer edge of a galaxy, something that is frequently seen in the more powerful of the FRI sources. Another point in this explanation is that these FRI galaxies often lie in clusters and so there is a cluster medium which interacts with any outward flow or wind from the active galaxy. A shock front is set up and when the jet encounters this, it is immediately disrupted and flares. An excellent example is Centaurus A, the closest FRI source to us and shown in plate 2. FRII jets, on the other hand, show strong laminar flow and are very much isolated phenomena, with the details of their structure depending on the jet itself. It is interesting to ask whether we would ever see any FRII sources if space were a perfect vacuum.

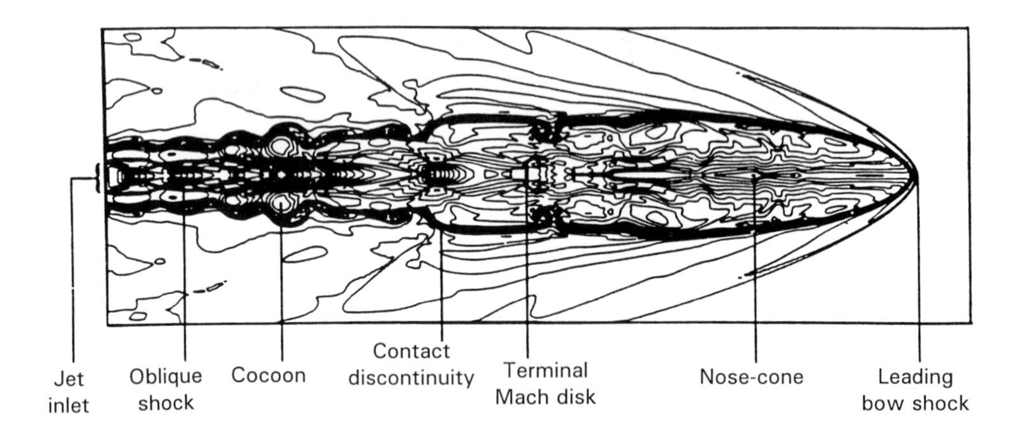

Fig. 8.3. Numerical simulation of a two-dimensional magnetically confined jet showing the relevant parameters. (From Clarke, Norman and Burns, *Astrophys.J.*, **311**, L63, 1986.)

It is tempting to suggest that the change from FRI to FRII could depend solely on the Mach number of the jet. If this were the case then the underlying cause might be the collimation and injection mechanism at the accretion disk interface. Something about the initial acceleration process produces different Mach number jets and hence we have FRI or FRII jets. We would expect therefore that the change from FRI to FRII jets would be directly dependent on the central engine. This might be because of differences in accretion disk or accretion rate, which might reflect gaseous conditions in the inner galaxy. Does this tell us something about the type of galaxy?

Optical studies of the parent galaxies show that the FRI and FRII classes separate out in the optical–radio luminosity plane. FRI galaxies have a much higher optical luminosity for a given radio luminosity. Another way of saying this is that the more optically luminous galaxies have weaker jets. One possible interpretation of this has far-reaching consequences for the jet and its production. It appears that for a fixed radio luminosity, which presumably translates to the power of the central engine energy generation, then as the optical power of the galaxy in the parent population is increased, a point is reached at which FRIIs switch to become FRIs. So what happens as the host galaxy becomes more luminous? It is possible that the more luminous galaxies (remember that for the FRIs we are thinking of the 'cD' type of galaxy in a cluster environment) have more central gas and at some point this has sufficient density to interact with the supersonic jet, decelerate it (say to a Mach number around 2) close to the nucleus before it can form a hot spot and hence we see a low velocity (FRI) jet source. This is an exciting idea and calculations of interactions and the required gas densities indicate that it has potential. In this picture, the jet production mechanism and central engine are unified for all sources, and it is the external factors of the closely surrounding medium which produce the observed differences between FRI and FRII sources. However, as we shall see later, this is not the only possibility.

In closing this section we should note that although it might be suspected that the radio lobes and jets are blasting out of a galaxy perpendicular to the plane of the galaxy, in fact there is no direct evidence to support this. The idea of a flattened system along with rotation naturally points to this simple concept; however, recent optical and radio studies strongly suggest that it is not borne out. The data show that the radio jets do not always lie along galaxy rotation axes, the normal paths of least resistance out of a galaxy. So we are left to wonder about the relation between the angular momentum of the central engine (determining the orientation of the accretion disk) and that of the surrounding galaxy. We shall return to this discussion in the final chapter.

8.2.4 Helical jets?

A further twist to the question of the opening angle of the jets forms a current topic of hot speculation. An intriguing statistical result from the first Caltech–Jodrell Bank VLBI survey of jet-sources was the so-called 'mis-directed jets' problem. When the relative angle between the directions of the VLBI parsec-scale jet and the kiloparsec-scale jet is plotted, it is found that although most sources cluster around zero degrees as might be expected (i.e. the two jets are perfectly aligned), there are a number of kiloparsec jets which are significantly mis-directed from their VLBI parsec scale progenitors. Furthermore, there is a significant clustering around an angle of ninety degrees, implying that the kiloparsec-scale jet has been bent through a right-angle in propagating from its parsec-scale core direction. Most pundits were content to attribute this to a fluke of small statistics (even though the numbers were not all that small), believing that with a larger sample the ninety-degree peak would disappear. Sure enough, the larger data set from the second survey revealed that the ninety-degree peak had disappeared. Most people breathed a sigh of relief. However, this was not the end of the story. When the degree of misalignment for the subset of those sources that have high optical polarization is plotted, the ninety-degree peak returns. The picture is definitely puzzling and requires a pause for thought.

If we assume that the distribution is not just an artefact of the statistical sampling, then a mystery remains to be solved: do the jets really bend through ninety degrees in travelling from parsec to kiloparsec distances? This seems highly unlikely and no satisfactory theory has been invented to explain it. However there is another intriguing possibility, the parsec-scale jet is travelling in a helical path, tracing out a spiral pattern in a much larger cone. Imagine drawing a pencil line around the inner part of an ice-cream cone from the vertex to the open end. The axis of the ice-cream large cone defines the true beaming direction, but at any time the pencil line makes an angle to this direction. Doppler boosting (see next section) then creates a preferential viewing angle for the helical jet and this turns out to be exactly ninety degrees. Hence the VLBI jet for these boosted sources does not reveal the true direction of the outflow and indeed, if we could monitor it for long enough it would trace out a helical path.

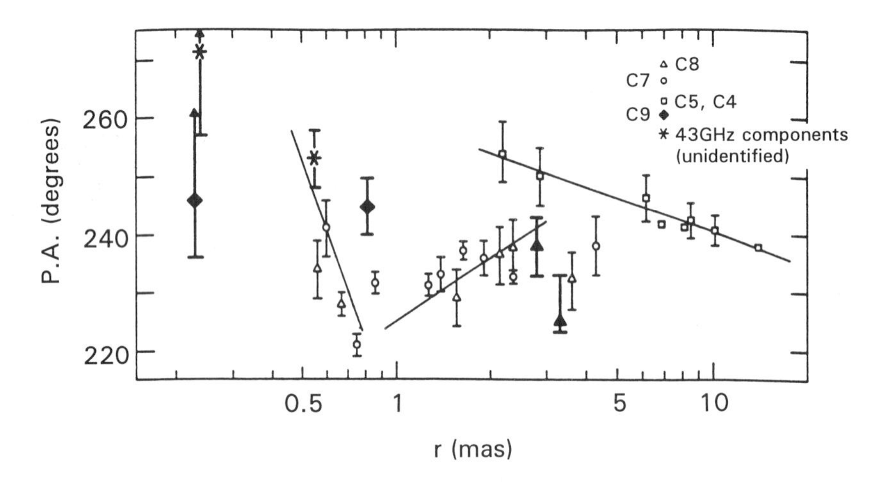

Fig. 8.4. The 'wiggle-cone' of the blobs of emission from millimetre VLBI observations of 3C273. Plotted is the position angle versus the core distance of the labelled components. (From Krichbaum, *et al., Astron.Astrophys.*, **237**, 3, 1990.)

There is some supporting evidence for this picture in that the blobs of emission observed by VLBI frequently do not follow a straight trajectory but rather, in the very first stages (of the order of parsecs), the jet appears to curve and even wiggle (or flap about). The locus of this wiggling traces out a much larger cone-angle as shown by the VLBI 'wiggle-cone' of 3C273 in fig. 8.4. We also know that in the case of 3C345 the blobs move in strongly curved trajectories (see fig. 8.7) and there is good evidence that they accelerate as they move down the jet. For 3C345 there is a monotonic increase with distance from the core, consistent with a model of a jet of constant Lorentz factor but which is bending *away from the* line-of-sight. The *apparent acceleration* away from the core might be crucial in terms of suggested explanations for the observed X-ray to radio 'problem' in the unification of BL Lacs and FRI sources (see section 8.4.2).

But we are now posed with the question of why a helical jet might form. Perhaps surprisingly, it turns out to be rather simple, although the choice of input parameters needs to be extremely carefully selected to explain the observations. Precession is an obvious candidate but we are then left with the question of what is causing precession for a supermassive black hole. The answer must be another black hole. Can we explain the existence of binary supermassive black holes? One possible answer lies in the theories of the formation of active galaxies and the supermassive black holes themselves. Galaxy mergers are believed to be common in the early Universe and calculations have shown that the possibilities of forming binary black holes are indeed reasonable.

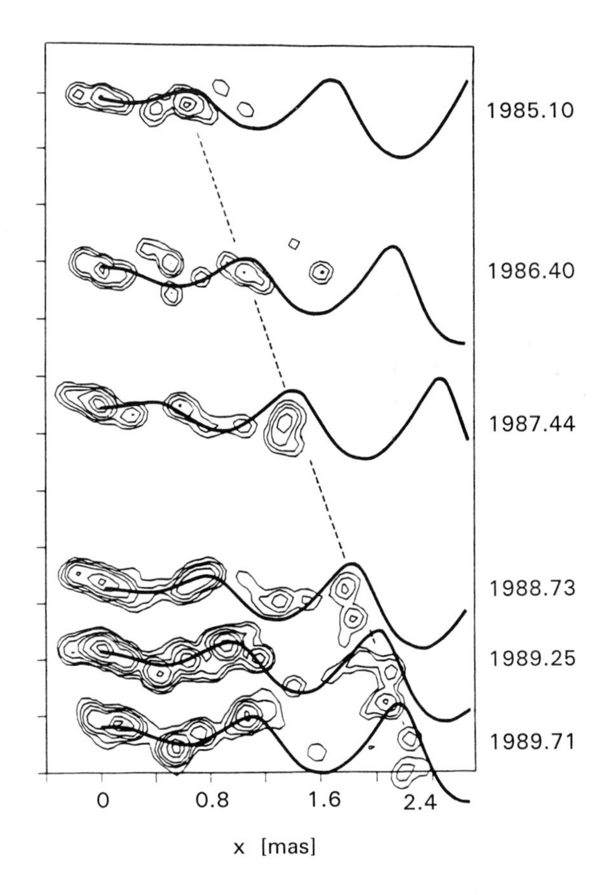

Fig. 8.5. A model of the 'sinusoidal' wiggle pattern claimed to be visible in the VLBI maps of the quasar 4C73.18. (From Roos, *et al.*, *Astrophys.J.*, **409**, 130, 1993.)

Is there any evidence to support the existence of binary black holes in active galaxy nuclei? Very little in fact, although we should remember from chapter 7 the HST and other data indicate a double peak in the brightness distribution which some have speculated might indicate that M31 possibly has two black holes in its nucleus. However, at the current time we are left with no direct evidence of binary supermassive black holes in galaxies. So we are left with indirect methods, and there have been a number of papers in the literature that claim that their data can be explained by such a phenomenon. Among these are the wiggles (fig 8.5) in the milli-arcsecond jet from a quasar (4C73.18). In this case calculations suggest a binary black hole with mass of order 10^8 $M_\odot$. Additional support is from claimed (although never conclusively substantiated) periodicities in the lightcurves of a small number of blazars. The most famous of these is the source OJ287. On the whole I tend to remain sceptical about claims of periodicities as the eye is very susceptible to finding periodic appearances in noisy observational lightcurves. Even some analysis methods (such as Fourier analysis) will home in and produce spurious periodicities if not used very carefully. We should note that binary black holes are not the only answer to helical motions. Perhaps a more mundane explanation is due to the effects of magnetic fields on blobs of material (see section 8.6) moving away from the central engine. Such models have been produced and do give a helical appearance.

We now leave the physics of the jets for the moment and turn our attention to the observational consequences of a relativistic jet pointing towards the line-of-sight. This is the realm of core dominated radio sources and the blazar phenomenon.

8.3 RELATIVISTIC BEAMING AND SUPERLUMINAL MOTION

8.3.1 Introduction

Relativistic beaming was introduced in section 3.5.3. We also saw in section 8.2 above that the lack of X-ray emission coinciding with radio brightness temperatures exceeding the 10^{12} K limit demand that relativistic beaming must be taking place.

Let us now take a more detailed look at the VLBI evidence for superluminal motion. We recall from chapter 3 that apparent motions in excess of the speed of light do not necessarily result in the motion of physical entities (such as radio components) at speeds greater than c. Let us expand on this. The evidence comes from observations of a number of extragalactic FRII radio sources by VLBI techniques and spanning a number of years. The parsec-scale jets in these sources are not smooth structures, but comprise blobs of emission. VLBI observations can identify these blobs and measure their separations from the core emission. When VLBI 'pictures' are taken over a number of years, the position and strength of the blobs are found to change. The blobs steadily move away from what is assumed to be the fixed 'core' and at irregular intervals new blobs appear in their place and in turn evolve away from the core. Evidence has now been obtained from millimetre monitoring campaigns linked with millimetre VLBI that the production of the new blobs of emission is linked with flares in the continuum energy output of the source. This is shown for the quasar 3C273 in fig. 8.6.

Before we can be totally satisfied with this overall concept, we really need to be certain that the core is a fixed entity so that we can know that the blobs are moving along the jet. One of the problems of VLBI is the astrometry with, for instance, the optical emission. Remember, an optical CCD image of a typical quasar reveals an unresolved blob the size of the seeing disk of the telescope facility, which might be 0.5

arcseconds. On the other hand VLBI investigates scale sizes of milli-arcseconds and less. So how do we know which features on a VLBI map are moving and which are stationary? The field of view of VLBI is very small and in virtually all cases there are no

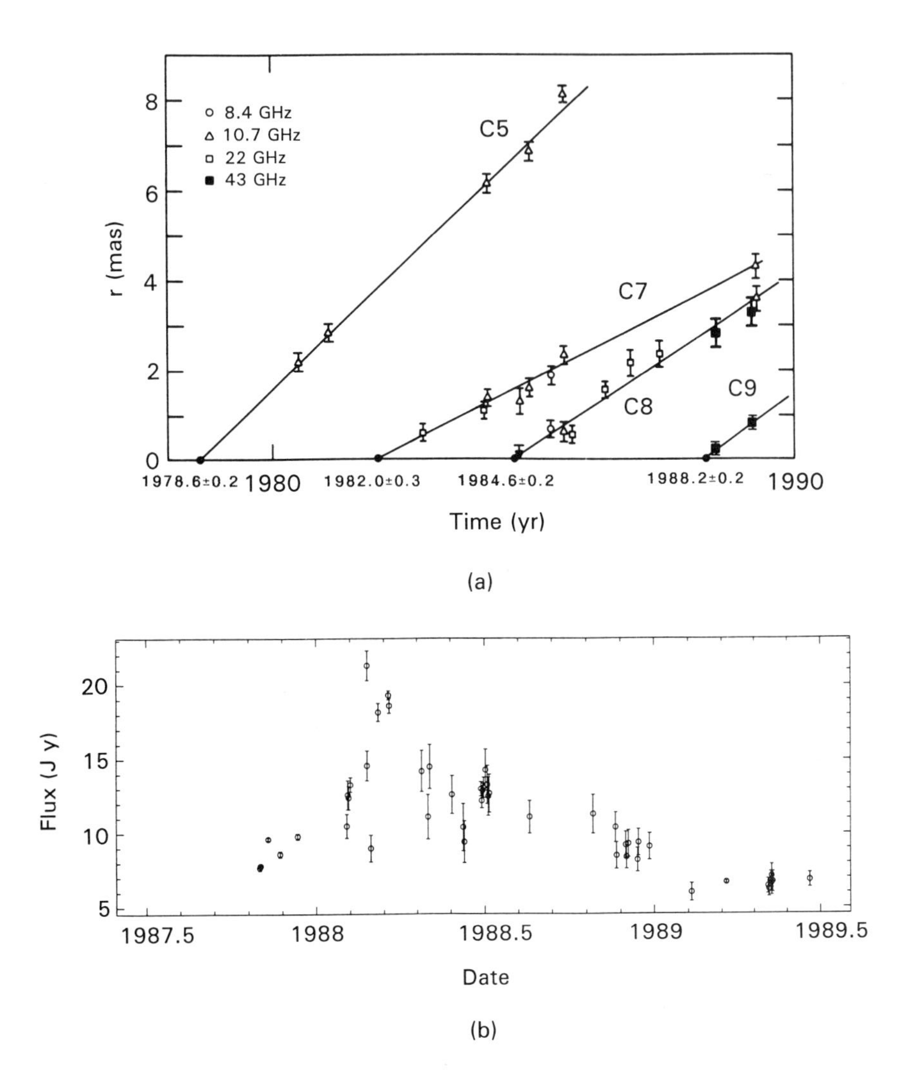

Fig. 8.6. (a) Data from millimetre VLBI of 3C273 showing the onset of a new blob and tracing it back to the time of its onset. (Krichbaum, *et al., Astron.Astrophys.*, **237**, 3, 1990.) (b) The 1.1 millimetre lightcurve showing the February 1988 flare associated with the new blob. (Robson, *et al., Mon.Not.R.Astron.Soc.*, **262**, 249, 1993.)

other sources of radio emission in the field. The main method of identification of the core component is the determination of the spectral indexes of the emission components, which requires two VLBI frequencies. The result is that the core tends to have a flat spectrum while the blobs have steep spectra. A distinct possibility is that the core is an optically thick photosphere or the innermost part of the jet.

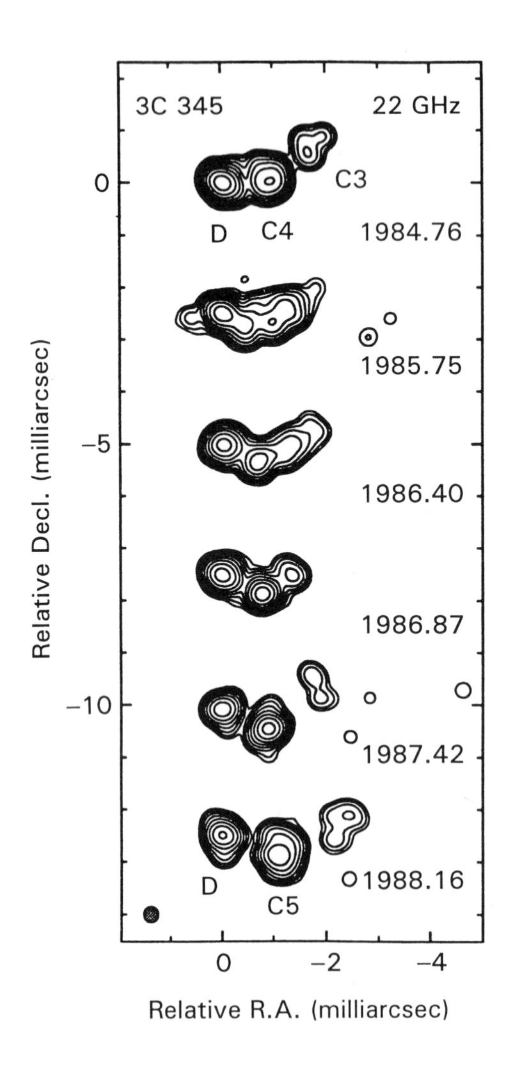

Fig. 8.7. Compilation of recent VLBI observations of 3C345 at 1.4 cm. The core is labelled D. Components C1 and C2 were observed at earlier epochs and have now expanded too far away from the core to be shown. The new component C5 appears in 1985.75 and gradually evolves outwards with an apparent speed varying between 4.8 and 10.6h. The jet ridge line as traced out by the superluminal blobs is strongly curved as indicated by the trajectory of C4. (From Zensus, *et al.*, *Astrophys.J.*, **443**, 35, 1995.)

Fortunately, for one source (the quasar 3C345, fig. 8.7), astrometry with respect to a nearby radio source has now firmly established that the core is in fact stationary and the blobs move outwards from this core. Therefore it is tacitly assumed that when we see cores and blobs the same is true for all other sources.

8.3.2 The geometry of superluminal motion

Let us now investigate the superluminal phenomenon of the parsec-scale jets in more detail. Consider 3C345, the latest high-frequency VLBI data are shown in fig. 8.7. This shows a separation of the blobs from the core seen *on the plane of the sky*. At this stage we have no information regarding the true expansion, we only see the projected effects. 3C345 has a redshift, $z = 0.595$, and using a Hubble Constant of $H_0 = 50$ km s^{-1} Mpc^{-1} a distance $D = 2{,}614$ Mpc is derived (from eqns. 1.6 and 1.3). The average angular separation of the blobs from the core as measured by VLBI over twenty years corresponds to 0.36×10^{-3} ″ y^{-1}. Translating this into seconds of time (1 y = 3.2×10^7 s) gives an angular separation velocity of 11.52 ″ s^{-1}. Because we know the distance of 3C345, this angular expansion can be translated directly into a distance. We elect to use the equation $x = r\,\theta$ (fig. 8.8(a)) where x and r are in the same units of distance and θ is in units of radians (1 arcsecond = 206,265 radians). So the angular expansion of the blobs is 5.45×10^{-17} rad s^{-1}, which translates into a linear expansion on the plane of the sky at the distance of 3C345 of 4.4×10^6 km s^{-1}. This is well in excess of the velocity of light (3×10^5 km s^{-1}), in fact it is 14.7c. This is the puzzle of superluminal motion; the blobs are apparently expanding away from the core at a velocity exceeding the speed of light. Changing the Hubble Constant does not help us out of this situation; even a value of 100 km s^{-1} Mpc^{-1} leads to an apparent expansion of 7.8c.

The solution is to be found in the geometry of the emission. Consider fig. 8.8(b), which shows the jet of blobs oriented at a small angle θ to the line-of-sight to the observer. Imagine a flare in the core emits radiation at a time t_0. This takes a time D/c to reach the observer. At the same time, a blob is ejected from the core and travels along the trajectory J with velocity v (close to c). A little time later, t_1, the blob is at the position shown. Radiation from this position reaches the observer only a short time after radiation emitted from the core at time t_0 because the blob is moving at almost the same velocity (c) as the radiation itself and in a direction almost towards the observer. When this geometry is projected onto the plane of the sky, the blob appears to be moving away from the core at a velocity greater than c. The two requirements to produce superluminal motion are (i) the blobs must be moving relativistically with velocities approaching the speed of light ($v \approx c$), (ii) the blob trajectory must be inclined at a small angle to the line-of-sight. These two parameters are related through the eqn. 8.1, which is relatively simple to deduce from geometrical arguments:

$$v_{sep} = \frac{v \sin\theta}{(1 - \frac{v}{c}\cos\theta)} \quad . \tag{8.1}$$

Using the standard format for $\beta = v/c$, eqn. 8.1 can also be written as

$$\beta_{sep} = \frac{\beta \sin\theta}{(1 - \beta \cos\theta)} \quad . \tag{8.2}$$

For example, if the plasma velocity $v = 0.9c$, and the angle to the line-of-sight $\theta = 10^{\circ}$ then the apparent separation velocity of the blob from the core $v_{sep} = 1.37c$; whereas for $v = 0.99c$ and $\theta = 5^{\circ}$, we have an apparent separation velocity of 6.27c.

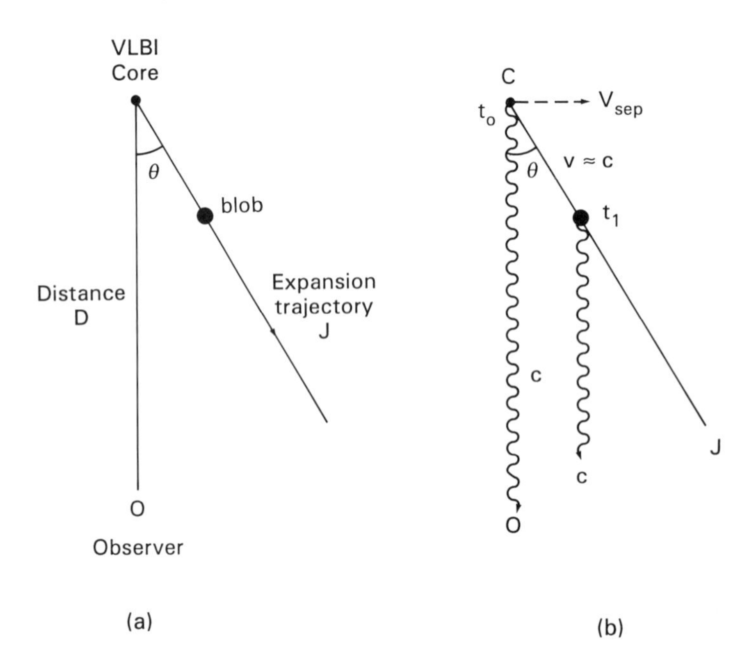

Fig. 8.8. (a) Projected apparent expansion of the radiating 'blob' from the VLBI core at an angle inclined towards the observer. (b) Showing the small difference in time between the arrival at the observer of radiation from the core and from the blob with the result that the blob appears to be separating from the core at a speed exceeding that of light.

8.3.3 Relativistic boosting

A further complication arises in the fact that because the blobs are moving relativistically, their output emission (flux density) is changed by relativistic time dilation. It turns out that for blobs of plasma moving towards the observer, the observed emission, S_{oi}, is boosted in energy over that emitted in the rest frame, S_{ei}, as shown in eqn. 8.3.

$$S_{oi} = S_{ei} \, [\gamma \, (1 - \beta \cos\theta)]^{-3} \, . \tag{8.3}$$

The Lorentz factor, γ, is given by $\gamma = (1-\beta^2)^{-1/2}$ (eqn. 4.22). Taking values of $\beta = 0.95$ and $\theta = 10^{\circ}$, then for blobs heading toward the observer the boosting factor is ~115. The flux density for the blobs receding from the observer is reduced by the same factor.

Therefore we see that the contrast between the approaching and receding blobs in the jets is large.

In fact eqn. 8.3 is an oversimplification in that it ignores the spectral energy distribution of the emitted radiation flux, which must also be taken into account in order to derive a precise value. In effect the flux densities in eqn. 8.3 are at different frequencies and there is a frequency shift to be taken into account. When this is included the exponent changes to $-(3-\alpha)$, where α is the spectral index of the radiation flux (which we define by $S_\nu \propto \nu^\alpha$). An equation relating the ratio, R, of the flux at a given frequency expected from the approaching and receding continuous jets can then be derived: this is given by

$$R = \left(\frac{(1+\beta\cos\theta)}{(1-\beta\cos\theta)}\right)^{(2-\alpha)} . \qquad (8.4)$$

Taking the standard value of $\alpha \approx -0.7$, we immediately see the drastic difference in brightness between the opposite jets. Using the same values as in our example above, i.e., $\beta = 0.95$ and $\theta = 10^\circ$, the ratio of the approaching to receding jets is $R \sim 10^4$.

Finally, we must remember a further implication of beamed radiation: the emission is no longer isotropic but beamed in a narrow cone angle—as in a lighthouse, or a pulsar. Therefore, if we calculate the luminosity of a beamed source based on the usual assumption of isotropic emission (using the equation relating flux to luminosity, eqn. 2.3), we always overestimate its true luminosity. For emission beamed into a cone of opening-angle 20°, the overestimate due to beaming alone (no relativistic boosting) is the ratio of 4π sr to the solid angle corresponding to the 20° cone. This is a factor of 130. The narrower the cone angle, the more the overestimate of the luminosity, and for very narrow beams, say 5°, the overestimate exceeds 2,000. Including the relativistic boosting, we thereby see that the beamed component can easily be overestimated in luminosity by factors of many hundreds to many thousands. This is a very important conclusion. A corollary to this is that the emission is heavily suppressed for angles outside the beaming cone, leading to the statistical arguments about numbers of sources as we shall see later. Furthermore, remember from section 4.3.1 that the cone angle is approximately given by γ^{-1}.

The very observation of superluminal motion demonstrates that some sources must be ejecting plasma with bulk relativistic velocities and furthermore that these sources are also oriented at a small angle to the line-of-sight. There is also a selection effect at work. Because of the small inclined angle and subsequent relativistic boosting in energy, these superluminal sources are also bright synchrotron emitters, especially at radio wavelengths. We will expand on this selection effect in section 8.4 and consider what the same source would look like were it not aligned close to the line-of-sight.

8.3.4 Luminosities, Compton drag and jet models

How many of these dramatically beamed sources are there? At the time of writing, forty-four sources have been observed to show superluminal motion; all are flat spectrum radio sources and the vast majority of these fall into the blazar category. The range of superluminal values is large, the highest to date being just less than 30c (assuming $H_0 = 50$ km s^{-1} Mpc^{-1}).

What does a relativistic jet look like? Nobody really knows, but some progress has been made in modelling their emission characteristics, particularly when applied to blazars. The current picture is that the stationary core of the VLBI maps is the point at which the jet becomes visible, at least in the radio. This point is located some distance (perhaps a parsec or so) from the actual accretion disk. It is fed by a pipeline or funnel from the central engine by a process that is not at all understood but is an area of much activity.

In terms of the jet, there are two broad possibilities for the inner zone: the jet is optically thin (and hence not visible) at all wavelengths, or it can be a strong emitter at, say X and γ wavelengths (see chapter 6). In the former case, the high-energy photon emission comes from a region in space very close to the VLBI core, whereas in the latter it comes from a region closer in—the funnel. This latter aspect has dramatic implications for unification scenarios (see sections 8.4 and 8.6) and requires the initial jet opening angle to be significantly larger than downstream from the VLBI core. This also tends to assume that the material in the jet is accelerated at the core. This presents a problem because as we shall see later, the best physical description for the radio emission from the jet is in terms of shocks within the jet. If the core of the jet is a standing shock front, which seems very reasonable for the shocked jet models, then this should be a region of deceleration rather than acceleration.

Clearly we need to know a lot more about the core and the inner funnel, and at present we are at an impasse. There are only two sources that are sufficiently close so that the core–inner jet can be resolved. One is M87 (see later) and the other is 3C84. This lies in the radio Seyfert galaxy NGC1275 in the Perseus cluster of galaxies and shows blazar-like properties and a jet–counter-jet (see section 8.4). VLBI mapping of this source shows that the core region is a confusing picture of individual blobs of emission, devoid of any general alignment as might have been expected. As such it unfortunately fails to provide us with a clear-cut answer, but points towards a more chaotic picture of multiple components of plasma emission in the inner jet rather than a smooth linear flow pattern.

Assuming that the VLBI radio core is where the material from the funnel delineates the beginning of the collimated jet, then from this point outwards, we are in the realm of the true jet itself. The usual modelling assumption is that the jet is not constrained but is free, flowing at a constant Lorentz factor and an opening angle of ϕ as shown in fig. 8.9. At the core, the values of the magnetic field and particle density are at their highest. For a free jet with constant opening angle, both the density of material and the parallel component of the magnetic field fall off as R^{-2}, while the perpendicular (or transverse) magnetic field component falls as R^{-1}. Correcting for the beaming and luminosity factors indicates that typical values of the synchrotron luminosity in the VLBI core in the rest frame of the source are of the order of 10^{38} W (give or take a factor of ten), which is gratifyingly less than the total luminosity of the source. This is an excellent reminder of two things: even though the radio emission appears spectacular, the synchrotron jet is a long way from dominating the luminosity; and second that beamed sources will dominate any flux limited (but otherwise unbiased) sample of sources.

The Lorentz factor is a crucial parameter of the relativistic jet. A number of arguments have suggested that if the jet is produced and focused close to the accretion disk, there is a natural upper limit to the Lorentz factor of $\gamma \leq 10$. This is due to what is rather descriptively termed Compton drag. Here the newly injected and ultra-relativistic particles at the origin of the jet are decelerated by inverse Compton interactions with the photons of the intense thermal radiation field of the hottest regions of the inner accretion disk. What is happening is that in their rest frame the electrons see the photons (which are isotropic in the rest frame of the quasar) approaching them head-on, which produces relativistic aberration. At its extreme this mechanism leads to the production of particle showers, starting with electron–positron pairs! Indeed, it also appears to be the case that once a value of $\gamma \sim 10$ is exceeded, the magnetic energy starts to dominate the particle energy and the synchrotron losses escalate dramatically. This then pushes up the energy requirements of the source and, as we shall see in section 8.4, it poses problems in having too many beamed sources for the supposed parent population. Nevertheless, at this point we note the apparent upper limit to γ and that theories of jet production must take this into account (see section 8.6).

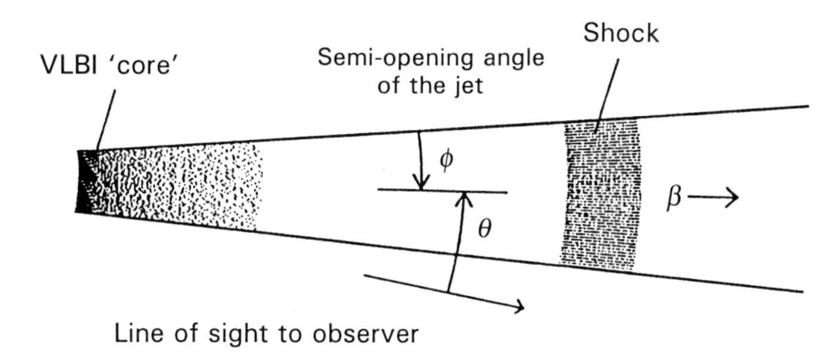

Fig. 8.9. Schematic of a shocked relativistic jet. In this picture, the flaring emission seen in the continuum synchrotron radiation along with new VLBI components originates in the shock rather than the non-shocked part of the jet.

So far we have been considering only a smooth jet that does not change over time. In that case, what are the blobs that appear on VLBI maps? In order to explain these features, new relativistic jet models were introduced which included shocks as origins of the VLBI blobs and the flaring behaviour observed in the output of blazars. The first of these was by Blandford and Königl in 1979, and the models have now been refined in a number of ways by a small group of workers. In most cases these models can explain qualitatively, and in some cases quantitatively, the temporal variations in the output emission at radio to X-ray wavelengths. The current 'best' model is that from Alan Marscher and his co-workers. The observations suggest that the timescale of the variability is a decreasing function of wavelength—the blazars vary more rapidly in the optical than the radio. The shock models explain this by postulating that the higher energy electrons responsible for the emission at the shorter wavelengths are confined to a thin zone next to the shock front while the lower energy electrons can propagate farther down the jet. When a shock forms, X-ray emission is seen first, followed by a rise in the millimetre–submillimetre part of the spectrum and finally a synchrotron decay phase when the submillimetre–radio emission declines. Multifrequency campaigns of blazar monitoring, such as noted in chapter 5, are carried out specifically to test these models.

Chapter 5 also introduced the recent discovery of gamma-ray emission from blazars and raised the question as from where the emission originates. Most models assume that the gamma-rays are inverse Compton scattered by the electrons in the jet; the question is from where do the seed photons arise. We know that the jet produces synchrotron emission, and that there is excellent evidence for X-ray production by the inverse Compton process due to these same electrons. Therefore it is a simple step to assume that the gamma-rays are produced by the same process, the synchrotron self-Compton (SSC) mechanism of the 'in-jet' scenario. On the other hand, as we saw in chapter 6, there will be an ample supply of photons surrounding the accretion disk. These could radiate the material in the jet and in so doing be upscattered by the bulk motion relativistic electrons. This is the 'external-jet' (or external inverse Compton) scenario. The photons could come from either the surface of the accretion disk, the extended halo of the accretion disk, or highly ionized plasma clouds in the super-hot corona.

At the moment the picture is not yet clear as to which mechanism is favoured. In chapter 6 we saw that there were general supporting explanations for the high-energy emission to derive from the very innermost parts of the jet. However, in the case of at least one blazar, 3C279, most observers lean towards the belief that it is the core region of the jet itself (rather than the funnel from the accretion disk) which is the source of the gamma-rays. This is based on two reasons: theoretical calculations predict quenching of gamma-rays in the funnel due to electron–positron pair production from the high photon number density of the X-ray radiation field; and secondly, the increased gamma-ray emission from 3C279 also coincided with a flare in the millimetre–submillimetre output emission (which is associated with the jet rather than the funnel).

With the new space-based high-energy observatories and ground-based facilities, such as the VLBA, the outlook for future advances in this field is very positive. Furthermore, much more work is now being done to ensure that ground and space-based multifrequency campaigns are well co-ordinated. In leaving this section we should refer the reader to some excellent reviews covering these topics. In particular, a detailed review of the physics of synchrotron radiation, the Compton catastrophe and their relations for jet sources is given by Phil Hughes and Lance Miller in *Beams and Jets in Astrophysics*.

8.4 BEAMING AND THE FIRST STEPS TO RADIO UNIFICATION MODELS

8.4.1 The orientation hypothesis

The beaming picture naturally explains, at least qualitatively, the one-sided nature of superluminal sources. What about the one-sided but non-superluminal sources, can these also be explained by the same picture? It is vital that we distinguish two factors at the outset, beaming effects due to relativistic motions and orientations and obscurations effects due solely to suppression of emission through orientation. It is also important to recognize that the determination of superluminal motion requires multi-epoch VLBI observations and the presence of radiating blobs at the epochs of measurement. Therefore there are undoubtedly many more superluminal sources than have actually been determined by observation.

For the beaming scenarios we conceptually have the overall picture of back-to-back relativistic jets ejected from an AGN core. The respective orientation to the observer can range from being in the plane of the sky to being viewed head-on. Where the jet presents a small angle to the line-of-sight to the observer, section 8.3 showed that the synchrotron emission from the jet moving towards the observer will be relativistically boosted while the receding jet will be correspondingly diminished. We can easily conceive of the case where it is boosted to such an extent that it dominates the emission from the rest of its host galaxy (such as the starlight or accretion disk emission). Such objects are not unknown to us by now; they are the powerful optically violent radio-loud quasars. Furthermore, depending on the precise orientation angle, there is also a distinct possibility that the source will show superluminal motion. There would also be an expectation that the sources would be entirely single-sided, due to the extreme contrast between the approaching and counter-jet.

Indeed, we can begin to paint a picture of the change of expected observable phenomenon with jet orientation. Starting with the jet pointing nearly towards the observer, we are more likely to discover superluminal single-sided radio objects, typically OVV quasars. As we increase the orientation angle so that the approaching jet is no longer pointing at such a small angle to the observer, we might expect superluminal motion to diminish until the point at which it is no longer seen but the object still appears as a single-sided core-jet source. The receding jet is still dimmed by relativistic effects to such an extent that it remains invisible to current detection techniques. However, as the angle to the line-of-sight increases further, the relativistic effects are reduced until the source appears as an unequal double-lobed radio source. Finally, for angles approaching 90° (i.e., lying roughly in the plane of the sky), the source appears as a true classical double with symmetrical lobes and jets.

The second effect we have to consider is the suppression of emission perhaps due to free–free absorption from intervening gas as was noted in section 8.2.1. This works in the same sense as Doppler boosting. For narrow viewing angles the difference in obscuration between the forward and receding jet is increased (being maximized for a source seen exactly head-on when no counter-jet is visible), whilst for jets in the plane of the sky the obscuration is equal and the jets look symmetrical. (This of course also assumes that the jets are aligned symmetrically with any symmetry axis of the parent galaxy to ensure equal passage of the jet through intervening gaseous material. We will subsequently see that this is not necessarily true).

We must further consider the effects of the molecular torus on the identification of the AGN (as we saw in section 6.5) particularly for the obscuration of BLR component and the radio core. In 1989 Peter Barthel brought together much of what was an on-going debate in the field and produced a picture in which all quasars are beamed towards the observer, while powerful radio galaxies are those sources that lie more in the plane of the sky and should not show superluminal motion. We will return to this general unification argument in chapter 9.

In the Barthel picture, all the jet-driven extragalactic radio sources have twin, back-to-back jets and the single-sided phenomenon is primarily caused by the orientation of the jets towards the observer, perhaps along with the bulk velocity of the jet itself. We obviously need to test this idea observationally and the only method is to revert to statistical tests. The orientation of jetted radio sources on the sky ought to be randomly distributed and from the known radio sources it should be fairly easy to calculate how many should be one-sided and superluminal sources. Unfortunately, this is much harder than it might first appear because of a number of complicating factors. Because an ejected blob of emission is a necessary but not sufficient condition to detect superluminal motion, sources need to eject blobs sufficiently often for them to be detected and subsequently monitored by VLBI in order to obtain the expansion velocity. Also, if there is a range of blob velocities, then this has a bearing on the degree of boosting for any particular blob and whether the same source always shows superluminal motion for all blob ejection. Finally, if the blobs slow down and become non-relativistic on the kiloparsec scale, then because VLBI observations are only possible for relatively bright sources, weaker doubles might have superluminal motion which is not currently measurable.

Nevertheless, bearing these caveats in mind we can make good progress. Consideration of jet physics and the orientation angle to the line-of-sight leads to interesting predictions about the relative values of apparent velocity for superluminal sources. For a randomly oriented sample of sources, less than 2% are predicted to show apparent superluminal velocities exceeding 10c, and no more than 8% greater than 5c. To obtain meaningful statistics requires a significantly larger sample of VLBI sources than is currently available, but nevertheless this points to a simple test for long-term studies. The detection of a large number of very fast ($\approx$ 5–10c) superluminals would pose severe problems for our hypothesis. But even here the game would not be up as there is a possible way out. This is by resorting to bent jets, i.e., jets that bend significantly as they move away from the core. The problem of helical jets and how to take this into account in the statistics clouds the picture even more.

A further difficulty in observationally testing the orientation scheme is to ask at what level we can differentiate a source with only a single jet from a double jet. This is back to the old story of the dynamic range of the radio maps. The dynamic range is the ability to detect very faint emission in the presence of a very strong source, or in other words, the ratio of the brightest to faintest flux levels in the map. Rearrangement of eqns. 8.3 and 8.4 show that the ratio of approaching to receding jet brightness $R \propto \gamma^5$. For any dynamic range on a radio map, the range of jet parameters at which a normal double-lobed source would appear only as a one-sided source can be calculated. For modest dynamic ranges, the answer is that for an orientation angle to the line-of-sight of around 30°, values of $\beta = v/c \geq 0.5$ are required.

Modern maps can have dynamic ranges in the thousands, and it is some of these observations that are pushing the twin-jet model to the very limit. For a number of jet sources that were firmly expected to show the counter-jet, none has yet been seen. The simplest explanation for this lack of counter-jet is to change the orientation of the source on the sky to produce a greater contrast between the approaching and receding jet. This then moves the orientation closer to the line-of-sight to the observer, which in turn drives the statistics dangerously close to requiring too many sources pointing directly towards us to be explained by a random orientation. Another way out of this dilemma is to resort to higher velocities in the jet. This is another uncomfortable, but perhaps necessary conclusion. A final possibility is to assume that the two jets are not equally bright (due to energy in the jet or the radiation mechanics of the jet).

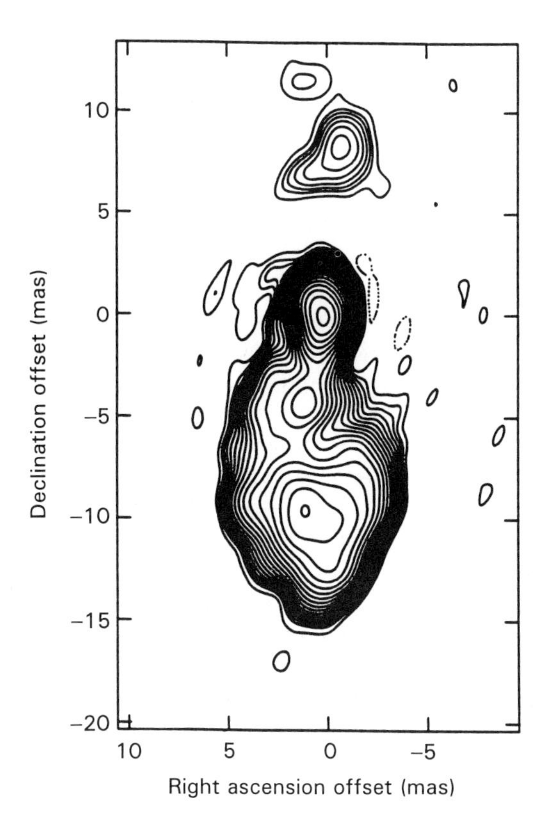

Fig. 8.10. Contour plot of the VLBA image of 3C84 at a wavelength of 3.6 cm. The core is located at a position 0,0 and the counter-jet lies about 8 milli-arcseconds north (top). The dynamic range is over 4,000. This is a persuasive diagram for unification pundits. The very bright forward beam is clearly unobscured as is expected, whereas as the gap between the core and the very faint counter-jet could be due to obscuration of the cm radio emission by free–free absorption from a molecular torus surrounding the central engine. (From Walker, Romney and Benson, *Astrophys.J.*, **430**, L45, 1994.)

Although this latter possibility seems to be true for one or two well-studied examples (see the case of M87 below), it cannot be the explanation for all the sources as it would always require the receding beamed jet to be the lower energy jet. This smacks of a major conspiracy as there should be no way that any source can know which jet should be preferentially brighter to suit our purposes. On the other hand, we should not be too downbeat about the counter-jet problem, because the long-sought-after counter-jet for the radio source Cygnus A has finally been detected (see chapter 9). Cygnus A has traditionally been the archetypal 'radio galaxy in the plane of the sky' source and so detecting the counter-jet is a very important result. Additionally, for 4C34.47, the single-jet quasar source with the largest extended radio emission, mild superluminal motion ($v_{sep} \sim 1$ to $\sim 2c$) has now been detected. This object should therefore be almost a unique quasar, being seen at the extreme angle of around 45° to the observer so that the large scale is still preserved and superluminal motion is also just apparent.

We therefore see that the testing of the simple hypothesis using statistical arguments is far from simple or conclusive. We can also revert to looking at individual sources and determine whether they support or oppose the model predictions. Our old friend 3C273 has been the prime target of detailed studies searching for the counter-jet. Dynamic ranges of 16,000 have now been obtained on this source, but the elusive counter-jet remains invisible. What does this imply for 3C273? At the VLBI wavelengths the spectral index is $\alpha = -0.8$, and so using eqn. 8.4 we obtain that $\beta\cos\theta \geq 0.91$. Although we do not know θ, the minimum condition of the true velocity for superluminal motion occurs when $\beta = \cos\theta$ and using this fact gives $\beta \geq 0.95$. When this is inserted into eqn. 8.1 it produces a superluminal motion requirement of 3.2c. The observed value for the superluminal motion of $\sim 12\, h^{-1}$ c ($H_0 = 50h$ km s^{-1} Mpc^{-1}) is sufficient to explain the lack of detection of the counter-jet.

Exciting data have been obtained from recent high dynamic range VLBI studies (including some of the first VLBA maps) which have revealed the first counter-jets. One of these is shown in fig 8.10 for 3C84 (in the Seyfert galaxy NGC1275). Even better data are now coming out of the VLBA but this early map (one of the very first taken with the VLBA) is sufficient to reveal the counter-jet. Based on spectral index information the picture is consistent with a mildly relativistic jet oriented between 30 and 50 degrees to the observer and the counter-jet is heavily obscured by free–free absorption from gas in what could be the molecular torus. So for 3C84 the previous need for a highly relativistic jet pointing almost towards the observer in order to explain the lack of detection of a counter-jet has been removed. Therefore superluminal motion is not required to explain the lack of all counter-jets, obscuration can suffice for some. Needless to say, both effects undoubtedly play a part in assembling the 'big picture'.

In terms of the overall number of radio sources, the radio sky is well surveyed down to a very low level and we are confident that there are enough sources already detected to ensure that there is no hidden population of extragalactic radio emitters that would confuse global statistics. Yet in being able to say for sure that the Barthel unification scheme is correct we eagerly await higher dynamic range maps of single-sided sources. The unification pundits expect these to reveal the counter-jets which have so far remained mostly elusive.

8.4.2 BL Lacs: the parent population analysis and FRIs

One of the ideas we met in section 3.6 was that the OVVs and BL Lacs making up the blazar category were demonstrating a phenomenon of relativistic jets. We will now look closely at the differences between the BL Lacs and the OVVs, are they the same or are they different? Although only about a dozen BL Lacs have both known redshifts and multi-epoch VLBI measurements, it is clear that the deduced speed measurements of the VLBI components is significantly less in BL Lacs than in quasars. Does this represent a difference in the jet-physics and/or the host galaxy or central engine? One of the old ideas was that BL Lacs were quasars viewed end-on, the beamed synchrotron emission dominating the underlying galaxy. Now that we have more of an idea of the central engine and broad-line regions, we suspect that this is far too simple. Indeed it is, and is not correct. We already have a very strong hunch that BL Lacs and OVV quasars arise from different objects, so this is a good place to start.

Let us consider the parent population of the objects that we are viewing preferentially as beamed sources. But what are they? The fundamental problem with answering this satisfactorily (in a statistical sense) is the need to take into account the redshift dimension (evolution) of the source population. This means we require to know how the population has evolved—i.e., the number of sources per unit volume of the Universe per luminosity range as a function of redshift. As we have seen in chapter 7, this is the luminosity function for the population in question and in order to determine it, deep and complete surveys at all wavelengths are required. Much work has been undertaken in looking at one class of object, seeking out the parent population and testing the result with respect to the beaming model and orientation. These are the BL Lac objects and we will now proceed to investigate their putative parent population.

In section 3.6.2 we saw that there are two types of BL Lacs, the radio and X-ray selected samples; the former we elected to refer to as 'classical' BL Lacs. [The reader will see that in astronomy 'classical' becomes a favourite word to describe the first sort of object found as soon as other and different examples are later discovered.] There are two ways to approach this problem. One can take the sample of observed classical BL Lacs, make assumptions about the beaming parameters and redshifts and hence determine their luminosity function, correct for beaming to determine the luminosity function of the unbeamed parent population, and finally test this prediction against observation. A second method uses the same principle but inverts the method by working backwards from a suspected parent population to arrive at the beamed population of BL Lacs and test against observation.

We know that BL Lacs are found in elliptical galaxies and an obvious starting question is to ask what other radio galaxies seem to fit the bill in terms of parent galaxy optical and radio luminosity. The answer turns out to be that FRI radio galaxies seem to hold out very good promise of being the same object as BL Lacs, but oriented at a sufficiently large angle to the line-of-sight to the observer that we see them as two-sided radio galaxies rather than BL Lac objects. This is a very fundamental proposition and therefore we need to investigate how well it stands up to tests. If it is correct, then what does it tell us about the physics of the jets?

Detailed studies have been undertaken on both the classical and X-ray selected BL Lacs, and although the determination of the luminosity function of FRI radio galaxies is far from simple, due to the limited number of sources, fig. 8.11 shows a compilation for the radio regime based on a sample containing 41 FRI sources. Because the lower two luminosity bins are very under-populated by FRIs (they contain only one or two galaxies), in order to obtain a good estimate of what the luminosity function might really look like, the authors added the elliptical galaxy radio luminosity function as it flows smoothly through the medium luminosity FRIs. Fitting two power-laws to the data shows strong evidence for the luminosity function falling into two distinct regions. It is clear that trying to work with luminosity functions based on small samples is always going to be difficult, but it is the best that can be done when the sample size is small.

The situation becomes more complex when, in seeking out the low-luminosity end of the luminosity function, we push the limits of what is a BL Lac. If we remember the basic definition of a BL Lac object (section 3.6.2) we note that a featureless optical spectrum is par for the course. As the luminosity of the radio (or X-ray) component declines, the strength of the active nucleus compared to the luminosity of the host galaxy will correspondingly decline rapidly. Hence at low luminosities it will be increasingly hard to determine whether the object is a BL Lac – or just an FRI radio galaxy with a bright nucleus. This is a very important statement and highlights the dangers of classification and the intrinsic selection effects which can take place. Because giant elliptical galaxies tend to have the same average luminosity, this places an immediate cut-off in the lower luminosity of the optical BL Lacs, which translates directly to a cut-off in the luminosity for radio and X-ray BL Lacs. Following this train of reasoning, the bottom line is that the number of BL Lacs is underestimated. Nevertheless, knowing this, steps can be taken to apply correction factors. Calculations predict a strong bias against identification of radio BL Lacs for redshifts $z > 0.04$. For the X-ray case, the situation is eased somewhat with the cut-off occurring around redshift $z < 0.4$. Observationally the only way out is to take the radio and X-ray data and then to undertake detailed optical-IR spectral-imaging studies to obtain more complete samples to increase the reliability of the sample base. In terms of analysis one can resort to more detailed statistical tests to attempt to counter the intrinsic biases.

With these difficulties in mind, let us now proceed to calculate the number density of FRIs. This is found by integrating under the luminosity function of fig. 8.11. Between the 5- GHz radio powers of 2.4×10^{20} to 6.3×10^{26} W Hz^{-1} the number density is of order 3×10^5 per cubic gigaparsec. We can now make use of eqn. 8.4 to work out the beaming ratio. However, a number of assumptions are hidden in this calculation, including the radio power of the underlying galaxy and the critical angle at which we see a beamed source. In its simplest treatment an estimate of the latter can be made from synchrotron theory (and seems to work well for pulsars) in that the size of the cone-angle of the beamed synchrotron radiation is linked to the Lorentz factor, γ, of the radiating electrons by γ^{-1}. But unless we know the Lorentz factors of the emitting blobs of plasma this is not very helpful. Unfortunately, determining γ is very difficult because in nearly all cases the beaming angle, θ, is unknown. Nevertheless, an estimate of the minimum value of θ can be made from the 11 BL Lacs which have been studied by VLBI (and all show superluminal motion). These lead to values of γ ranging from $\gamma > 2$ to $\gamma > 16$ (depending on the value of H_0) but with an average value tending towards smaller values. If we assume an 'average' value of γ then the predicted luminosity function of BL Lacs (the beamed FRIs) can be determined. The answer, perhaps amazingly, seems to agree with the observations. Detailed and sophisticated treatments have been applied to this concept and there is now general support for the hypothesis that BL Lacs are beamed FRIs. Nevertheless, it must be stressed that it depends on the relatively small samples of the original number count observations and we are not completely out of the woods yet.

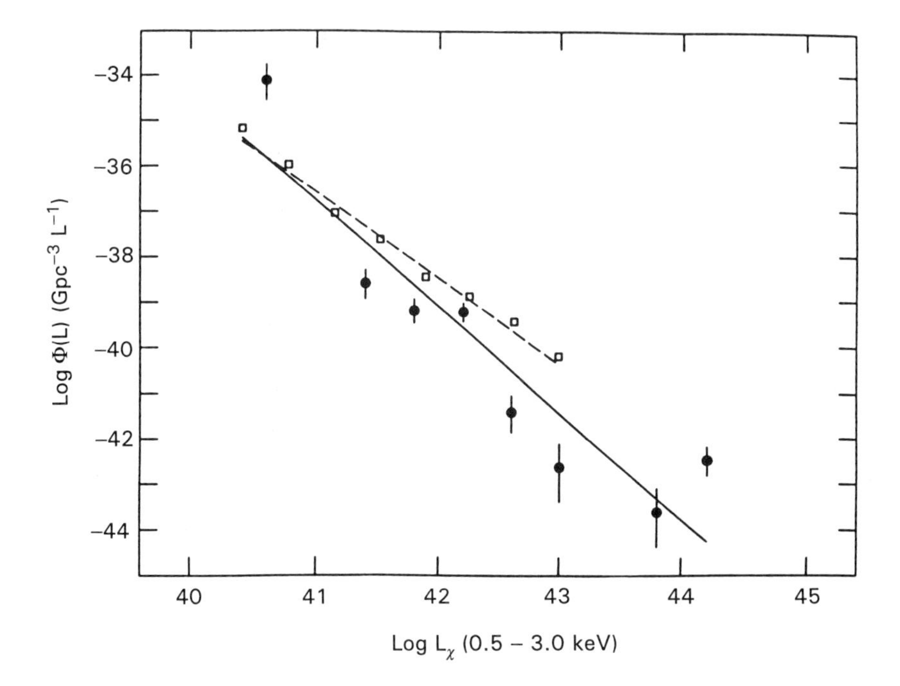

Fig. 8.11. Differential X-ray luminosity function for FRI radio galaxies. This shows the number (Φ) of FRI sources (filled points) per unit gigaparsec with luminosities, P, in a given range of luminosity, plotted as a function of luminosity. The line represents a power-law fit. Also shown for comparison (open squares) is the radio luminosity function for elliptical galaxies. (From Padovani & Urry, *Astrophys.J.*, **356**, 75, 1990.)

Observationally, further support for this type of hypothesis is gained from another approach. If BL Lacs are lobe-dominated radio galaxies seen more or less end-on, then the lobe emission should remain the same because it is not beamed and therefore unaffected by the orientation. A classic study in 1985 by Antonucci and Ulvestad paved the way. They obtained very deep radio maps of a significant number of blazars that showed that the 'core-dominated' and supposed 'point-like' radio sources did, in fact, show extended, albeit faint, halo emission. Furthermore, the OVV quasars in the sample had large radio luminosities in their extended components. This supported the idea that the FRII radio galaxies (although at the time the FRI and FRII terminology had not well permeated the language of the discipline) might be the parent population of the OVVs. This was at a time when the idea of a unified picture based on jets and beaming was just beginning to catch on. For the student of the subject, their paper is well worth a read. They were very convinced by the orientation arguments and since that time, the number of converts has grown.

We can take this analysis further and look at the optical and X-ray emission from BL Lacs in the same way. This allows us to make even more progress because once we believe we know the parent population luminosity function, we can then determine jet parameters and specifically the Lorentz factor, γ. Depressingly, it turns out that the luminosity function is very hard to determine in the optical because there are no complete samples available. An estimate of the luminosity function can be made from the number of known optical BL Lacs and the lack of BL Lacs on deep sky survey plate searches. Perhaps surprisingly, the X-ray luminosity function is better determined but the results from this analysis are relatively controversial or at least unexpected. They seem to point to the requirement that the Lorentz factors producing the X-ray emission from the BL Lacs are significantly less than for the radio emission. In fact only a narrow range of γ's (i.e. $\gamma \approx 2$–5) is required to explain the X-ray emission, compared with the large distribution ($8 < \gamma < 20$) to explain the radio sample. What does this mean?

The simplest explanation is that the cone angle for the radio emission is much less than that for the X-rays (which also explains the much higher number density of X-ray BL Lacs than radio BL Lacs). This means that the material producing the radio emission has higher boosting factors than that for the X-ray. We tend to assume that the higher energy (X-ray) photons come from the innermost parts of the jet, while the radio photons come from farther out. Hence because the radio emission is produced by material moving with higher velocities we reach the conclusion that the material is accelerated down the jet. This is surprising and is not supported by most models of jet physics, although some do predict possible acceleration. Nevertheless, this conclusion is controversial and still remains to be tested by further observations.

There are a number of ways out of the accelerating jet hypothesis. The most obvious solution is for the X-rays not to be produced in the jet. Another way out is that the X-rays are not produced by the same synchrotron process as the radio photons but are inverse Compton scattered from regions farther down the jet. However, both of these explanations have problems in that the continuum spectrum seems smoothly continuous from the radio to the X-ray, strongly suggesting a single and connected component. Perhaps the X-rays come from the funnel region and some form of re-collimation produces the narrow- radio jet? This sounds reasonable. Perhaps the acceleration occurs in the pinch at the core of the jet – but then we hit the problem of the standing shock which is a deceleration for the material flowing through it. A final thought, to which we will return later, is to ask whether we are measuring the speed of the jet when this is derived from VLBI multi-epoch observations, although this tends to push things in the wrong direction to help our puzzle.

Clearly, there is still some way to go to complete this picture, but one gets the feeling that it is all starting to come together, from the accretion process of the central engine, via the high-energy processes of the funnel, to the collimation of the jet, the jet itself, the beaming aspect and hence to the parent population. In spite of the limitations of the samples and some of the surprising conclusions, most pundits in the field are now quite comfortable with the belief that classical BL Lacs are FRI radio galaxies pointed closely towards the observer, while the X-ray selected BL Lacs are oriented at much larger angles, some even approaching 90 degrees.

In looking at this general separation of jet populations we should note another facet, that of polarization. Very difficult to undertake measurements (but which the VLBA should revolutionize) has shown that there is a distinct difference between the OVV quasars and the BL Lacs in terms of the polarization of the blobs. The VLBI polarizations for the BL Lacs are aligned along the VLBI jet direction, indicating that the magnetic field is perpendicular to the jet direction. An obvious explanation for a strong transverse magnetic field is that there are strong shocks in the jet and these compress and enhance the transverse component. For OVV quasars the fields lie along the jet direction, which tends to support the idea that this represents the direction of the underlying jet magnetic field in the absence of strong shocks. This is a very recent field and the observations are still in their infancy. Nevertheless, already a number of fascinating results have been produced. In a small number of notional BL Lacs, the field is found to have an appreciable longitudinal direction. Sure enough, for one of these sources, 1308+326, closer inspection of a range of other data reveals it is probably a mis-directed quasar rather than a BL Lac and for a second, 2007+777, there is the suggestion that it is an FRII source. This is helpful because this object has a higher jet-speed than the 'typical' BL Lac.

A second area of this research, and closely linked with millimetre and submillimetre monitoring and polarization is that for a couple of sources it has been found that the core polarization can change dramatically in terms of strength and position angle. Again, these are early days, but there is a possibility that when a new component is forming, due to a strong shock in the jet, the polarization is preferentially along the jet. For OVV quasars it has been found (albeit for only two objects so far) that the millimetre polarization can actually decrease during a flare. This also fits in with the shock model because the underlying jet field is parallel to the jet and the shock introduces a transverse component which thereby reduces the overall level of polarization. So shocks seem to be a common factor to all jets, even though the jets themselves are different. However, a note of caution. When we talk of the speed of a jet as measured by multi-epoch VLBI, we actually measure the speed of the shock evolution along the jet. This is more correctly referred to as the *pattern speed*. (We use the term speed because we generally do not know the angle the jet is making to the *line-of-sight*.) There is no *a priori* reason why the pattern speed should be the same as that of the underlying bulk jet and there is observational evidence to suggest that for at least one object the pattern speed is less than the underlying flow. Given that BL Lacs show a much higher degree of variability than OVV quasars and have generally tangled magnetic fields rather than strong longitudinal fields, perhaps the shocks form much easier in their jets and the jets are therefore shock dominated. Perhaps the quasar jets are much less shock dominated because shocks are harder to form in the strong fluid flow of the jet.

Have we now grasped the Holy Grail of unification? No, and indeed our discussion of BL Lacs has posed almost as many questions as answers. Nevertheless, it has pointed the way forward and it has demonstrated the methodology by which all unification scenarios are tested. In summary we can say that we have now obtained a consistent picture in which BL Lacs and FRIs are elliptical galaxies with a supermassive black hole at their centres, but showing very little evidence for a broad line region of highly excited gas. The jets are comparatively weak and have tangled magnetic fields, or at least a small longitudinal component, and because the flow in the jet is relatively low (compared with quasar-FRIIs), transverse shocks are easily formed which explains the frequent flaring seen in these objects and the predominant transverse magnetic field found in the jets. What causes the difference in the jets is still an unanswered question but we will return to this topic in chapter 9. Now let us turn our attention to the FRII radio sources, the OVV quasars and the steep spectrum radio sources. For the reasons described above, these sources have different jet-physics and probably represent different conditions in the central engine.

8.4.3 The FRIIs and quasars

In what is now regarded as a classic study, Peter Barthel in 1989 brought together into a single paper many of the thoughts that were in vogue at the time. He made some very clear statements concerning unification, beaming and the presence of obscuring molecular tori in the nuclear regions of AGNs. For the moment we shall concentrate on the beaming part only, reserving the presence of lines and obscuring tori to the final chapter. Barthel pointed out the great attractions of unified models based on beamed radio emission, but highlighted and then tackled the two main problems posed for the models. The first was that a number of quasars which were believed to lie close to the line-of-sight to the observer (because they showed superluminal motion) possessed extremely large projected linear dimensions (which was unexpected as the dimensions should appear small). This culminated in the finding (also by Barthel) of superluminal motion in the core of the quasar with the largest dimensions on the plane of the sky. The second problem arose in the degree of length asymmetry between the jet and counter-jet, the difference being too high to reconcile with most models of randomly oriented sources.

In answering the problems he proposed that the parents of the radio-loud quasar population are the FRII radio galaxies and argued that all radio quasars are indeed beamed towards the observer at some angle. If they happen to be oriented towards the observer within an angle of anything between 40° to 50° and directly along the line-of-sight (Barthel quotes a very precise value of 44.4° from statistical arguments but it is not to be taken literally) then they are quasars. Hence, because no quasar is ever seen close to the plane of the sky, the jet and counter-jets are never expected to be equal due to the relativistic effects. Also the size problem disappears, and the de-projected quasars fit in well with the sizes of the class of FRII radio galaxies—the quasars in the plane of the sky. This paper brought together the work of many astronomers and was seen to be a turning point in terms of radio unification scenarios. This overall picture is now widely accepted as being basically correct in principle. Therefore we will assume that the radio-loud quasars and FRII sources are united and comprise powerful elliptical galaxies producing strong and fast jets which are characterised by longitudinal magnetic fields and in which shock flaring is less frequent than for the BL Lacs. Why some galaxies make FRIs and why some make FRIIs will be addressed in the final chapter when we also look at the question of why powerful jets are a relatively rare phenomenon in AGNs. Before we take our first look at mechanisms and processes for forming the jet we will illustrate what can be learned about jets and the jet environment by examining a particularly well studied source, the nearby FRI source M87.

8.5 A SPECTACULAR JET SOURCE—M87 (VIRGO A, 3C274)

M87 is a giant elliptical galaxy with a mass around 10^{12} $M_{\odot}$. It lies at a distance of ~16 Mpc (plus or minus about 2 Mpc), close to the centre of the Virgo cluster of galaxies. We have met M87 in a number of places in this book, particularly in chapter 6 when we discussed the evidence for the presence of supermassive black holes. We will now concentrate on another claim to fame for M87, the presence of a spectacular synchrotron jet blasting out from the nucleus. Taken on the largest scale, M87 is a FRI radio galaxy, immersed in a giant halo extending about 30 kpc (7 arcminutes) from the centre in a mostly NE–SW direction but with a strong eastern plume shown in fig. 8.12. This is fed by a single-sided jet (fig. 8.13) which appears to begin to flap around just before it ends and opens out into the amazing plume that sweeps back on itself to a position 180 degrees away from its start. The full east–west extent of this emission is about 6 kpc, the bright jet being about 1.6 kpc long.

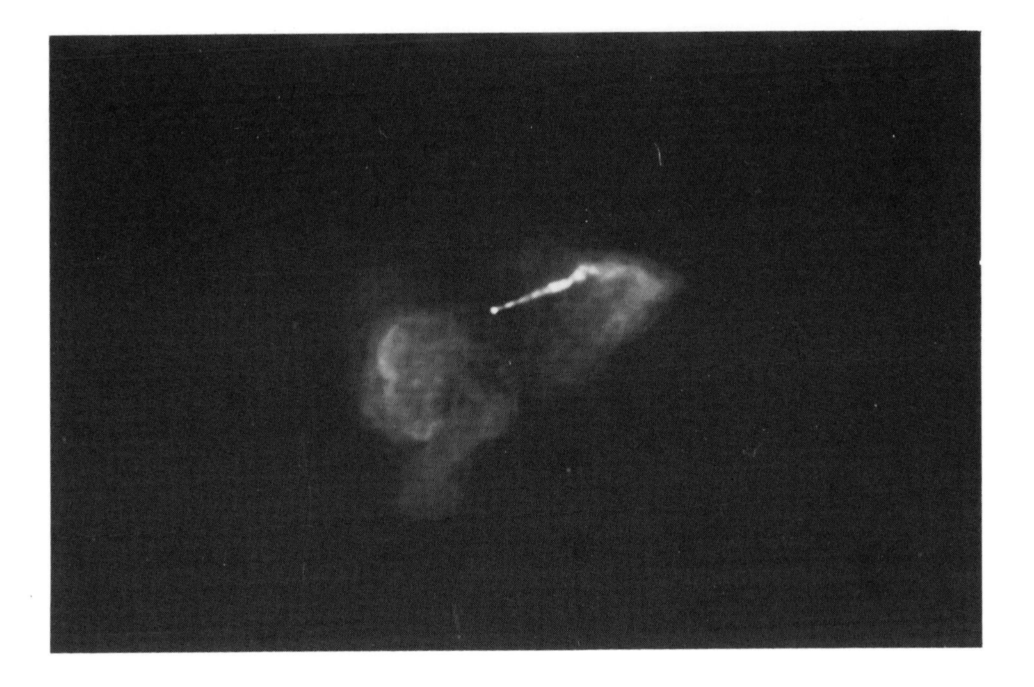

Fig. 8.12. A 6-cm VLA map showing the spectacular radio jet and plume of radio emission from M87. (Courtesy of NRAO/AUI.)

The radio jet has been extensively observed at all size scales; indeed the jet of M87 is the best observed of all jets. It is subluminal, with expansion velocities ranging from 0.3c on the parsec scales to at least 0.7c on the kiloparsec scales. It is highly collimated from the closest distance to the core of about 0.1 pc to its outer end at about 1.6 kpc. Indeed

VLBI observations show evidence for alignment (although not good evidence for collimation into a narrow beam) of blobs on scales as small as 0.01 pc. This is just over 2,000 AU. These VLBI maps also allow the opening angle of the initial jet to be determined, the result being a value of less than 10°. As well as showing knots, the innermost jet is clearly edge-brightened in some regions, although when this occurs both edges are not always equally bright. Moving away from the core, the north and south edges brighten alternatively with centre-brightened regions also being visible. Even on these scales the behaviour is seen to be complex and some interpretations have resorted to a helical jet structure. The inner jet can be traced out to around 14 pc from the core and over this range it has a width of just less than a parsec. The overall appearance is that of a filamentary structure.

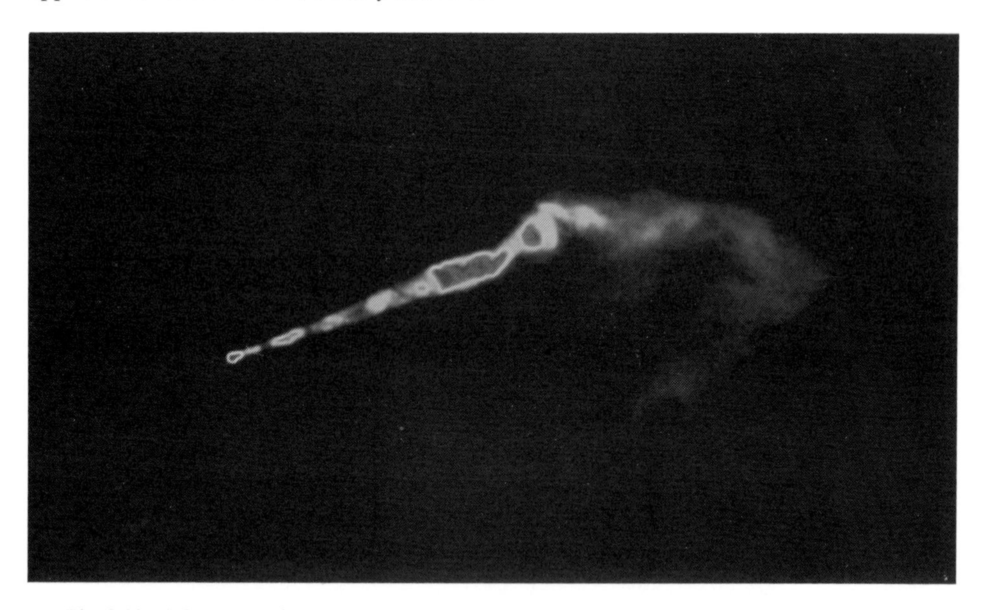

Fig. 8.13. A 2-cm VLA image of the M87 jet. (Courtesy of NRAO/AUI.)

Moving to the VLA (and HST) images of fig. 8.14 we see remarkable and excellent agreement between the two. Indeed even the polarization seems to map on a one-to-one basis between the two wavelength regimes. The VLBI filamentary structures are present but tend to merge into the knotty structure which is very apparent on both images. The overall direction of the jet remains remarkably straight, with no sign of bending and the opening angle of this part of the jet is just less then 7°. Nevertheless, when the structure of the jet is examined in more detail, the most striking features are the knots. The jet appears to show evidence of bending following each of these knots and indeed, the features seen on these scales have also been interpreted as being due to a helical jet. (The reader might be forgiven for thinking that when in doubt and a jet is not perfectly straight, resort is made to a helical jet.)

Although the images are spectacular we must remember that our aim is to understand the physical processes at work. What are the knots? Examined in more detail in both the radio and optical the knots are clearly resolved and possess structure, varying from circular to elliptical emission. Two knots in particular show a slab-like shape, oriented at a slant angle to the jet axis. The radio spectral index of the synchrotron emission from the jet is nearly constant at about $\alpha = -0.6$, which might invoke us to wonder why the electrons do not show signs of decay along the jet. Does this point to a lossless process or re-acceleration? This lack of variation in the spectral index is mirrored in the optical-UV, where a value of $\alpha = -0.6$ in the inner kiloparsec changes smoothly to -0.8 beyond about 2 kpc. Perhaps surprisingly there are no major changes seen in the optical spectral index between the knot and inter-knot regions; however, the knots are regions of local maxima.

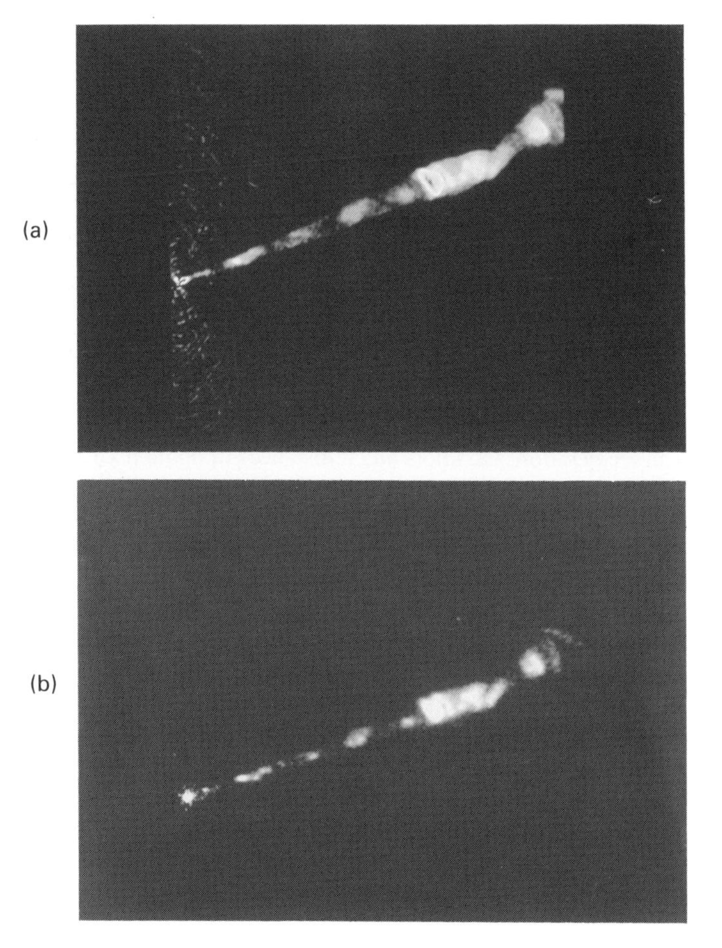

Fig. 8.14. (a) A grey-scale image of the high resolution and high dynamic range 2-cm VLA map of the jet of M87. (From Owen, Hardee and Cornwell, *Astrophys.J.*, **340**, 698, 1989.) (b) HST UV image. (From Boksenberg, *et al.*, *Astron.Astrophys.*, **261**, 393, 1992.)

The polarization data are striking and although not shown here, the reader is referred to the excellent reviews and figures of M87 in Röser and Meisenheimer's *Jets in*

Extragalactic Radio Sources. The radio polarization reveals that the magnetic field is parallel to the jet throughout its 1.6 kpc length. The exception is in the two brightest knots toward the end of the jet, where the field is perpendicular to the jet axis and indeed seems to follow the transverse slab-like shape seen in both these knots. On the scale of the finest detail, the parallel component is found to follow the lines of the filamentary structure, being tilted at various small angles to the axis of the jet. Furthermore, in two prominent twists in the emission within the jet, the magnetic field is found to match faithfully the emission morphology.

We now turn to one of the most critical aspects of the observations of the jet, the synchrotron emitting electrons and the physical processes taking place. The striking similarity between the radio and optical images poses an immediate question. Remember that the emission we are seeing is due to synchrotron radiating electrons, and we know from eqn. 4.42 that there should be a large difference between the lifetimes of electrons radiating at radio (2 cm) and optical wavelengths. Application of eqn. 4.24 tells us that the electrons must have energies differing by a factor of 200, and this translates through eqn. 4.42 into lifetimes that differ by the same value. So at first sight we would not expect to see such very similar features; we would predict that the radio emission should appear smoother due to the electrons lasting longer and so travelling farther and smearing out the finer detail. Evidently this is not the case.

To progress further we need some idea of the actual lifetimes. These depend on the magnetic field strength as given by eqn. 4.42. The assumption of equipartition leads to an estimate of the magnetic field in the jet of around 30 nT. Although this is not a strong field, it nevertheless leads to synchrotron lifetimes for the radio emitting electrons of less than 30 years. This immediately demands that the electrons must have undergone local re-acceleration; the knots are at distances of about a kiloparsec from the nucleus, and even assuming velocities of the speed of light gives travel times of thousands of years at the very least. We note that the knots are also a much shorter distance apart than this and so either the magnetic field must fall rapidly outside the shock-acceleration region (the knot) or we must think again.

Are there ways out of this dilemma? Possibilities have fallen into three general categories. The first resorts to lowering the magnetic field values by invoking a hollow-field jet, while the second seeks to use regions of highly relativistic flow and hence invoke relativistic beaming along with energy and time reductions. Neither of these categories have managed to explain the observations, and so for the jet of M87 we are left with the third and only remaining possibility, that the electrons are accelerated throughout the jet length.

The problem for this model is to explain why there is so little variation of the radiation spectral index along the jet, and even through the knots. The knots are obvious places where re-accelerations of the electron population are probably taking place. The observations that the radio spectral index is roughly the same in the knot and inter-knot zones strongly suggest that not only is the same physical process of acceleration taking place in both places, but the power-law electron distribution retains some fundamental form, or at least the acceleration produces the same output spectral form. The physical process must be precisely the same throughout the jet.

Indeed, the puzzle increases when HST observations of other jets sources are taken into account. On a larger scale (because the sources are farther away) the physics again seems to be the same along any given jet, but individual jets are very different. So whatever the mechanism, it is not a universally unique process giving the same output spectrum at all times. Instead it seems to depend on either the galaxy in which it occurs, or the central engine that generates it. Astronomers are eagerly awaiting the results from the refurbished HST to provide even more spectacular images for study. Clearly with these matched new high resolution multi-wavelength images the challenge will have been laid down to the numerical modellers. The next step for them may be to incorporate magnetic fields within their frameworks in the requirement to explain the detailed features of the amazing jet of M87.

Because the entire radio structure is so well studied, the lack of a counter-jet is well established for this source. Exciting new results have come from the optical on the other hand. These revealed a compact synchrotron emitting blob of plasma located 24 arcseconds to the SE side of the nucleus, in a position identical to that of a region of bright radio emission. Because this is exactly symmetrical to the NW extending jet and it is produced by synchrotron emission, it would seem reasonable to assume that its origin lies in an unseen jet. The observations of the blob of radio plasma were not a clinching argument because this still allowed a flip-flop jet argument to hold. But because the radiative lifetime in the optical is less than 1,000 y, then the idea of a flip-flop jet requires the last flop to have occurred within this time period, otherwise the optical feature would have decayed. This smacks of being far too contrived, and jet-feeding is a much more appealing concept. This then begs the question of why we do not see the counter-jet. Is it because of relativistic beaming? Dynamic range arguments require a jet orientation of less than around 30° to 40° to the line-of-sight with a corresponding bulk particle velocity of $> 0.8c$. At first sight this looks difficult to sustain because the velocities observed in the jet are $< 0.5c$. But these are for the inner jet and the velocities in the outer jet are much higher which still allows the jet in M87 to be oriented generally more towards the observer than 45 degrees.

We are therefore left with the conclusion (ignoring the flip-flop argument) that M87 has symmetrically spatial, back-to-back, twin jets, but in radiation emission they are very asymmetric. This is due either to differences of intrinsic power, or to how the jets interact with the surrounding medium, or to their structures. Interestingly, more than one radio astronomer has commented in published works that the NW side of M87 is a typical FRI source, whereas the SE is a FRII. With this tantalizing thought we will leave further discussions of the jets and lobes of M87 until the final chapter.

8.6 FORMATION OF THE JET

VLBI observations of M87 show that the jet originates only 0.01 pc (2000 AU) from the stationary core. With this in mind we conclude this chapter with a brief overview of how the jets might be formed. This will extend our discussions of accretion disks from section 7.5.2. As we have gathered, the mechanism by which the jet is produced is a hot topic of theoretical research, and X- and gamma-ray observations are very important in deciding which of the competing models might provide the best description. Again it should be pointed out that the observed width of the VLBI jet in M87 is just less than a parsec, but this is still tens or hundreds of thousand times larger than the size of the accretion disk. Indeed, it should be borne in mind that there is no real evidence for the collimation of a jet into a narrow cone until we get to VLBI scales, which is, for most sources, addressing parsec length scales. This is a crucial point to bear in mind.

We can make a start by asking about the composition of the jet. So far we have only discussed electrons. What else might be present? Presumably the jet material originates in matter from the accretion disk which is normal material, albeit in a plasma (ionized) form. The possibility of electron–positron components in the funnel zone has been a subject of intense speculation over recent years, although no direct evidence for the positron existence has yet been found from observation of a gamma-ray annihilation line (but see section 8.5). Nevertheless, the jet must remain neutral and so ions (e.g., protons or positrons) must also be present in the material. In any event, by some process as yet undetermined, material from the accretion disk has to be funnelled into an outward directed plasma flow, which at some point becomes collimated into the narrow beams observed in the radio. Making life even harder for the theoreticians is the knowledge that in some cases this plasma flow is relativistic. Magnetic fields provide the clue and current modelling now looks towards complex and detailed magneto-hydrodynamic solutions, including special relativistic treatments where possible. This is probably one of the most challenging areas of current theoretical astrophysical research and the reader is referred to the review by Max Camenzind in *Jets in Extragalactic Radio Sources* for detailed discussion and references.

The thick accretion disk model provides an obvious mechanism for funnelling of the stream as shown in fig. 7.2(b). Remember that thick disks are formed either from high accretion rates in the case of radiation pressure supported tori, or very low accretion rates for ion tori. The actual process of jet production is too complex for this book, but broadly speaking it consists of turbulence in the funnel zone between the accretion disk and the Schwarzschild radius. There is a corresponding poloidal pressure gradient which ultimately forces the plasma to be ejected along the polar axis. The inner walls of the accretion disk provide a natural boundary, defining an initial wide-angle collimation of the order of 30°. As this is much greater than the opening angle of the thin VLBI jets, secondary collimation is required. One of the methods by which this might occur is to make use of a magnetic field and some form of magnetic pinch, focusing the beam before it becomes completely free and blasts out into the galaxy. This focusing might be at the 'core' seen in the VLBI maps, sometimes called the throat of the jet. However, even here we should remember the very different size scales. The hot and radiating parts of an accretion disk for a supermassive black hole are of order 10–40 AU, while the VLBI core components seen in the jet of M87 have a width of around 1 pc. We again stress that getting from the accretion disk to the parsec scale with some form of secondary collimation is far from simple.

In terms of accretion disk models, an important general conclusion from the modelling studies is that radiation pressure supported disks cannot produce outflows which are relativistic, nor can they provide narrow collimation unless the radiating luminosity is in some way super-Eddington. This limits their usefulness for AGNs unless ways are found to support super-Eddington luminosity for adequate times (see section 8.4). Recent work suggests that the thin accretion disk is a better bet for providing well collimated and relativistic jets. We saw that the thick torus provides a natural medium angle collimation, but how does a thin disk fare better? The situation is even more complex for thin disks as the material forming the jet derives from the wind emitted by the inner surface of the accretion disk—a form of super-coronal wind. The basic collimation mechanism is no longer the geometry of the inner disk (the funnel) but comes from the interaction of the streaming particles and a magnetic field. The outward particle flow comes from the pressure of the extremely hot corona and this is threaded with rotating, magnetic, current-carrying field lines. The accretion disk is effectively a rotating magnetized plasma zone; the magnetic lines of force extend outwards until at some point there is a shear and a resulting magnetic re-connection as the field lines can no longer co-rotate with the disk. (This is similar to scenarios describing emission from pulsars and at very low energies even to explanations of solar flares.) The jet is then collimated downstream by these re-connected poloidal magnetic sheaths in some form of a pinch effect. This might be the core of the VLBI jet.

It is encouraging to note that a prediction of these models is that the accretion rate and the jet velocity are correlated. This is due to an overall balance (or at least a limiting ratio) between the magnetic and turbulent energy in the accretion disk, leading to higher accretion rates producing higher jet velocities. Following on from this picture, we can now see from our discussions of accretion disks in section 7.2 that quasars would be expected to have strong relativistic jets and FRI sources to have lower or just sub-relativistic jets. We have see in this chapter that observations generally support this prediction but with the caveat that unification restrictions (for BL Lacs for example) limit the lower speed allowed for FRI sources.

So far so good, but much work remains to be done and this is a field of intense activity. Interestingly enough, the models can produce relativistic jets with a very large range of possible Lorentz factors, much higher than is observed from superluminal motion studies. So we need some mechanism to place a natural limit on the Lorentz factor, and of course this could well be the Compton drag effect of section 8.3.4. Therefore we end this section with the conclusion that serious modelling work on magnetized accretion disks shows promise in terms of explaining the formation of the jets of energy which give rise to the narrow-collimated beams, which in turn are the power conduits for the extended extragalactic radio plumes and halos with which we began the chapter.

8.7 SUMMARY

Extended radio galaxies are spectacular phenomena with plumes and halos extending many times greater than the outline of the galaxy as defined by stars. Theoretical work showed that back-to-back jets produced by an accretion disk surrounding a supermassive black hole could provide the mechanism for transporting energy from the core of the galaxy to the radio emitting lobes. These narrow parsec-scale jets were eventually found by VLBI techniques, and subsequent observations led to the discovery of superluminal motion. This shows that bulk relativistic motion must be taking place in the jets of the radio sources. The extreme variability and lack of observed X-ray emission independently demonstrated the need for bulk relativistic motion in order to explain the predicted but not observed Compton catastrophe.

Relativistic effects can explain the lack of counter-jets for superluminal and many other radio sources. The current working hypothesis is that all radio galaxies which are jet sources have twin, back-to-back jets. The approaching jet which is oriented toward the line-of-sight is significantly enhanced, the degree of enhancement depending on its bulk relativistic velocity and the closeness of the angle to the line-of-sight. The receding jet is correspondingly dimmed to such an extent that it often remains invisible on the deepest radio maps. Orientation which does not depend on beaming can also enhance the contrast between forward and backward jets. In this case the culprit is obscuration in the galaxy, perhaps very close to the central engine, such as the torus, which preferentially dims the backward jet by free–free absorption. Such an example is seen in the nearby active galaxy of NGC1275 (=3C84). Both orientation and beaming therefore explain the core-jet sources, the OVV quasars and BL Lac objects (the blazar phenomenon) lying closest to the line-of-sight but not necessarily very close to the line-of-sight as is often quoted. Superluminal sources should all be oriented towards the observer (rather than closer to the plane of the sky) and are therefore quasar-like or BL Lac-like. Likewise all beamed sources should be quasars and BL Lacs. Radio galaxies are those mostly wide-angle radio doubles whose jets lie close to the plane of the sky.

Extension of this concept of beaming and a search for the parent populations of the beamed objects led to the proposed unification of BL Lac objects with the lower radio luminosity FRI radio galaxies, and of radio-loud quasars with the higher radio luminosity FRIIs. We then find that the OVV quasars fall into this class. The blazar phenomenon of beaming suggests orientation closer to the line-of-sight but this does not need to be very close to the line-of-sight, and in fact even for BL Lacs this is the case. The beaming and orientation hypothesis applies to the FRIs and the FRIIs giving us the BL Lacs and the OVVs. A surprising outcome from the studies of the luminosity functions for FRI radio galaxies, radio and X-ray selected BL Lacs is that in order to explain the relatively higher number density of the X-ray selected BL Lacs, the Lorentz factor of the emitting material must be significantly less than for the radio emission (assuming that the X-ray emission is synchrotron radiation from the jet). This leads to the conclusion that the material is being accelerated down the jet. This is still controversial and whether it will stand the test of time is an open question.

The differences between FRI and FRII jets is now thought to be understood in terms of the jet dynamics, with FRI jets being lower speed than the FRII jets, the latter showing strong relativistic pattern flows. The polarization data indicate that the FRII jets are smooth flows with strong longitudinal magnetic fields whereas the FRI jets are weaker with more tangled fields and the jets are more susceptible to shock formation. This explains the higher degree of temporal variability of the synchrotron emission than is seen in the OVVs and also the predominantly tangential magnetic fields observed in BL Lac jets. There is substantial evidence to show that either the trajectories of the early stages of the jets are curved, or the shocks within the jet are occurring at different orientations to the general direction of the kpc jet.

The spectacular wealth of detail in the VLA and HST images of the jet in M87 shows that by studying this nearby source significant progress in jet physics can be made. The refurbished HST and the VLBA will provide a new generation of images with even greater detail for the jet modellers to tackle. Nevertheless, the idea of jets forming very close to the nucleus is firmly supported, and new modelling of hydrodynamic jets from thick and thin accretion disks shows excellent promise. Clearly, theoretical studies can now form jets of the required size scale and velocity. Although the precise collimation mechanism which gets us from the scale sizes of the accretion disk (of order 50 AU) to the width of the jet of around 1 pc is still an area of dispute, the future looks extremely promising.

So we leave this chapter with firm support for jets and beaming models. Unification of jets and parent galaxies is a burning issue. There is now much evidence in favour of unification, and a major band-waggon of support is up and rolling and in the final chapter we shall consider the parent galaxies and conditions of the central engine or fuel which might explain the differences between the FRI and FRII jets. We shall also extend this difference to see if it will allow us to solve one of the fundamental problems of AGNs, why are powerful jets such a rare phenomenon, given the observations that the overwhelming majority of AGNs are distinctly radio-quiet and so jet-less.

8.8 FURTHER READING

General and reviews

Begelman,M.C., Blandford,R.D., & Rees,M.J., 'Theory of extragalactic radio sources', *Rev.Mod.Phys.*, **56**, 255, 1984. (A classic paper, a must for the student of extragalactic radio astronomy.)

Burns,J.O., 'Chasing the monster's tail: new views of cosmic jets', *Astronomy*, p39, August 1990. (An enjotable review of radio jets and sources for the general reader.)

Hughes,P.A., Ed., *Beams and Jets in Astrophysics*, Cambridge University Press, 1991. (An excellent series of review of a wide range of extragalactic jet phenomena.)

Kundt,W., *Astrophysical jets and their engines,* Reidel, 1987. (Jets galore.)

Miley,G.K., & Chambers,K.C., 'The most distant radio galaxies', *Scientific American*, p54, June, 1993. (Intriguing.)

Pacholczyk,A.G. *Radio galaxies*, Pergamon, 1977. (A physical and mathematical treatment of radio galaxies, detailed.)

Röser,H.J. and Meisenheimer,K., Eds., *Jets in Extragalactic Radio Sources*, Lecture Notes in Physics, 421, Springer–Verlag, 1993. (An excellent series of reviews and specialised papers covering a wide range of extragalactic jet phenomena.)

Roland,J., Sol,H., and Pelletier,G., Eds., *Extragalactic radio Sources - from Beams to Jets*, Cambridge University Press, 1992. (Wide ranging set of reviews and papers.)

Urry,C.M., & Padovani,P., 'Unified schemes for radio-loud active galactic nuclei', *Pub.Astron.Soc.Pac.*, Sep.,1995. (A detailed review and excellent source references.)

Specialized

Double radio galaxies and jets

Blandford,R.D. & Rees,M.J. 'A 'linear exhaust' model for double radio sources', Mon.Not.R.Astron.Soc., **169**, 385, 1974

Blandford,R.D. & Konigl,A. 'Relativistic jets as compact radio sources', Astrophys.J., **232**, 34, 1979

Bridle, A.H. *et al.*, 'Deep VLA imaging of twelve extended 3CR quasars', Astron.J., **108**, 766, 1994.

Burns, J.O., Norman, M.L. & Clarke, D.A., 'Numerical models of extrgalactic radio sources', Science, **253**, 522, 1991.

Hardee,P.E. & Norman,M.L. 'Asymmetric morphology of the propagating jet', Astrophys.J., **365**, 134, 1990

Pearson,J & Readhead,A. 'The milliarcsecond structure of a complete sample of radio sources II first epoch maps at 5 GHz', Astrophys.J., **328**, 114, 1988

Scheur,P.A.G. 'Models of extragalactic radio sources with a continuous energy supply from a central object', Mon.Not.R.Astron.Soc., **166**, 513, 1974.

Wilkinson, P.N. *et al.*, 'Two-sided ejection in powerful radio sources: The compact symmetries objects', Astrophys.J., **432**, L87, 1994.

Zensus, J.A., Cohen, M.H. & Unwin, S.C., 'The parsec-scale jet in quasar 3C 345', Astrophys.J., **443**, 35, 1995.

Vermeulen, R.C., Readhead, A.C.S. & Backer, D.C., 'Discovery of a nuclear counterjet in NGC 1275: A new way to probe the parsec-scale environment', Astrophys.J., **430**, L41, 1994.

Relativistic beaming and superluminal motion

Blandford,R.D., Begelman,M.C. and Rees,M.J. 'Massive black holes binaries in active galactic nuclei', *Nature*, **287**, 307, 1980.

Davis,R.J., Unwin,S.C. & Muxlow,T.W.B., 'Large-scale superluminal motion in the quasar 3C273', *Nature*, **354**, 374, 1991.

Roos, N., Kaastra,J.S., & Hummel,C.A., 'A massive binary black hole in 1928 + 738 ?' *Astrophys.J.*, **409**, 130, 1993.

Sikora, M., Begelman,M.C. & Rees,M.J., 'Comptonization of diffuse ambient radiation by a relativistic jet: The sources of gamma-rays from blazars ?', *Astrophys.J.* **421**, 153, 1994.

Conway,J.E. & Wrobel,J.M., 'A helical jet in the orthogonally misaligned BL Lacertae object Markarian 501 (B1652 + 398)', *Astrophys.J.*, **439**, 98, 1995.

Relativistic beaming and unification

Antonucci,R.R., and Ulvestad, J., 'Extended radio emission and the nature of blazars', *Astrophys.J.*, **294**, 158, 1985.

Antonucci,R.R., & Barvainis,R., 'The cores of lobe-dominant quasars', *Astrophys.J.*, **325**, L21, 1988.

Bartel,N., *et al.*, 'The nuclear jet and counterjet region of the radio galaxy Cygnus A', *Proc. National Academy of Sciences*, **2**, in press, 1995.

Barthel,P.D., 'Is every quasar beamed?', *Astrophys.J.*, **336**, 606, 1989.

Barthel,P.D., *et al.*, 'Superluminal motion in the giant quasar 4C 34.47', *The American Astronomical Society*, **336**, 601, 1989.

Baum,S.A., Zirbel,E.L., & O.Dea,C.P., 'Toward understanding the Fanaroff-Riley dichotemy in radio source morphology and power', *Astrophys.J.*, **451**, 88, 1995.

De Young, D.S., 'On the relation between FRI and FRII radio galaxies', *Astrophys.J.*, **405**, L13, 1994.

Gabudza,D.C., *et al.*, 'Evolution of the milliarcsecond total intensity and polarization structures of BL Lacertae objects', *Astrophys.J.*, **435**, 140, 1994.
Ghisellini,G. & Maraschi,L., 'Bulk acceleration in relativistic jets and the spectral properties of blazars', *Astrophys.J.*, **340**, 181, 1989.
Jackson,N., Browne,I.W.A., Warwick,R.S., 'The soft X-ray spectra of quasars and X-ray beaming models', *Astron.Astrophys.*, **274**, 79, 1993.
Jackson,N., & Browne,I., 'Optical properties of quasars II: emission line geometry and radio properties', *Mon.Not.R.Astron.Soc.*, **250**, 422, 1991
Laurent-Muehleisen,S.A., *et al.*, 'Radio morphology and parent population of X-ray selected BL Lacertae objects', *Astron.J.*, **106**, 875, 1993.
Padovani, P., 'Is there a relationship between BL Lacertae objects and flat-spectrum radio quasars ?', *Mon.Not.R.Astron.Soc.*, **257**, 404, 1992.
Padovani, P., 'A statistical analysis of complete samples of BL Lacertae objects', *Astron.Astrophys.*, **256**, 399, 1992.
Rawlings,S., & Saunders,R., 'Evidence for a common central-engine mechanism in all extragalactic radio sources', *Nature*, **349,** 138, 1991.
Urry,C.M., Padovani,P., & Stickel,M., 'Fanaroff-Riley I galaxies as the parent population of BL Lacertae objects. III. Radio constraints', *Astrophys.J.*, **382,** 501, 1991.
Urry,C.M. & Padovani,P. 'Unified scheme for radio-loud active galactic nuclei', *Pub. Astron.Soc.Pacific,* Sept. 1995.
Wills,B.J., & Browne,I.W.A., 'Relativistic beaming and quasar emission lines', *Astrophys.J.*, **302**, 56, 1986.

M87

Biretta, J.A., Stern,C.P., & Harris, D.E., 'The radio to X-ray spectrum of the M87 jet and nucleus', *Astron.J.*, **101**, 1632, 1991.
Biretta, J.A., Zhou,F., & Owen,F.N., 'Detection of proper motions in the M87 jet', *Astrophys.J.*, **447**, 582, 1995.
Boksenberg,A., *et al.*, 'Faint object camera observations of M87: the jet and nucleus', *Astron.Astrophys.*, **261**, 393, 1992.
Owen,F.N., Hardee,P.E., & Cornwell,T.J., 'High-resolution, high dynamic range VLA images of the M87 jet at 2 centimeters', *Astrophys.J.*, **340**, 698, 1989.
Reid,M.J., *et al.*, 'Subluminal motion and limb brightening in the nuclear jet of M87', *Astrophys.J.*, **336**, 112, 1989.
Stiavelli,M., *et al.*, 'Optical counterpart of the east lobe of M87', *Nature*, **355**, 802, 1992.

Forming the jet

Blandford,R.D., 'Accretion disc electrodynamics—a model for double radio sources', *Mon.Not.R.Astron.Soc.*, **176**, 465, 1976
Blandford,R.D., & Levinson, A., 'Pair cascades in extragalactic jets I. gamma- rays', *Astrophys.J.*, **441**, 79, 1995.

9

Unification

9.1 INTRODUCTION

In the previous chapter we saw how astronomers have built up an impressive understanding of the workings of radio-loud active galaxies. Radio galaxies and radio-loud quasars are twin-jet phenomena with the jets being produced and collimated by the processes associated with the rotating supermassive black hole. The power of the jet, and whether or not it is relativistic, depend on the details of the black hole and accretion processes. These in turn probably depend on the mass of the host galaxy, or at least on the fuelling capacity of the central regions of the galaxy. However, believing that the basic power source of all AGNs is the same, namely a supermassive black hole, we need to explain why so few active galaxies are radio-loud with jets.

Moving on to the more general unification scenarios, we remember that the detection of broad-line emission from some Seyfert 2 galaxies when observed in polarized light produced a breakthrough in unification. This provided overwhelming support for the orientation of the spin axis of the supermassive black hole with respect to the observer being a critical parameter in AGN classification. In the resulting unified model, the lack of broad-line emission is explained by the central engine of the galaxy being surrounded by an extended torus of dust and molecular gas. This effectively provides a shielding cocoon for events in the vicinity of the supermassive black hole.

In this scenario, broad-line (type 1) objects are those whose core can be viewed directly, whereas many objects that show only narrow lines (type 2) possess a broad-line region hidden from view by the obscuring nature of the torus due to its orientation with respect to us. In this chapter we shall refer to some of the most recent research on selected AGNs to support the scenario and to provide greater insights into active galaxies. We return to one of the favourite sources of AGN study, the Seyfert 1 galaxy NGC4151 and we will consider HST imaging data.

We will then move on to investigate the ultraluminous far-infrared galaxies and in the process we shall discover that some of these are better described as buried quasars. We will also discover that one of them is gravitationally lensed, amplifying its apparent luminosity. This is a relief because it now more easily fits into the big picture. The data revealing this startling conclusion concerning lensing have only been possible because of the latest generation of telescopes, specifically the 10-m Keck on Mauna Kea. Unification continues to be in good health and future studies are confidently expected to reveal more buried quasars, hidden behind 50 to 100 magnitudes of extinction. This means that infrared and even infrared polarimetric studies are required in order to detect the heavily obscured and in some objects scattered BLR.

We will then home in on the complexities of the physical processes taking place in the central regions of AGNs. This is the realm of the X-ray observations and we will pull together the continuum and line observations of previous chapters. We will then step back and present a pictorial overview of the big picture in terms of how we see it today. Finally, we will consider some of the problems that remain and some key studies to be undertaken in our on-going quest: to find the ultimate answer to AGNs—how they form, evolve and die—the stellar evolution picture for active galaxies.

9.2 UNIFICATION—PROBING THE BIG PICTURE

9.2.1 Introduction

Let us assume that our picture of the central kiloparsec of a radio-quiet AGN consists of a supermassive black hole surrounded by (moving radially outwards) an accretion disk, broad-line clouds an obscuring torus, and narrow-line clouds, all of which might be embedded in a much larger disk of gas, dust and hot stars (the starburst disk). Not all AGNs will look the same, some may lack the starburst disk, but all are expected to reveal the other components to some degree. Why we observe such different objects as type 1 and 2 is explained by orientation and whether or not we can directly see the BLR. Therefore the orientation of the black hole spin axis with respect to the observer is a key parameter.

Although this overall picture has gained general consensus, many details are yet to be determined. We will now consider a particular source as an example of tests of the unification model. This will highlight the overall success of unification but will also expose some of the remaining weaknesses of the detailed picture.

9.2.2. NGC4151—a Seyfert on the edge?

We briefly consider an AGN that has attracted much attention over many years, the Seyfert galaxy NGC4151 (fig. 1.7). In section 5.4.2 we saw that over time it dramatically changed from a Seyfert 1 to at least a Seyfert 1.8 (although the broad Hα lines never fully disappeared) and then back again, at least according to the usual Seyfert classification based on its emission lines. This was also accompanied by a change in the shape and strength of the continuum. In terms of our unification model and obscuring torus, one simple interpretation of this change could be that we are viewing NGC4151 at a glancing angle to the edge of the torus; sometimes we can see directly into the BLR, at other times it is much more obscured. Another, and more probable reason is that the changing strength and shape of the continuum ionizing radiation spectrum directly alters the make-up of the BLR, in the process changing the line strength and shape of the emission lines and the subsequent classification of the source.

Of course we do not know the exact geometry or precise scales of the torus and inner accretion zone, and therefore calculating the range of inclination angles to the line-of-sight contains a number of assumptions. On the other hand, following up our earlier description of ionization cones and an extended narrow-line region, we can simplistically say that if we see twin ionization cones and an ENLR, then the rotation axis of the supermassive black hole must lie more in the plane of the sky than out of it. In NGC4151 the ENLR is very anisotropic, being extended along a NE-SW direction, providing supporting evidence for radiation from the central source being directed into a cone by an obscuring torus. Can we start to put all these pieces together to come up with a detailed picture of the accretion zone for at least one AGN?

The answer is that we are getting there, but perhaps we are not quite there yet. This is due to the blurred optics of the HST, and as the new data come out from the corrected optics (expected imminently) the true story may be told. At the moment we have a conflicting description of the geometry of the emission from NGC4151, the best images coming from the Wide Field Planetary Camera obtained through a filter centred on the [OIII] 500.7 nm line in the galaxy. As we saw in section 6.4, [OIII] is one of the best diagnostics of the NLR and we see that in NGC4151 it does not trace out a uniform distribution. Instead the gas is resolved into a number of knotty and string-like regions extending over scales of order 10-20 pc and is distributed in a cone-like shape (similar to the structure seen in NGC1068 in plate 5).

A very important result is that there is essentially no line emission in the expected species (such as [OIII], Hα, [NII]) in a direction perpendicular to the cones. This gives very strong support to the concept of an isotropic central ionizing radiation field being roughly collimated by a torus of gas and dust. Closer investigation, and some degree of faith, shows that the emission is in fact bi-conical, the apexes of the two back-to-back cones coinciding with the nucleus of the galaxy, delineated by unresolved emission in Hα. The cone angle of the [OIII] emission is $\sim 70^{\circ} \pm 10^{\circ}$ and has an NE–SW orientation.

NGC4151 also has anisotropic core radio emission and a small VLBI jet. Although these are aligned in the same general direction as the optical radiation cones, there is a definite misalignment that is puzzling (see later). Nevertheless, the presence of ionizing cones and a VLBI jet would seem to be powerful evidence for the presence of an obscuring torus and a supermassive black hole. Furthermore, the [OIII] data from the HST argues for the line-of-sight to the nucleus of NGC4151 to fall outside the ionizing cones. However, the fact that an unresolved BLR is clearly seen along with a relatively unobscured UV nuclear source argues that we are looking down the radiation cones. How can these two apparently contradictory facts be reconciled? The HST data also reveal that the central engine, delineated by strong and unresolved Hα emission, can be seen directly, and further evidence suggests that there is little obscuration in the line-of-sight. So this poses a problem: how can we see into the torus yet also see well-delineated twin cones of NLR ionization?

At the moment there are two competing viewpoints. The picture presented in fig. 9.1(a) argues that we cannot see into the torus and therefore the simple AGN unified picture of a thick molecular torus breaks down. On the other hand, the authors of the diagram depicted in fig. 9.1(b) strongly believe that once the position and orientation of the supermassive black hole with respect to the plane of the parent galaxy are taken into account, the unified picture emerges unscathed. The key to their argument makes use of the misalignment of the radio jet to the ionization cones. They believe that this is revealing a local density effect in the galaxy, the ionization cones appear stronger where there is more gas to be ionized. The shape and strength of the ionization cones therefore

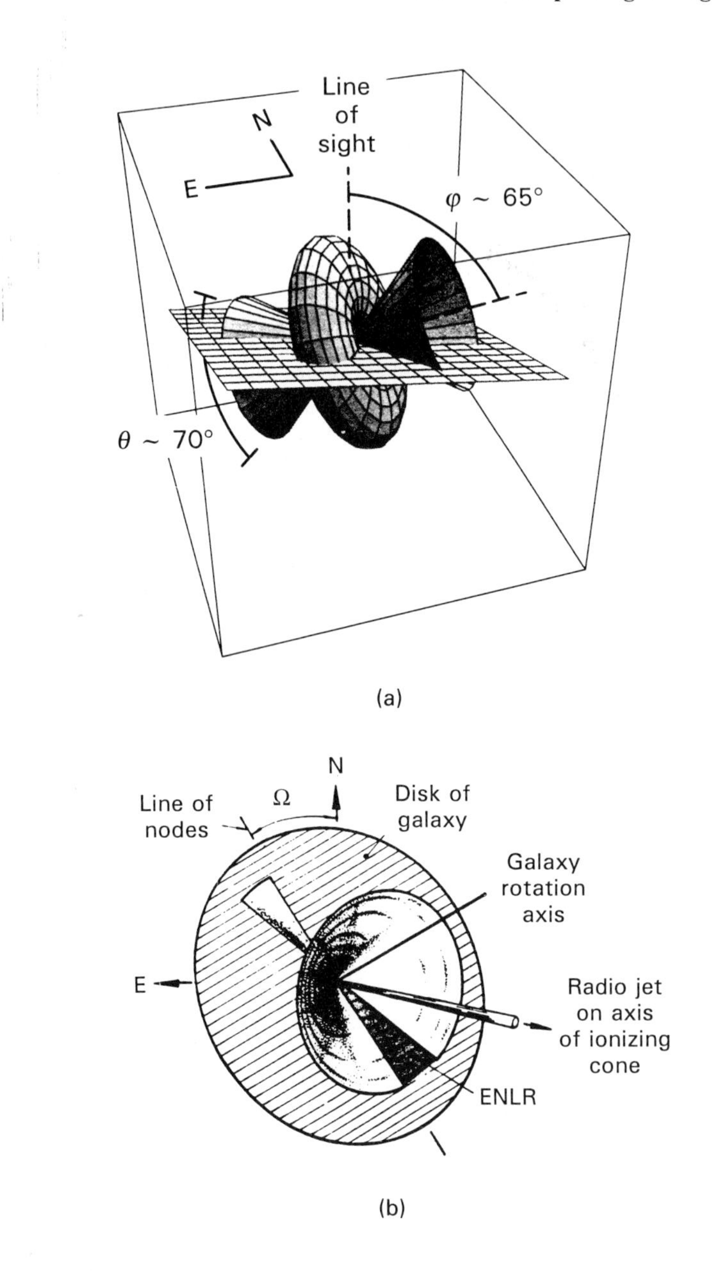

Fig. 9.1 (a) Schematic of the geometry of NGC4151 taken from Evans *et al., Astrophys.J.*, **417**, 82, 1993 showing that in their picture, the line-of-sight to the nucleus must be intercepted by the torus, which seems to be incompatible with the data. (b) Cartoon (from Boksenberg *et al., Astrophys.J.*, **440**, 151, 1995) showing that the misalignment of the spin axes of the black hole and the galaxy can explain the results due to a gas density gradient as seen by the ionizing radiation.

depend on the gas density in the parent galaxy. The model depicted in fig. 9.1(b) assumes that the direction of the radio jet delineates the spin axis of the black hole, which is found to be tilted some 25° from the rotation axis of the galaxy. The optical ionization cones appear to be bent with respect to this direction due to an optical illusion, the ionizing radiation intercepts differing density regions of gas in the disk of the galaxy. In support of this inhomogeneous interpretation is that we can see the cones are far from perfect shapes.

The case of NGC4151 demonstrates that we are now in a position to undertake very detailed studies of the central engine of AGNs. Although the dust has not yet settled on the story of NGC4151, future HST imaging and spectroscopic data will certainly be able to test whether the picture of 9.1(a) is correct, or, as most unification pundits expect, something more like 9.1(b) explains what is happening. It is probably the case that given the other evidence of NGC4151, we will find that we are relatively close to the correct viewing angle of a cone which is modified by a gradient in gas density in the parent galaxy, and furthermore that the torus is not as sharply bounded as in simple models. More clumpy structures for the ionized gas and the molecular torus are most probably the order of the day when sufficiently high spatial resolution observations enable these structures to be probed adequately. So we will leave NGC4151 with the tantalizing prospect of new HST observations clinching the puzzle.

9.2.3 Radio-quiet AGNs, ultraluminous far-infrared galaxies and buried quasars

The IRAS satellite discovered a new class of galaxy known as far-infrared galaxies (FIRGs). Of the 324 galaxies in the IRAS bright galaxy survey, ten were found to be ultraluminous, with far-infrared luminosities in excess of 10^{12} $L_{\odot}$. The source of the radiation is clearly heated dust, but what fundamental power source is responsible for heating the dust? Quasars are the only known objects which have comparable luminosities, and an obvious possibility is that these ultraluminous FIRGs harbour quasars that are hidden from view by the same dust that is responsible for the objects being so prominent far-infrared emitters. One of these galaxies became the focus for extensive investigations, and we shall use it as an example of the studies that have been applied to many members in this class.

The object is Arp 220 and the name immediately tells us that this galaxy is not an ordinary run of the mill spiral or elliptical galaxy. Anything that appears in Halton Arp's catalogue of peculiar galaxies has, by definition, a strange morphological form. The photographic image of Arp 220 is disappointing, revealing only a smudge of light with an irregular appearance. Deep optical CCD images show what appears to be two bright nuclei on either side of a dust lane, leading to the suspicion that Arp 220 is the result of a merger between two galaxies. Indeed, very deep imaging of Arp 220 reveals a highly disturbed morphology, including faint tidal tails, clear signs of a recent merger event. An alternative viewpoint is that these two bright putative nuclei are actually a single nucleus bifurcated by a dust lane, giving a similar appearance to the galaxy NGC5128 (plate 2).

The solution to the controversy came from one of the best results recently obtained from the now somewhat long in the tooth Palomar 200-inch telescope. High spatial resolution infrared imaging showed that there are indeed two nuclei and these are separated by only 300 pc. Neither nucleus is coincident with the bright optical peaks but the brighter of the two IR nuclei is located centrally to the obscuring dust lane. This is shown in fig. 9.2 which spectacularly demonstrates the ability of high spatial resolution infrared imaging for detailed studies of dust enshrouded nuclei.

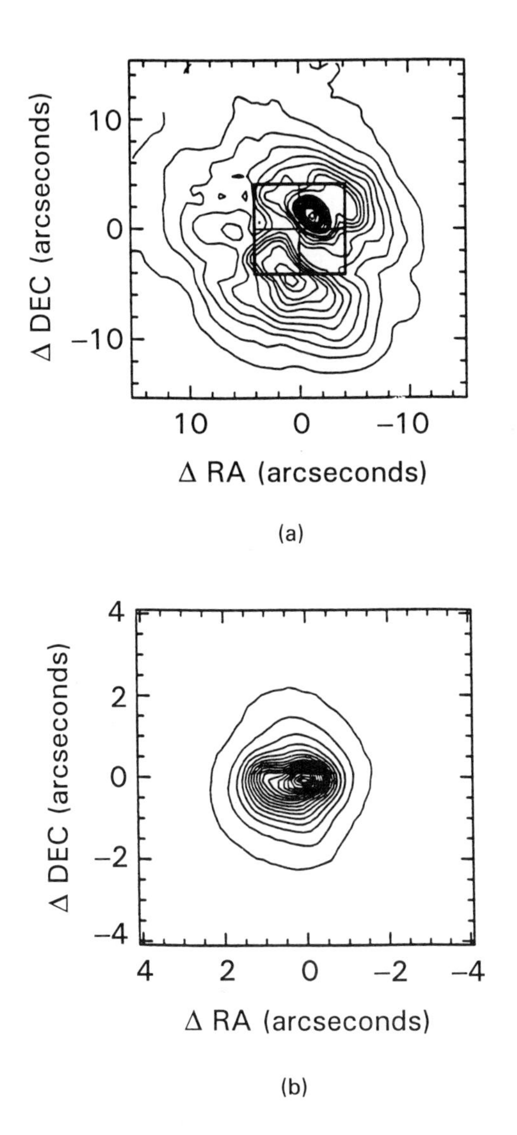

Fig. 9.2. (a) An optical R-band image of Arp 220 showing the double optical 'nuclei' separated by a dust lane. The box represents the size of a single infrared frame which is shown in (b), a 2.2 μm image showing the infrared nucleus which is coincident with the dust lane. The infrared nuclear emission nucleus is extended but both [FeII] and Paβ emission is found to be strongly concentrated on the western infrared nucleus at 0,0 in (b). (From Armus *et al.*, *Astrophys.J.*, **440**, 200, 1995.)

Arp 220 is one of the original members of the 'megamaser' class, in which the OH maser line radiates with a strength nearly a hundred million times that of a normal galactic maser. This emission is concentrated in the nuclear region of Arp 220 and a number of observations point to a high level of star formation taking place in the central kiloparsec of the galaxy. This results in an intense infrared radiation field in the nuclear region, which provides a good explanation for the pumping source of the very strong maser emission. (Further support for the infrared pumping comes from the excellent correlation found between the infrared and maser luminosities, albeit in the relatively small number of megamaser sources discovered so far.)

Strong CO emission from Arp 220 gives mass estimates of up to 20 times the amount of molecular hydrogen found in the Milky Way galaxy, leading to values of the molecular gas masses of the order of 10^{10} $M_{\odot}$. An enormous quantity of neutral hydrogen absorption, with a line width of over 700 km s^{-1} is also seen against the bright nuclear source. It is now believed that this HI absorption arises from two components; a compact and cold disk about 1.5 kpc in size with a mass of order 4×10^8 $M_{\odot}$, and a higher velocity component coming from a second, more extended region that is not oriented in the same plane. This again is strong evidence for a merged system.

In fact all the ultraluminous FIRGs show strong evidence for current merging activity as displayed by tidal tails, plumes, separate nuclei and multiple disks. So the suggestion that merging triggers star formation would seem to be very reasonable. But what about an AGN core? All the ultraluminous far-infrared galaxies are emission line objects and all but one have linewidths in the range 1,000–2,000 km s^{-1}. In spectral line classification terms, most are classed as Seyfert 2 or LINERs, but one of them, the well-known galaxy Mkn231, is classified as a Seyfert 1.5. Therefore, from what we have already learned in this book (section 3.2), this provides evidence for a broad-line AGN core. Further evidence began to point to the idea that the ultraluminous objects were more likely dust- enshrouded quasars than super-starburst (non AGN) galaxies. The main reasons were that the infrared colours are at the very extreme for starburst galaxies while the bolometric luminosity is significantly higher than for any pure starburst galaxy.

Another powerful reason has come from mid-infrared studies of Arp 220 and Mkn231. The nuclear 10 and 20 μm emission is far more compact than can be explained purely on the basis of a star formation model. For at least one group of astrophysicists, these facts were enough to convince them that many of the ultraluminous FIRGs galaxies were dust-enshrouded quasars. Furthermore, merging provided a distinct possibility of triggering the AGN activity. We should note that this is not to say that a strong starburst is not occurring simultaneously, and we know that in many objects (such as NGC1068), this is certainly the case.

As even more evidence has accumulated from spectroscopic imaging studies, the notion of buried quasars has been strengthened. Indeed, for objects such as Arp 220, models of starbursts and subsequent evolution of the starburst fail to satisfy all the observations and fall short in terms of luminosity. For Arp 220 and NGC6240 (another extreme IRAS superluminous galaxy) it is agreed that another source of energy is required and a buried AGN core certainly fits the bill. But what is the acid test for identifying a buried quasar? So far, although persuasive, the arguments have been almost of a secondary nature. Detecting a BLR, as we have consistently maintained throughout this book, would provide the clinching argument. Let us pursue this further.

Because of the large IRAS beamsize it was some time before all of the newly discovered sources could be unambiguously associated with an optical counterpart. The number of ultraluminous FIRGs has been increased significantly as a result of ground-based follow-up studies. Some of these galaxies were promptly catapulted to fame as they were found to be the most luminous objects so far observed in the Universe. One of these identifications, amazingly obtained on the first night of a concentrated search, led to perhaps the most spectacular of these objects being discovered. Its unmemorable name is IRAS FSC 10214+4724, but it is now more often referred to as the Rowan-Robinson object after its discoverer, or just IRAS10214 for short. (For the curious, the FSC stands for Faint Source Catalogue.) This object is at first sight amazing. It has a redshift of 2.286 and a far-infrared luminosity of 3×10^{14} $L_{\odot}$, easily beating the bolometric record of the previous holder, a quasar. The Hα and [NII] lines fall into the infrared K-band window for an object at this redshift, and as shown in fig. 9.3, these structures are clearly extended over a few kiloparsecs indicating the presence of an excited gas rich disk.

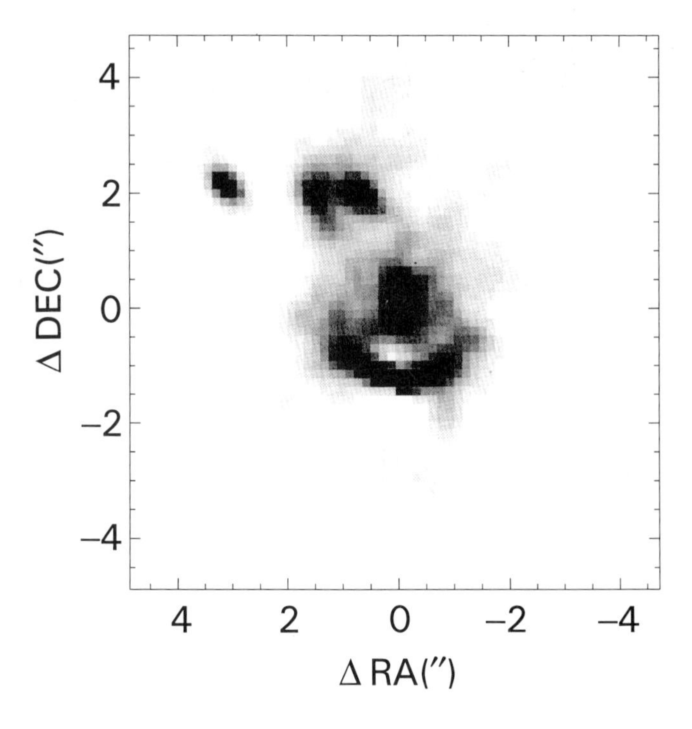

Fig. 9.3. A spectacular 2.2 μm image of IRAS 10214+4724 after image restoration techniques have been applied. The image was taken on the 10-m Keck I telescope in 0.4″ seeing. The nucleus lies at 0,0 on the plot and the arc-shaped feature is indicative of gravitational lensing. (From Graham & Liu, *Astrophys.J.*, **449**, L29, 1995, courtesy of James Graham.)

If IRAS10214 is powered by a starburst then this must be very vigorous in order to account for the far-infrared re-radiated luminosity, and furthermore, the models require these stars in the starburst to be massive. Therefore the lifetime for this phase is extremely short. However, there is a problem in that the sum of the starburst luminosity appears too low to account for the overall luminosity of the object, although a number of starburst models managed to reproduce most of the observations. On the whole there was a general feeling that IRAS10214 was most probably a dust enshrouded quasar. (In either case, the far-infrared dust emission vividly demonstrates that large quantities of dust, and hence heavy elements, must have formed by a redshift of 2.)

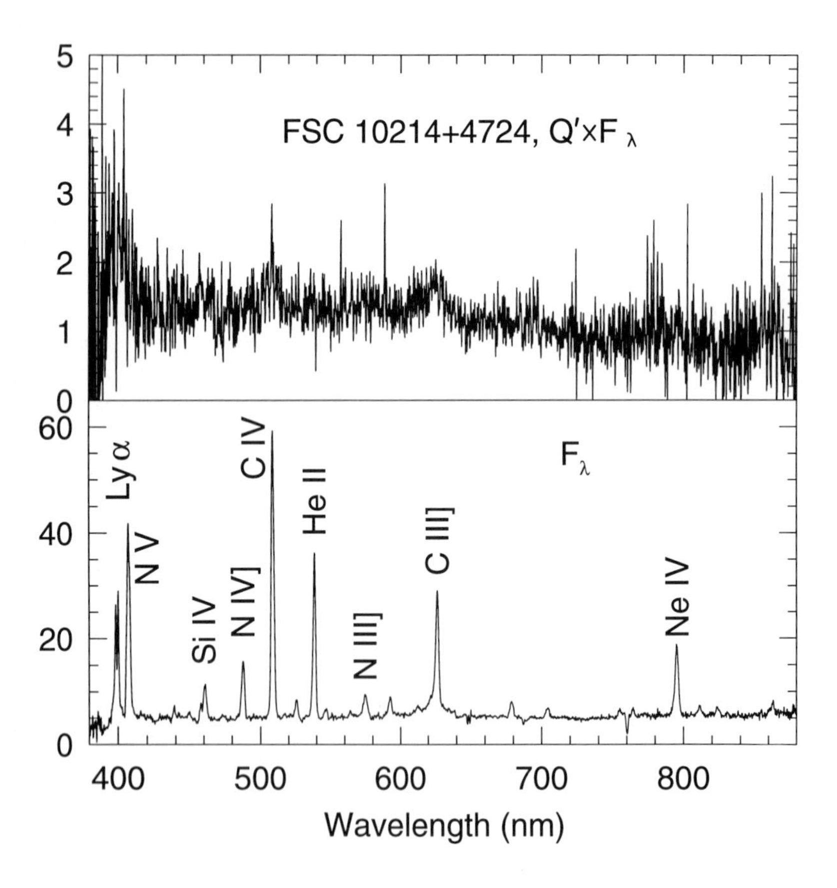

Fig. 9.4. The total flux (bottom) and polarized flux (top) of IRAS FSC 10214+4724 obtained using the Keck I telescope. The broad emission lines of Lα/NV, CIV and CIII] can be clearly seen in the polarized spectrum. (Courtesy Robert Goodrich.)

The story took another dramatic turn with the production of a spectacular 2.2 μm image from the Keck telescope (fig. 9.3) which showed a suspicious arc-like feature. Arcs are a usual indicator of gravitational lensing by an intervening cluster of galaxies (see section 3.9) and spectroscopic data confirm this interpretation. Therefore, at the time of writing, it is strongly suspected that the luminosity of IRAS10214 has been significantly magnified by this 'lensing' and it is probably more like a 'normal' ultraluminous FIRG. (In passing we can note that another ultraluminous galaxy, that also happens to be a radio-quiet quasar, 1413+117, the 'cloverleaf', is also believed to be enhanced by lensing because of its multiple [cloverleaf] image.) So is 10214 a buried quasar? With the reduction in the luminosity due to the lensing is this no longer a requirement?

Further detailed and deep optical and infrared studies of IRAS10214 (and some other ultraluminous far-infrared galaxies) have finally swung the balance in favour of a buried quasar hypothesis. Very broad emission lines have now been detected (sometimes in polarized light—fig. 9.4), clearly revealing the presence of a hidden BLR. This has changed the classifications from putative Seyfert 2s to type 1 objects, i.e., Seyfert 1s or quasars depending on the luminosity of the galaxy. In our unified approach we take this as evidence for the presence of a supermassive black hole and do not worry too much about the Seyfert/quasar sub-classification. We will conclude that IRAS10214 is indeed a buried quasar (which also happens to be lensed by an intervening galaxy or cluster) and that it is our orientation with respect to the central engine that makes it appear different from its obvious quasar counterpart of 1413+117 (which also happens to be lensed).

The concept of buried quasars has been given further support from recent observations of the FRII radio galaxy Cygnus A (fig. 3.12). The first optical observations in 1954 revealed a prominent bifurcated appearance for the central (radio faint) object. As for Arp 220, this led to speculation that Cygnus A was either two galaxies in collision or a single galaxy crossed by a dust lane. For many years this puzzle remained unresolved. New infrared imaging observations provide strong evidence that the central optical galaxy in Cygnus A possesses a single infrared core. This is unresolved on scales of an arcsecond, is located at the centre of the galaxy, and is coincident with the radio position of the centre of the Cygnus A jets. Towards the centre of the galaxy (a giant elliptical or a cD), but not evident at the centre itself, is a source of blue continuum luminosity that contributes well over 50% of the continuum emission from the galaxy. The question is whether this blue continuum extends right to the centre and is not observable because of dust obscuration, or is it in some form of ring.

Near infrared spectroscopic observations have provided the answer by revealing that the central source is hidden behind at least 50 magnitudes of optical extinction. This extensive reddening would make a blue central continuum source totally invisible. Further support for a central point source comes from polarization measurements of the central regions of the galaxy. These demonstrate centro-symmetric patterns, strongly suggestive of radiation scattering from dust illuminated by a central point source. CCD images also show that the emission from the central regions is not smooth but irregular in structure.

If the central continuum source and the BLR are obscured by a molecular torus, we can say that the axis of the torus should be aligned almost in the plane of the sky (because we see two symmetric radio jets). Furthermore, the CCD observations restrict the size of the putative torus to less than $\sim$ 800 pc. The anisotropy of the line emission in the inner zone gives further support to the buried quasar scenario and combined with the radio data on the VLBI jet, leads to a picture of twin back-to-back beams of ionizing radiation with their axes oriented between 50° and 60° to the line-of-sight. The observer is not able to see directly into the ionization cone due to the orientation, and hence there are no observations of the central continuum UV source or a broad-line region.

Therefore we see that the current data for Cygnus A strongly suggest that this FRII radio source has jets oriented away from the line-of-sight, in agreement with the jet beaming unification models proposed in section 8.4, and furthermore, is powered by a buried quasar in the central galaxy. Conclusive proof that the galaxy powering the FRII radio source harbours a buried quasar should be obtainable with future increased instrument sensitivity to detect broad lines in the polarized scattered light.

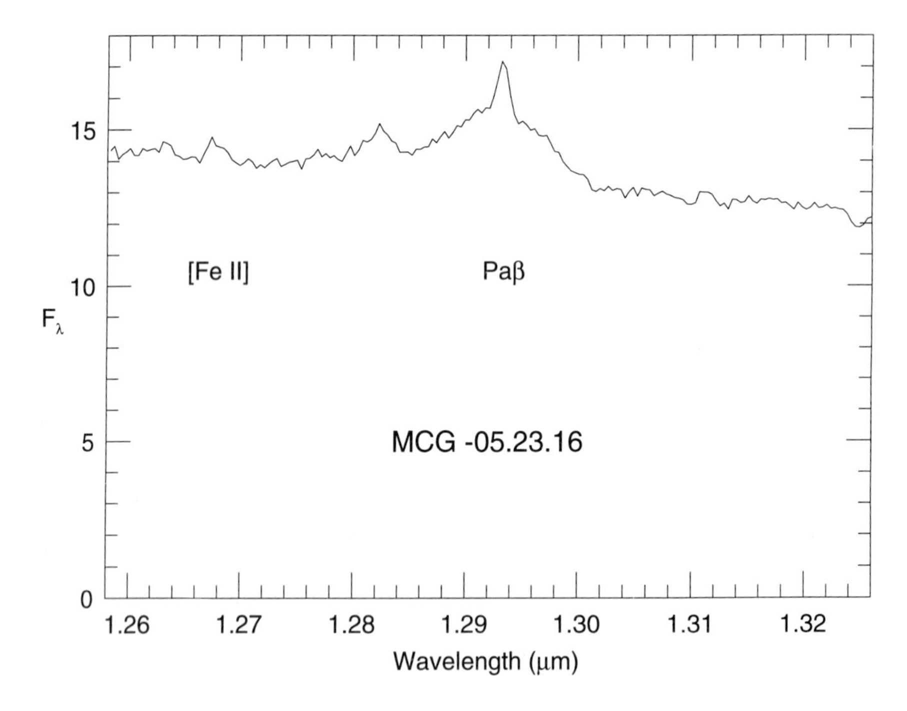

Fig. 9.5. Near infrared spectropolarimetric observations of the hidden BLR in the Seyfert 2 galaxy MCG -05.23.16 revealing the broad component of Paschen β, on top of which sits a narrow core. This is an example of the ability of IR observations to reveal details which are forever hidden to the optical. (Courtesy of Robert Goodrich.)

These searches for buried BLRs are moving to the infrared in order to minimize the extinction properties of the obscuring torus (e.g., fig. 9.5). The data from these early days are strongly suggestive that the extinctions in these objects are extremely high, and that when lines are detected, even in the infrared we may be seeing them scattered from a mirror rather than direct. Therefore we are left with the growing suspicion that the central regions of AGNs are incredibly complex, and the path of a photon from the central engine to us can follow a complex scattering route.

9.2.4 A note about evolution

We should note that in the above discussion we have only considered the ultraluminous FIRGs; the source of luminosity for the more run-of-the-mill FIRG is still not certain. It is clear that they possess heated dust because that is what we observe, but whether they harbour buried quasars will have to await the observations of ISO. There is no *a priori* requirement for an AGN core, and models of starbursts can explain satisfactorily the observed spectral energy distribution and luminosity. The few deep infrared spectral observations have shown, albeit for only a small number of objects studied so far, that the FIRGs are entirely consistent with a starburst power source. I shall assume that the vast majority of FIRGs are starburst galaxies and not AGNs.

Furthermore, it is unlikely that all the ultraluminous FIRGs ($L \sim 10^{12}\ L_{\odot}$) possess an AGN core, in fact consideration of the luminosity functions (where available and probably far from complete) suggests that the space density of these ultraluminous FIRGs exceeds that of quasars of similar luminosity by a factor of ten. However, this gap shrinks with increasing luminosities and may be comparable for $L \sim 10^{13}\ L_{\odot}$. Therefore, in general it can be argued that the ultraluminous galaxies do not have AGN cores, and perhaps only one in ten houses an AGN.

This is important as it strongly suggests that quasars cannot have evolved from ultraluminous FIRGs. On the other hand, there is smooth trend in the comparison of the luminosity of a Seyfert to a starburst component, and this hints at a very similar physical process taking place in the two systems. This further suggests that both might have a common origin, and the obvious one is galaxy interactions and mergers in the early Universe. For the ultraluminous FIRGs that house an AGN core, the ratio of the power radiated in the starburst to that in the optical–UV (the AGN) is around 0.3, which might just reflect the efficiency with which a black hole can convert mass into radiation compared with stars (both ultimately using gravity as we saw previously.)

The evidence for merging continues to build up as more data become available, especially the very deep imaging by the HST and the ground-based infrared studies of the host galaxies of quasars. Disturbed systems appear to be common–place and currently we believe that interactions and merging are responsible for both starburst and AGN activities. Both of these are driven by gravitational processes, in one case gas collapsing into a condensed cluster of massive stars (the starburst), or alternatively, onto a supermassive black hole (the AGN). Why should they be different, and why do all mergers not result in a supermassive black hole? We saw the reasons in chapter 7—a black hole will form only if certain conditions are met. Perhaps what is surprising is that this seems to occur so often, rather than so infrequently.

9.2.5 The radio picture

We have covered much of this topic in the previous chapter, the burning question remains as to why so few AGNs possess strong radio emission via jets. Let us recap. The picture we now have is that there is unification of radio sources in that they are all jet-powered and how we see them and what we subsequently classify them as depends on jet parameters, orientation and obscuration effects. Nevertheless, we are left with the simple result that there are only two fundamental types of jet, those powering the FRII sources and those powering the FRIs.

We have also seen that the FRI radio galaxies, which themselves are giant ellipticals, are most probably the parent population of the BL Lacs. The BL Lacs are oriented so that the jets are beamed towards the observer, this gives the classical BL Lac its dominant radio emission and the synchrotron spectrum effectively drowns out all other emission components from the galaxy and the BLR and blue bump of the central engine, which in many cases are known to be much less luminous than in their quasar counterparts. Perhaps this is due to a lack of gas supply surrounding the black hole. Maybe this is a clue as to why there is radio emission. We will return to this later.

The FRII sources are generally powerful quasars. Here the precise classification depends again on the viewing angle, obscuration and jet parameters. For those sources beamed towards the observer, we have the synchrotron-dominated OVVs, but not sufficiently so that the big blue bump is masked; in fact this is usually very apparent. As we reduce the synchrotron component, the prominence of the other components is increased, notably the big blue bump and the X-ray emission, revealing the radiation from the central engine.

There is one type of object which has escaped much scrutiny so far in the book, the narrow-line radio galaxies (NLRG). As the name implies, these are strong radio galaxies but are not quasars, Seyferts or BL Lacs. We have met them in other guises, Cyg A is a prime example even though they have been labelled otherwise (FRII radio sources). Why are they not called quasars? Although they reveal narrow lines, and hence a NLR, they are devoid of an obvious BLR. But wait, we have been here before. If our unification scenario is correct, then these objects should merely be quasars on the plane of the sky. They should house a BLR, which is deeply hidden from view by the obscuring torus. I will confidently predict that within the next few years, broad lines will be detected from what we currently refer to as narrow-line radio galaxies. This will close the circle on unification, the only thing we are then left with is why some galaxies manage to make powerful jets from their central engines. This is still an unsolved problem, and we shall touch on it again in the next section.

9.3 THE CENTRAL ENGINE REVISITED

9.3.1 Introduction

We are now in a position to bring all the factors together to come up with the really big picture of what makes an AGN the way it is; the grand unification scenario. In this we will assume the unified model provides a basis for the power generation. Let us determine those fundamental parameters that make an AGN what it is. We shall start by listing what these parameters might be and then consider whether they can provide the answers.

9.3.2 The fundamental parameters

We should start by considering the fundamental parameters of the primary component of AGNs, the supermassive black hole. What are its fundamental parameters? They are the mass, spin and charge. Let us begin with the mass. Although it is obviously important, the precise relationship between mass and AGN luminosity is unclear due to the very small number of black holes masses measured so far. However, we do know that a low-mass black hole cannot give rise to an extremely luminous AGN.

What about the spin? The black hole could be spinning either slowly or rapidly and this is expected to have an important effect. Furthermore, the spin would be expected to change with time due to the transfer of angular momentum from the accreted material. The simple picture is that the accretion disk is rotating in the same direction as the spinning black hole and therefore the hole will be spun up with time. But what if there is severe misalignment between the accretion angular momentum and the black hole? Could this end up with a slowly spinning black hole?

The electric charge is really an unknown quantity, but it is likely to be neutral. However the extended corona is an obvious region of charged particles and the interaction of a rotating charged region with any magnetic field (or vice versa) will inevitably produce electromagnetic particle acceleration of some form. Rotation in the presence of a magnetic field has obvious scope for jet production and a number of models have determined potential jet production as we saw in the last chapter.

So we are left with the idea that bigger means better as far as supermassive black holes and AGN luminosity is concerned. Spinning gives greater potential for jet production via magnetic induction processes, but how this happens is still a matter of speculation.

9.3.3 The radio properties

Once the black hole parameters are set, the next important topic for AGN luminosity is the fuelling of the black hole, i.e. the accretion rate. This was discussed in some detail in chapter 7 with regard to quasar lifetimes. We will now consider another aspect, the galaxy in which the black hole is located. We can now turn to the radio emission to provide us with significant clues as to what might be happening. We know that AGNs including quasars occur in both spiral and elliptical galaxies. But we also know that the powerful radio galaxies and radio-loud quasars are *only* found in ellipticals and never in spiral galaxies. This must be telling us something important.

In considering the radio emission we should refer back to section 3.7.1, where we noted that the extended and easily detected radio emission was in fact the smoke of the 'smoking gun' which pointed inexorably to the need for a collimated jet of electrons and a central engine to produce the jet. Therefore, radio emission crudely translates into the presence of a jet and supermassive black hole. But we also know that there are low-power, small-scale radio jets in Seyfert galaxies. Therefore the type of the galaxy (spiral versus elliptical) cannot be the sole determinant of the presence or absence of a jet. But given these two pieces of evidence, could it nevertheless be the requirement for the separation between a powerful and a weak jet and hence a radio-loud and a radio-quiet object?

What are the differences between spiral and elliptical galaxies? There are two main differences: the high rotational angular momentum of a spiral galaxy versus the three-dimensional orbital motion for the stars in ellipticals; the presence of a significant interstellar medium in spirals (and the associated events such as star formation and supernovae) compared to the general lack of an interstellar environment in an elliptical. Additionally, we should note that there is also a difference in the external factors of a galaxy, i.e. the habitat. Ellipticals tend to be found in dense clusters of galaxies, whereas spirals tend to be either field galaxies or are located in the outer extremities of clusters.

Staying with the internal differences rather than environment we know from chapter 8 that the presence of a radio jet must be determined in the immediate surroundings of the accretion disk. Rapidly spinning and well-fed black holes might be the requirements for powerful radio jets. But do these conditions only occur in ellipticals? Presumably yes, at least for the powerful cases, but they must also occur to some extent in spirals to explain the low-luminosity jets. So now we can ask what makes a rapidly rotating black hole. Because not all giant elliptical galaxies are AGNs, not all elliptical galaxies possess a rapidly spinning black hole with an adequate fuel supply. And here we see the problem, how can we separate the fuel supply from rotation? This is as yet an unsolved problem.

Indeed, fuel supply is an obvious method of lowering the luminosity but not obviously a method of preventing radio emission. To some extent this mitigates against the interstellar medium being a key parameter in determining whether a jet can form, although it might be important in the *extent* of the jet, as we noted in section 8.2.3. Also, because radio-loud and radio-quiet quasars are found in giant elliptical galaxies, the interstellar medium cannot have a fundamental bearing in terms of fuel supply. So what about angular momentum?

In considering this aspect, we need to take a step backwards and ask what is different between spirals and ellipticals in terms of their formation. It is now widely believed that many ellipticals were formed by interactions and mergers of galaxies in a dense cluster environment. Mergers can be a method of forming massive black holes in the resulting remnant, and if giant ellipticals are the result of multiple mergers (the aptly named cannibalism model) then a supermassive black hole becomes a distinct possibility.

The question of the orientation of the spin axis of the black hole and the axis of the parent galaxy becomes interesting. In a spiral galaxy it is easy to conceive that the spin axis of a supermassive black hole is more likely to be closely oriented with that of the parent galaxy. This is due to the conservation of angular momentum on the large-scale formation process and the subsequent lack of galaxy merging or interactions. Even though this sounds appealing, there is evidence from the radio jets that the spin axis of a black hole can be severely misaligned from that of the parent galaxy, and so it is certainly not true all the time.

For ellipticals on the other hand, the situation is much more complex. Although there is an overall tri-axial symmetry in elliptical galaxies, the question of global angular momentum and a spin axis is much less clear. Furthermore, merging and multiple merging would rapidly remove any correlation between the spin axis of the supermassive black hole, and the potential axis (or infalling direction) of the fuel supply and/or the accretion disk. This could translate into a misaligned or warped accretion disk. Eventually, the angular momentum of the accreted material will cause the axis of the black hole to become re-aligned with the disk. However, calculations point to this requiring perhaps as much as 10^8 y to take place.

Therefore we have found another possible difference between spirals and ellipticals, and also between ellipticals themselves: the orientation and angular momentum direction of the fuel and the spin axis of the black hole. If this difference is important, then we have identified another lifetime of AGN activity, the relaxation of the black hole spin axis.

Unfortunately this is where this particular story currently concludes. Is it the spin orientation of the black hole that determines radio emission? This fits the requirements that all radio-loud AGNs are in ellipticals. On the other hand, there is something contrived about getting the numbers to balance. To show an example of this possibility, however, we can take the case of NGC5128 (plate 2). It is clear that this elliptical galaxy is a merger remnant and the radio jet is significantly misaligned from the normal to the dust lane, and this presumably defines the axis for the fuel supply. We also can see that the extended emission direction is not aligned with the current parsec-scale jet, there is a definite angular shift, perhaps due to a precession of the black hole spin axis? An observational test might be to investigate the correlation between the orientation of the fuel supply with respect to the radio jet axis (the latter presumably indicates the spin axis of the black hole). This is clearly work for the future.

What about the interstellar medium? If we look to the local interstellar medium (ISM) in the vicinity of the black hole, then for our purposes this can only have an effect if it acts as a quenching agent because ellipticals and spirals both form luminous AGNs. Furthermore, it might be thought that a powerful jet should make light work of propagating through an ISM. However, if there were interactions in the region of the collimation, then perhaps only very slow jets are formed and these can rapidly dissipate when they impact a local ISM. So there is still hope for much further work along these lines. The ISM has also been used as a collimation mechanism for the jet. The idea is that a powerful wind is emitted from the central engine, a thick torus helps in this context, and as it expands into the low pressure medium recollimation is possible. A cartoon giving an example of this possible process is shown in fig. 9.6. As we saw in chapters 7 and 8, it is highly likely that magnetic fields are crucial, and so the picture is undoubtedly more complex than the simple view of fig. 9.6, nevertheless, this is a pictorially appealing diagram.

9.3.4 The FRI–FRII divide

What about the FRI–FRII divide? Observational studies now allow a number of global statements to be made about these objects. Perhaps the first question is in relation to the jets. Do the jets start out the same but the FRI jets are decelerated over a short distance due to a dense ISM, or are the jets intrinsically different? Let us recap what we know about FRI and FRIIs. The transition from FRI to FRII occurs with increasing optical luminosity of the galaxy but at a fixed radio luminosity. We know that FRII jets are not radio luminous in themselves, suggestive of low interaction with the surrounding medium, and there is often a corresponding low luminosity 'gap' seen in FRI sources before the jet becomes very pronounced. Perhaps this is indicative of the fast flow before the deceleration sets in. What scale-sizes are we talking about? The answer is the order of tens of hundreds of parsecs and some calculations have shown that a dense ISM could decelerate a fast jet over this length without requiring excessive gas densities. The outcome of this hypothesis would be that FRI galaxies have much higher ISM densities than FRIIs. Although this test sounds easy to undertake, in fact it is extremely difficult and the observations have not yet been carried out with the required degree of precision on sufficiently large samples. Although the idea that FRIIs reside only in powerful elliptical galaxies, which on average might have lower central ISM densities, is appealing, it is not conclusive.

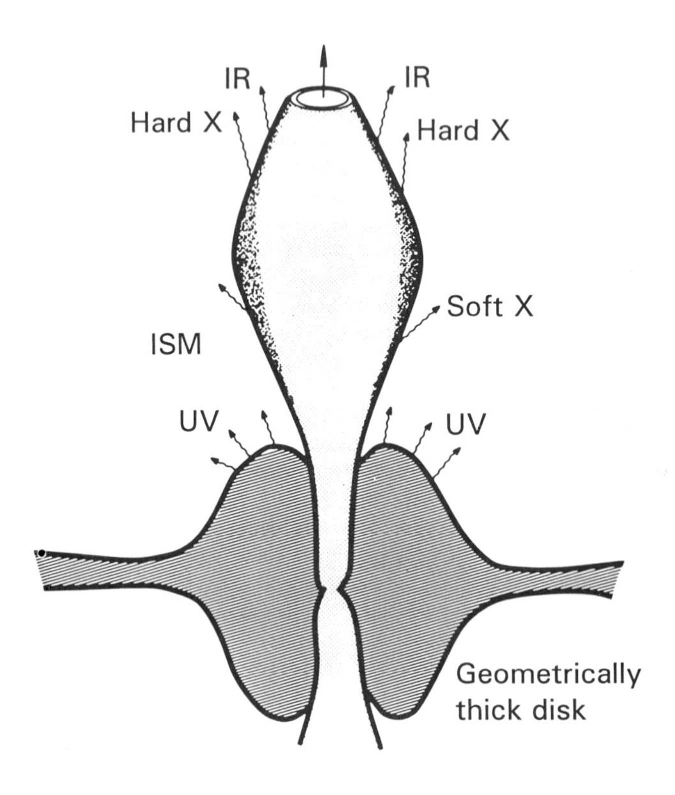

Fig. 9.6. Schematic of the collimation of a powerful wind by an external interstellar medium into a shock-emitting jet. (From Courvoisier & Camenzind, Astron.Astrophys., **224**, 10, 1989.)

Although we know that there is an excellent correlation extending over 10 orders of magnitude between the total radio luminosity and emission line luminosity, to explore differences further we need to dissect the radio emission into core and extended, and we will concentrate on the core radio luminosity. Once this is done, the divide between the FRIs and FRIIs reduces, but does not close, the difference falling by a factor of about ten (from 40 to ~ 4), but is still very significant. Observations then show that the line emission (assumed to be centrally photoionized) from FRIIs exceeds that of FRIs when measured at either the same host galaxy luminosity or radio luminosity by a large margin. For example, at a given radio core luminosity FRIIs have 10–50 times the emission line luminosity of FRIs, while at the same emission line luminosity, the FRIs produce 200–300 times as much core radio power. This can be interpreted in a number of ways but some conclusions are inescapable. We know that for radio-quiet AGNs and FRI sources there is a good correlation between the emission line luminosity and luminosity of the host galaxy—bigger means more. So we might think that one inference is that FRIIs have more cold gas than do FRIs perhaps from larger host galaxies, but in fact there is no correlation between emission line luminosity and host galaxy for FRII sources. CCD studies also point to the fact that FRIIs are found in smaller galaxies than are FRIs.

Another interpretation is that the gas excitation is much more efficient in the FRII sources, perhaps the jet-production or central engine photoionization is different in the two cases. let us now briefly explore the latter idea. Because the FRIs are much weaker AGNs than the FRIIs at a given radio power, this opens up the possibility that the central engine is weaker in FRIs, having a lower photon output but perhaps being more efficient at converting energy into the radio jets. We are then left with our earlier question regarding the fuelling process. A speculative picture is that the central engine in FRIIs is fed at a high accretion rate and that a significant fraction of this energy is converted into radiation, whereas the FRIs are fed at a lower rate and the jets are preferentially powered. In this picture, although it is not absolutely necessary, it is generally assumed that the jets start out different due to the differing properties of the central engine.

9.3.5 The link to the gamma-ray production

Where we have a powerful jet of relativistic electrons, the possibility for photon–particle interactions is highly probable. We used this idea to explain the gamma-ray emission described in chapters 5 and 8. Figure 9.7 shows a schematic of the various processes which might be occurring. The clouds in the BLR or black hole corona provide a ready scattering medium to deflect photons back into the jet where they are upscattered. The outflowing wind of particles expected to be present from the throat of the accretion disk also has the potential to scatter photons from the central engine back into the jet for upscattering by the relativistic electrons. So this provides a neat arrangement which most workers feel comfortable with. In this case the gamma-ray emission is from photons external to the jet rather than synchrotron photons upscattered from within the jet. As we saw, which of these is correct is clearly testable by observations of variability. This is something on which I am currently engaged and the next year promises to be very exciting and should reveal the answer.

The progress that can be made is shown by the multiwavelength observations of a TeV/X-ray flare in Mkn 421. The key factors of the observations of this 1994 flare were that although the flare was very pronounced at TeV and 2-5 keV X-ray energies, it was either very weak or not apparent at other wavebands, especially from UV to longer wavelengths. A key factor is that the gamma-ray flux measured by the EGRET detector on the CGRO ($E > 100$ MeV) did *not* show a dramatic rise. So what mechanisms can explain and TeV gamma-ray and X-ray flare but not a few hundred MeV gamma-ray flare? It seems apparent that the TeV and X-ray emissions are correlated, and so a synchrotron-self-Compton origin would look to be an obvious possibility given that we know the X-ray emission is of synchrotron origin. But what can cause just the high-energy tail of a synchrotron spectrum to increase? The only obvious answer is for an increase in the number of high-energy electrons (by some unknown mechanism). This would produce a flare in the X-ray region with a corresponding inverse Compton flare at extreme gamma-ray energies, while the radio to UV synchrotron flux would change little, as would their inverse Compton scattered counterpart in the few hundred MeV gamma-ray region. So inverse Compton emission could be the origin of the intrinsic power law, at least for the radio loud objects. The rapidity of the flare and subsequent decay poses further problems for more exotic models which we will not discuss further.

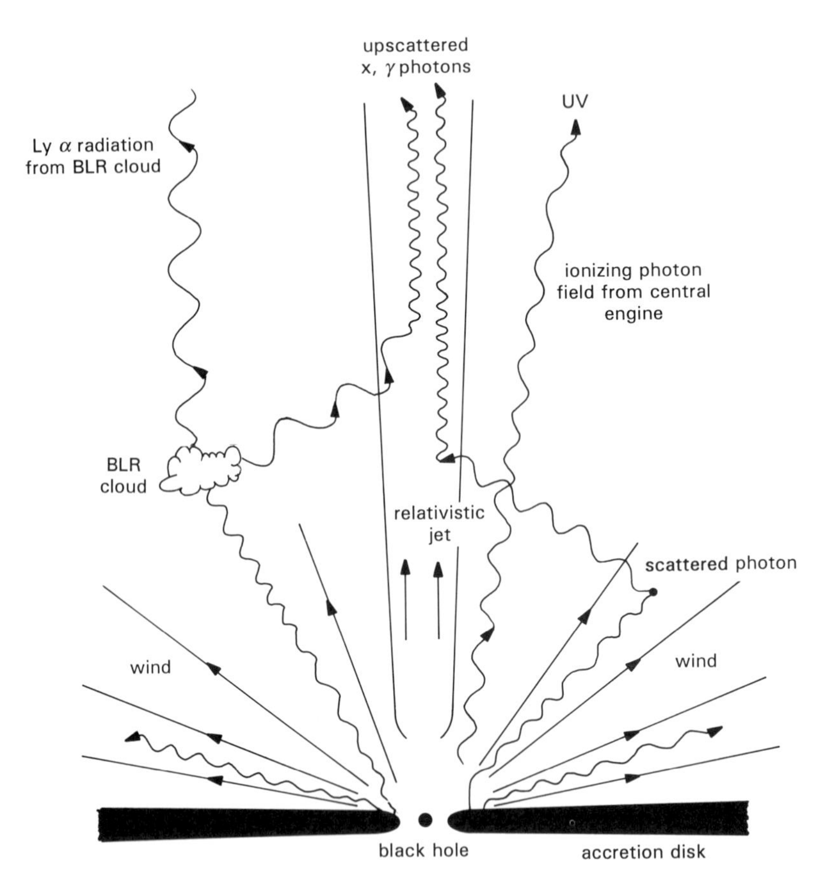

Fig. 9.7. Schematic of gamma-ray production from photons upscattered by the electrons of the relativistic jet. (Adapted from Sikora, Begelman & Rees, *Astrophys.J.*, **421**, 153, 1994.)

9.3.6 The central engine

We will bring this section to a close by taking another look at the central engine, and in particular the energy generation processes and interactions. We assume that gravitational energy is the prime driver for the output luminosity. The mediator of this is the supermassive black hole through its attendant accretion disk. Gravitational energy is released by material accreting onto the disk and being converted into radiant energy by viscous dissipative coupling. The precise details and physical processes taking place in this energy conversion are not yet completely clear, but good progress has been made.

Let us remember that although we have talked about an accretion disk throughout this book, the evidence for its existence is definitely on shaky grounds. We saw in chapter 7 that accretion disks come in two flavours, thin disks and thick tori. Thick tori help the luminosity by allowing higher accretion rates, and provide steep-walled funnels for rough collimation of photons or winds (as shown in fig. 9.6—but are unhelpful in powerful jet production by themselves without spin and magnetic interactions). Although it is difficult to fit the emission expected from a thick torus to the observations there is more scope and parameter space available for model 'tweaking' and so the picture is far from lost. The thin disk is the simpler case, but even this comes up against a number of observational difficulties.

Let us summarize the problems for accretion disks. These include: the difficulty of fitting the observed spectral energy distribution with that expected from a thin accretion disk; a thick torus fares even worse; one would expect a correlation between the spectral shape of the big blue bump and the luminosity of the central engine, but this is not observed; hydrogen continuum absorption shortward of 91.2 nm is expected from an accretion disk or torus and this has never been detected; and finally, the variability lag between the UV and optical is expected to be of order weeks for the accretion disk sizes anticipated to exist around supermassive black holes, but observations are consistent with around zero lag.

This latter aspect is very important. If the inner face of the disk is energized, say by direct heating due to photon irradiation, then the outer regions only become energized as this additional energy is transferred through the body of the disk. The speed at which this can happen depends on the sound speed in the disk, and all models predict that delays between the UV and the optical should be of the order of weeks, certainly not of the order of zero days. This latter figure is much more indicative of a 'prompt' response, which points towards a thin gaseous medium where information is transmitted by photons at the speed of light rather than atoms at the speed of sound.

In fact the weight of evidence tends to suggest (disappointingly perhaps) that for the lower luminosity AGNs we might not be seeing an accretion disk at all. In this case the big blue bump could still be extended gaseous emission surrounding the black hole, but not in the form of a disk, rather in more distributed clouds. There is no agreement on which precise model best describes this picture, although there are a number in the literature. A closely related origin has been proposed, that of optically thin free–free emission from gas in the immediate vicinity of the supermassive black hole. Free–free emission fits the shape of the 1 μm to 100 nm spectral energy distribution and explains many of the observations, but not all. The free–free emitting region must be sufficiently small for its predicted output spectrum to match the observations but cannot be too small otherwise it cannot remain optically thin and would no longer be able to fit the optical–UV spectrum.

Note that none of what we have just described argues against the presence of an accretion disk, it is just that we may not be seeing it in the form of the big blue bump. At the moment, the jury is out on the precise origin of this important component but we shall assume that much of it is some form of reprocessing. In any event we will no longer assume that in all cases the soft X-ray excess is part of the big blue bump. We will now investigate what this might be and paint our big picture of the central engine.

Within the close vicinity of the black hole will be a plasma field and it is now thought likely that the primary photon emission from the central engine, demonstrated by the X-ray power-law (as we discussed in section 5.2.5 and 5.4), is due to the thermal electrons Compton scattering the lower energy (UV, soft X-ray) photons. But from where do these soft photons originate? This is, of course, the million dollar question and where everything becomes very complex. Let us step back and review what we know about the central engine.

In Seyfert 1 galaxies the X-ray observations of the last decade or so have revealed that there is K-α emission from fluorescent iron, an iron K-shell absorption edge and a high-energy spectral component. These all tell us that the primary X-ray source must be illuminating material which is both neutral and partially ionized. These give us our 'cold' and 'warm' absorbers respectively. Let us take a look at the latter.

The requirement for the material to be partially ionized comes about because of the absorption features in the X-ray spectra. This means that the material cannot be too shielded from the central radiation ionizing continuum spectrum and neither can it be too exposed (else it would be totally ionized.) As we saw in chapter 6, this material is likely to derive from an outflowing wind (e.g., fig. 6.4) and provides one potential explanation for the scattering 'mirror' which is crucial in our unification picture. This wind could in turn come from the inner face of the molecular torus due to intense irradiation from the central engine or from the accretion disk itself. The latter is unlikely for two reasons, first it would be too hot and second, it makes the 'mirror' extremely small. Hence the former is generally favoured.

What about our cold absorber, or cold reflector? This is required to explain the iron emission and the high-energy reflected (scattered) component. Calculations of the respective luminosities in the regions requires that this cold absorber is seen by a substantial fraction of the primary X-ray emission and that of the order of 10% of the primary X-rays are reflected while the rest are absorbed and thermally re-radiated from the material. It is this component, with a pseudo-blackbody emission spectrum from a range of temperatures, that neatly explains the big blue bump. The gas in this case is believed to be in the form of clouds some distance from the black hole, possibly the broad line clouds. As an aside, gas at a temperature of $\sim 3 \times 10^4$ K is expected to have its ionization state dominated by the hard X-ray radiating photons, hence the use of the 'cold' concept as far as the irradiating photon field is concerned.

But does this really work, can reprocessing explain the big blue bump? One obvious test is to consider the energetics and it is to this that we shall look next. For a number of reasons discussed in previous chapters, the power-law extending from a few keV to the gamma-ray region (in some instances) is taken to be the signature of the central engine and the primary photons. Because this power-law extends without deviation down to a few keV, then in general it cannot be mixed with colder gas otherwise as we noted above it would rapidly become optically thick to bound-free absorption which would be immediately noticeable below ~10 keV. Although there is always the possibility of reverting to geometrical effects, this is just another way of saying that the power-law component and the thermal gas may co-exist is the same region of space, but are physically separate and do not mix. We have a separation, or a two-phase model.

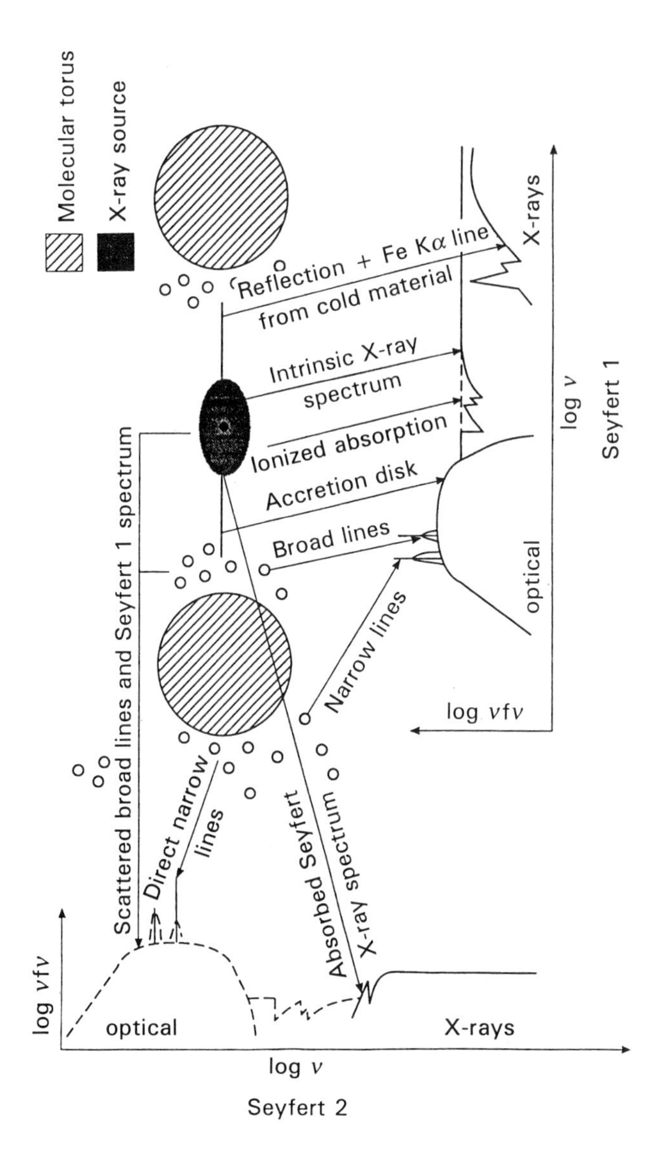

Fig. 9.8. Schematic of the inner parsec of a radio-quiet AGN showing the main constituents and the interactions processes. The hard X-ray primary source is produced in the immediate vicinity of the supermassive black hole and irradiates the surrounding medium, which in turn absorbs, reflects and scatters the X-rays producing secondary emission in the process. The model depicts what we would see looking along various lines-of-sight assuming unification based on orientation of the obscuring molecular torus. (From Mushotzky, Done and Pounds, *Ann.Rev.Astron.Astrophys.* **31** 717, 1993.)

What about the luminosities in the power-law component and the big blue bump. To answer this all we need to do is compare the two luminosities. This is nothing like as simple as it sounds. While in many objects the big blue bump–soft X-ray is much less luminous than the hard X-ray tail, it is clear that in some cases the scenario runs into difficulties because there is insufficient energy in the hard X-ray photons to explain the thermal luminosity. On the other hand, the ratio of the thermal to non-thermal luminosity seems to lie in a band spanning only about a factor of ten, from ~ 0.3 to ~ 3. For cases where the ratio is around unity or greater, then we assume any reprocessing is minimal and the most likely cause of the thermal emission is either gravitational or extraction of rotational energy of the spinning black hole. Gravitation would favour the accretion disk scenario. So some objects should show evidence for accretion disk emission and this should be the big blue bump.

But what about the case where the ratio of thermal to non-thermal luminosities is clearly much less than unity? Here we might expect appreciable reprocessing, and as we saw above, this should be manifested by Compton scattering and atomic absorption (bound–free transitions). Both are seen in some AGNs and so reprocessing seems on firm grounds, and in these cases we anticipate that the big blue bump originates mostly, or solely, due to reprocessed hard X-rays.

So how can we differentiate? Temporal correlations of variability between the UV and the hard X-rays is an obvious answer. If there is significant reprocessing, they should be strongly correlated, with the hard X-rays leading the UV, but not by very much as these regions are closely co-spatial. Lack of correlation would indicate no, or very little reprocessing. Also, the entire big blue bump should respond as a single entity, with no appreciable delays. An accretion disk should show easily measurable lags from the UV to the optical as the energy is transported outwards through the disk. Clearly, there are exciting times ahead to test these scenarios and determine how this fits in with central engine and galaxy luminosities.

There is an entire industry at work calculating the complex radiation processes involved in the potential reprocessing and attempting to match observations of the continuum spectra. Clearly, the situation is going to be complex when one considers there will be a range of interactions, from zero reprocessing to perhaps complete reprocessing and whether there is an accretion disk, an accretion torus, or a thermal plasma. We can finally note that if the thermal and non-thermal luminosities are about the same, and there is no reflection component seen in the X-ray spectrum, then there cannot have been much interaction and so the big blue bump must be thermal in origin (e.g. due to gravity and an accretion disk) and not reprocessing.

The next phase of AGN research will be filling in the details and refining the picture even more. The latest X-ray satellites are hard at work on the details of the spectra and the line absorption and emission. Figure 9.8 shows a highly illustrative and useful cartoon by Ken Pounds showing how the various processes in the central engine produce the differing output spectra.

In closing this section we remember that we still have to explain the primary photon spectrum, the hard X-ray power-law as we refer to it. We saw in section 6.6 that there are a number of ways to produce this, and we addressed the two most popular models, These involve electron–positron pair cascades along with an intense photon field and Compton scattering from thermal electrons. We discussed the latter and saw that a quasi-equilibrium situation could be set up with equipartition existing between the high-energy photons and the low-energy photons and gives good agreement with the observations. Future high-energy observations are required to test these theories.

9.4 OVERVIEW - THE BIG PICTURE

We now come to the final picture, and we will describe it as such. At the outset of starting on this book one of my goals was to produce a 'picture' of the scale-sizes of the workings of AGNs, from the kiloparsec-scale radio lobes to the fine details of the accretion disk. However, Roger Blandford beat me to it and fig. 9.9, adapted from his excellent treatise on AGNs in the Saas Fee Lecture series referenced in section 3.11, provides the answer. This is by far the best example I have seen of the 'big picture' in terms of the relative sizes—something that is all-too-often not fully appreciated.

9.5 FUTURE WORK

In many respects this can be said to be more of the same but better and with higher signal-to-noise on larger and better selected samples. The new facilities and instruments being constructed or planned will provide the means to acquire these data. Below I list a selection of topics and observations which I have picked out as being required in order to understand better the limitations of the current unification models and subsequently continue to refine, or discard them and move on. (I do not believe the latter to be likely but one must always hold out this possibility in science, as we saw at the very beginning in chapter 1.) The list is far from comprehensive and tries to select a number of topics from the various wavelength regimes. In this aspect, I have moved away from the physical processes approach, and targeted it towards new techniques and developments.

I will start with the radio and work downwards in wavelength One of the key series of observations is to determine better the jet parameters for FRI sources. This requires a much larger sample, more VLBI and much higher dynamic range imaging of the jets. A requirement for the unification models of radio sources and AGN parent populations is that the jet starts out from the core with relativistic velocities. To date there is some (weak) evidence that the jets may be non-relativistic, and this causes major problems for unification. However, as we noted in section 8.4 these VLBI observations typically relate to standing shocks in the jet and as such derive the pattern speed which might be significantly less than the true flow speed. The differences between the FRI and FRII jets need continued investigation, especially in two areas, the flow and shocks within the jet, and the deceleration of the FRI jets from the assumed relativistic to non-relativistic velocities. Again, the VLBA and space VLBI are crucial in this context. Also, multi-epoch polarization measurements, both in total flux at millimetre and submillimetre wavelengths and on the parsec scale using VLBI to investigate the knot–interknot regimes are needed. The suggestions alluded to in section 8.4 that the jet differences between FRI and FRIIs are manifest in the order and strength of the magnetic field along the jet, with FRII sources showing a strong longitudinal component and smooth flow must be followed-up. Showing how this is moderated by the transverse fields from shocks will be a powerful probe of the jet dynamics and structure. Needless to say, wiggles and different orientations of shocks in the jet will undoubtedly make any simply interpretation difficult.

The VLBI maps are also anticipated to reveal the expected single-sidedness and/or superluminal motion, the standard tests of relativistic jets. In the week before this book closed for press I was at a meeting in Hawaii where I saw a new VLBA map for Cygnus

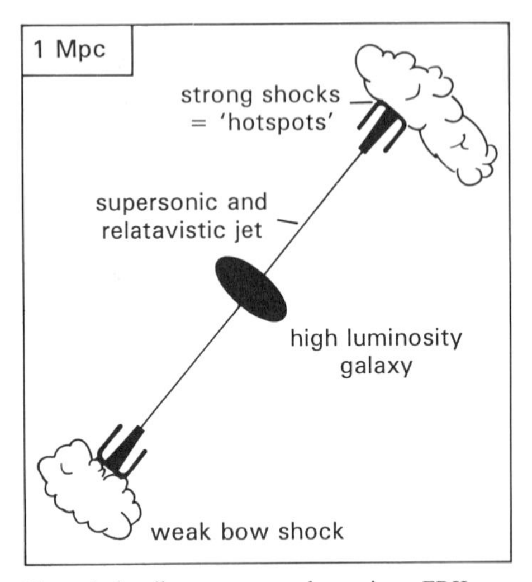

Extended radio sources — shown is an FRII source with an edge-brightened structure. The FRIs have lower jet velocities and fade-out to the ends.

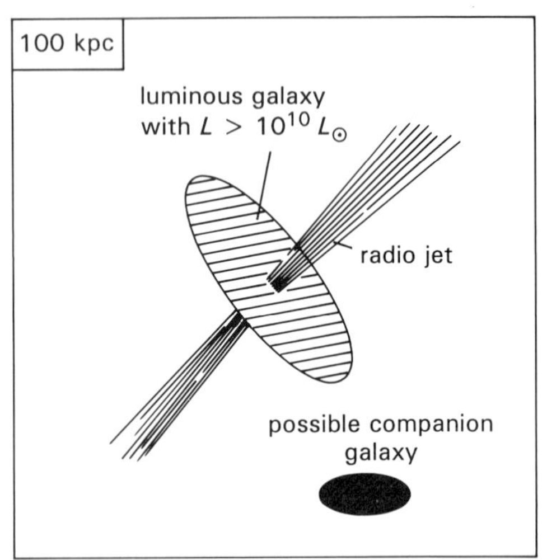

The host galaxy. Although shown as an early type galaxy with a smooth profile, it could also be highly irregular with multiple nuclei as a result of merging.

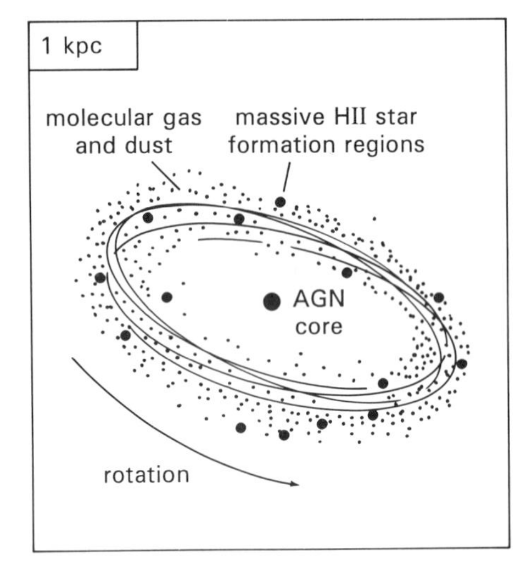

The central kpc star formation disk. This strong far infrared emitting zone might be fed by a bar structure, as seems to be the case for NGC1068.

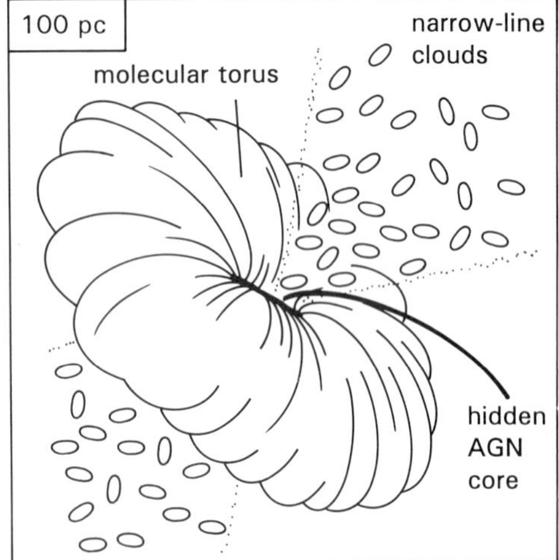

The narrow-line region comprising small but numerous clouds of the interstellar medium ionized by the central AGN core.

Fig. 9.9 Cartoon of the representative scale sizes of an AGN. How we eventually see the object depends on a number of parameters, the main one being the orientation of the obscuring torus with respect to the observer. (Adapted from Blandford, *Active Galactic Nuclei,* Saas-Fee Advanced Course 20, Springer–Verlag, 1990.)

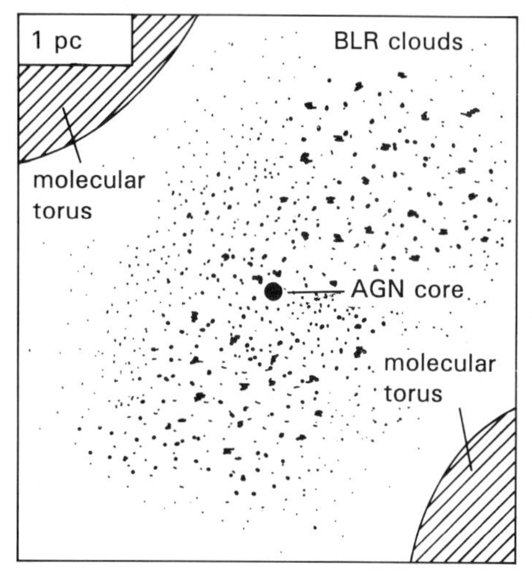

The outer extent of the broad-line region and the deep-walled molecular torus which can provide an effective shield of the central AGN, depending on the relative orientation of the observer.

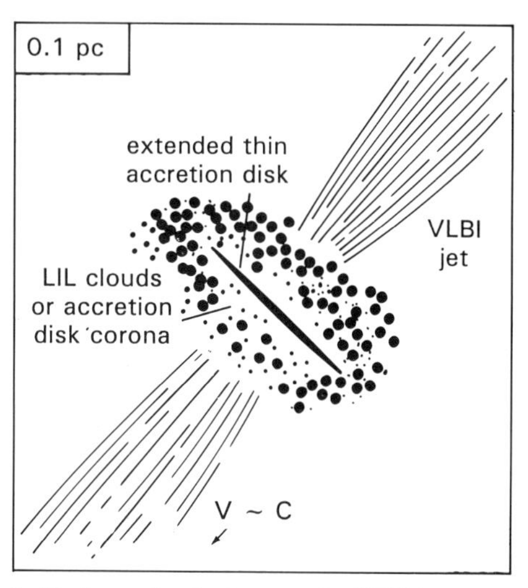

Inside the molecular torus — the VLBI jet becomes self-absorbed closer in, and the low ionization lines of the BLR, which might be the corona of the accretion disk.

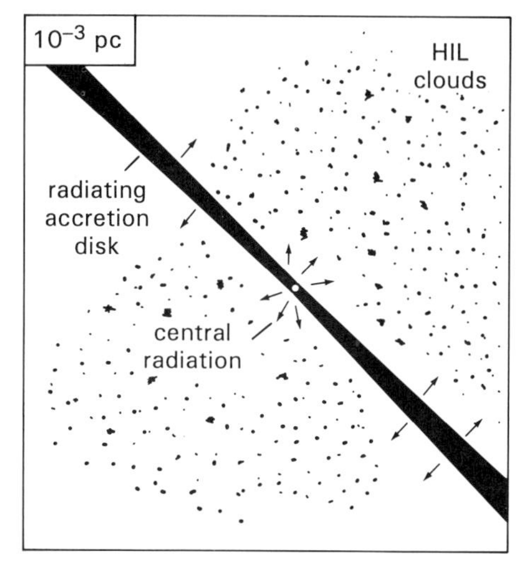

The accretion disk which radiates strongly at UV and optical wavelengths. The high ionization clouds of the BLR are excited by the central continuum radiation field.

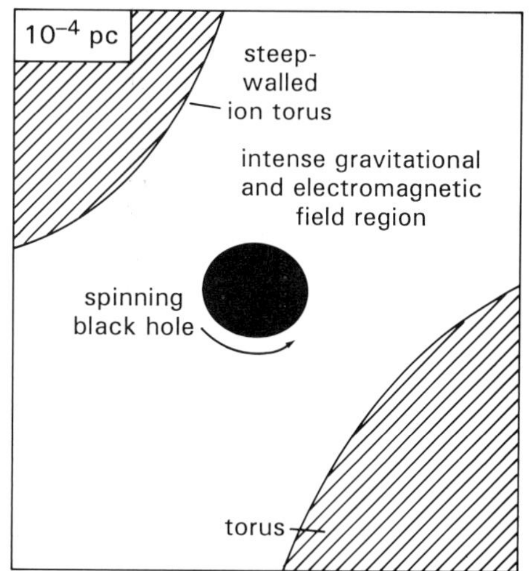

The black hole. The Schwarzschild radius for a $10^8\ M_\odot$ black hole is 2 AU (10^{-5} pc). The spin will introduce twisted magnetic field lines and particle acceleration.

A in which the evidence for the long-sought-after counter-jet looked very convincing. The author, Norbert Bartel, e-mailed me a postscript version of the image which is reproduced in fig. 9.10. The results fit in very well with all we have said about twin relativistic jets and unification.

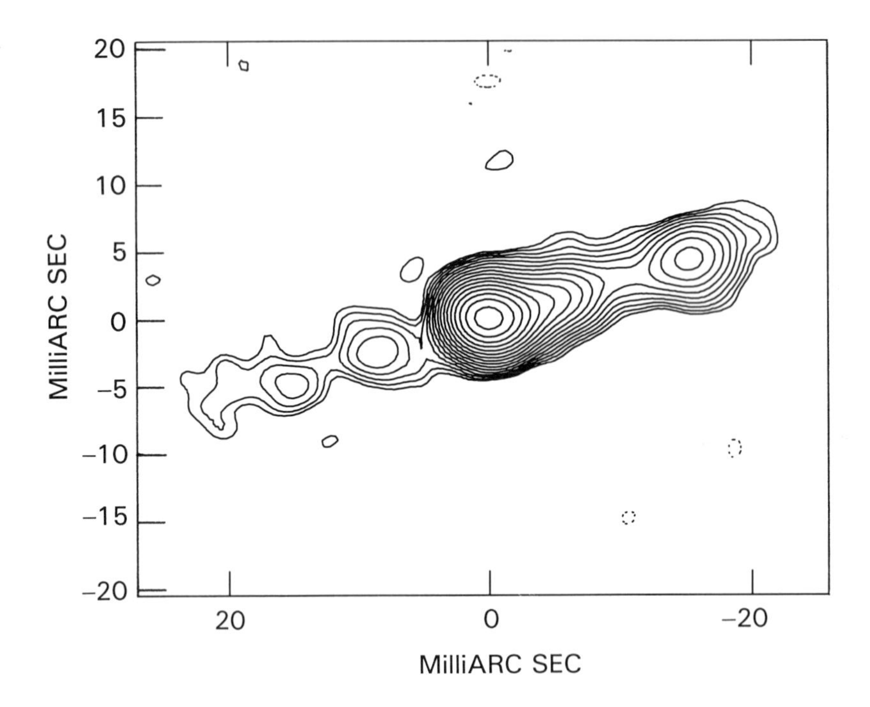

Fig. 9.10. The counter-jet in Cygnus A (3C405) revealed by a 5 GHz VLBA map at a spatial resolution of about 3 milli-arcseconds. This should be compared with fig. 3.12 which shows the full extent of this FRII source. (From Bartel,N., *et al., Proc. National Academy of Sciences*, **2**, 1995, courtesy of Norbert Bartel.)

The entire area of searching for the counter-jets must be stepped up in terms of dynamic range and sensitivity. Only recently have observations started to bear fruit and so it looks like we are on the edge of a breakthrough. Larger samples with counter-jets will test the statistics of the beaming and put limits to the beaming factor, which is something that is important for gamma-ray production mechanisms (through Compton scattering) and also the X-ray BL Lacs. The extension of the polarization studies on the FRII sources could do with much more work, especially with high dynamic ranges. Furthermore, another area which will benefit from the introduction of the VLBA is that of the detailed investigations of the small-scale (parsec-scale) jets in the nuclei of non-FRI or FRII galaxies, the Seyferts for example. What can we discover about the differences in these jets and can this help us with clues as to what makes a radio-loud AGN? We will return to this later.

Millimetre and submillimetre spectroscopy will become increasingly important. For the nearer AGNs (and NGC1068 is the obvious example) high spatial resolution, high dynamic range and high signal-to-noise mapping of various molecular species are needed in order to identify the sources of obscuration and gas motion streaming, especially within the inner 0.5 kpc. This project will be a prime target for the proposed millimetre arrays, pushing downwards in wavelength from the VLBA. Of course these will also open up the mapping of the molecular ISM in the host galaxies. The idea that the FRIs are embedded in regions of higher gas densities can also be probed by this technique and extensive surveys of FRI–FRII sources are needed.

Moving to the submillimetre and infrared regimes. The ISO satellite and detectors like SCUBA will allow statistically significant samples of all kinds of AGNs to be studied to show that (revealing my bias) the entire far-infrared radiation comes from heated dust, which is extended on scales of about a kiloparsec, and that there is excellent continuity in dust properties and mass between the ultraluminous FIRGs and powerful radio-quiet quasars. We hope to be able to determine whether there are measurable differences in the dust parameters (temperature, composition) as a function of galaxy type and redshift, and this will be especially important for high-z objects to look for changes in metallicity.

In the mid-infrared regime, ISO and new instruments on UKIRT and Gemini North will allow the thermal radiation from the molecular torus to be better separated from the more extensive and cooler dust and with large samples this should place restrictions on the possible geometries of the torus. Moving to the near infrared, the search for more broad lines in embedded objects is an obvious must, with a special emphasis on narrow-lined radio galaxies. Separating out the ultraluminous FIRGs with AGN cores from those (and the ordinary FIRGs) that are not AGN-powered is important and will require large samples. Can we get a solid handle on why some have a black hole and stars while others only make stars? The infrared should also throw up very reasonable statistical samples of AGNs to test the requirement that we should see a BLR relatively easily in all pole-on objects, even when buried by dust in the host galaxy. The degree of reddening and the host galaxy and torus obscuration should come out of these studies. Similar investigations of the narrow-lined radio galaxies are also anticipated to reveal the hidden broad-line regions which should be present if our unification ideas are correct.

Investigations of the host galaxy of a wider range of AGN should be extended. This will require larger panoramic detectors in the infrared and higher resolution imaging, both of which should be available on the UKIRT telescope with its larger camera (with finer pixels) and the new tip-tilt secondary mirror to produce almost diffraction limited images at 2 μm. This will open up the way for definitive studies on a large sample of radio-loud and radio-quiet AGN of all varieties. This has dramatic bearings on what makes a radio-loud object. Will our result that radio-loud activity is only found in ellipticals (or at least spheroidal systems) stand and be strengthened or not? The new HST infrared camera will provide lots of new information, but will probably not make as much impact in terms of spatial resolution as in the optical due to its relatively small aperture size. On the other hand, because of the absence of the atmospheric absorption features and sky emission, it will be much superior in producing contamination-free wide wavelength coverage.

In the optical–UV regime, adaptive optics will enable much higher spatial resolutions to be obtained from the large flux collectors on the ground. The investigations of the cores (and hosts) of AGNs at higher redshifts with imaging and spectroscopy is something for the new 8 and 10 m telescopes, the HST cannot match the shear collecting area and these are investigations with which the HST probably cannot compete. There is a need for detailed spatial investigations of the make-up of the narrow-line regions, especially the region closer-in to the core. This will remain a problem due to the relative small number of close-by AGNs, but progress will undoubtedly be made in getting a handle on the general structures. Further extensive optical studies are required to test the hypothesis that the jets in FRI and FRII sources start out the same but the former suffer rapid deceleration over scale sizes less than a kiloparsec due to a dense interstellar medium. Accurate imaging and spectroscopy of the central kiloparsec is required to decide this crucial question.

Another hot topic is the availability of gas in the FRI and BL Lac sources. Where is the broad-line region? Orientation can explain the situation for many FRI sources, but what about BL Lacs? Is it totally absent or just very weak and if it is the latter, what is the cause, lack of gas or conditions of the central engine? Very deep infrared–optical searches for buried broad-line regions, or reflected broad lines using infrared–optical polarizations must be undertaken for a range of BL Lac and FRI sources.

The HST comes into its own in the ultraviolet, where it is unique. Here there are a number of very important observations to be undertaken. One of the outstanding questions that still limits knowledge of the central engine is the precise shape of the continuum radiation making up the big blue bump. We have see that this might be from an accretion disk or reprocessed X-rays, or a combination of both and is probably different for different AGNs. Precise UV measurements of this for a range of objects which have different ratios of X-ray to 'thermal' luminosity are crucial to test out the reprocessing model, and to refine the details if it is shown to be applicable. The puzzle of the FeII lines needs resolving and is already the subject of much further work. There is scope for seeing how these lines correlate with other aspects of the AGN, but good samples are needed for this.

In the general optical regime there is further work to be done on the blazar class as a whole. These all ought to show a BLR at some level of strength, although for the BL Lacs it may be difficult unless they are caught at a time when the synchrotron jet emission is at a low level. The need for rapid response programmes to capitalize on these events is an obvious requirement and hopefully much more will be made of this in the future. In this light, multiwavelength and reverberation mapping studies are crucial. Remember, these provide spatial resolutions of order micro-to nano-arcseconds. Indeed, it is only the latter that can reveal the details of the central engine in terms of the line emission and the BLR sizes. The simultaneous multiwavelength monitoring campaigns are absolutely crucial, and becoming more so as we home in on definite models for the central engine. These must include full X-ray and UV coverage to allow the components of the central engine to be disentangled. For the radio-loud objects, the link between the synchrotron jet and the gamma-ray emission is of prime importance and the answer seems tantalizingly close. But is there a single answer, or do different processes take place in different AGNs? This is something that we need to determine.

The entire X-ray region is awaiting ever-further exploitation and a number of questions can be addressed. These basically focus on the geometry depicted in fig. 9.8 and how this changes with AGN luminosity, or not. Better determination of the 'primary' 10 keV X-ray power-law is still required, is this really constant in all AGNs? This immediately leads us into the determination of which of the two current hard X-ray power-law production mechanisms is correct (if either). This can be achieved by investigating the shape of the radiation spectrum in the vicinity of the electron–positron annihilation line of 511 keV. The two models predict significantly different strengths of the 511 keV emission line and the shape of the continuum spectrum.

There is still much to be learned about the soft X-ray excess, both in terms of better spectral shape and correlations with other wavebands. The key question is whether it is part of the big blue bump or not. The multi-wavelength variability studies have thrown up one or two examples where it clearly is not and therefore must be a separate radiating component. Is this the case for all AGNs? And if not, why?

Finally, there is always scope for more work on the theoretical side. As a long-past instrument-builder, now an observational astronomer, I find theoreticians come up with more suggestions or models to explain data than I could ever conceive, and some of these are even applicable in the real Universe! Following this tongue-in-cheek remark it must be stressed that the impact of theoretical studies for AGNs is absolutely outstanding. Indeed, this is a field in which many of the predictions, that perhaps seemed outlandish when they first appeared, have been shown to be absolutely spot-on. There are always fashionable topic areas to be worked on, and these come in waves; supermassive black holes, accretion flows, the BLR, the molecular torus, and the production of gamma-ray jets. At some point the theories or models run out of testable observations, they are far ahead of their time. The theorists then move on to other challenging areas while we experimentalists build the new facilities and instruments to test the theories and so advance scientific progress. Clearly we have not heard the last word on most the intimate details of the physical processes taking place in AGNs, and perhaps we never will, but it will not be because of a lack of imagination and brilliance on the part of the theoreticians.

9.6 SUMMARY

The unified model whereby all AGNs are powered by a central supermassive black hole appears intact. The overall predictions of this model satisfy most, if not all of the observations. The idea of orientation being a crucial parameter has found strong supporting evidence and this is now the generally accepted picture of AGNs. However, there is still no agreed picture as to why some galaxies are AGNs and others are not. The fact that all powerful radio sources lie in giant elliptical galaxies and none lie in spirals is a clear clue. The presence of an outflow of material in the form of a jet (usually indicated by the presence of luminous radio emission—the smoking gun pointing to the black hole) does not depend on the mass of the black hole alone. Other parameters must also be taken into account, namely the accretion rate and also, quite possibly, the spin of the black hole. This latter aspect, linked with the angular momentum of the fuel supply, which is well ordered in a spiral galaxy as opposed to being potentially misdirected with respect to the spin of the black hole in an elliptical, may also house clues as to the big picture. The formation of jets is presumably linked to some electrodynamic or magnetohydrodynamic process close to the black hole horizon.

There is now extensive evidence to show that at least some of the IRAS ultraluminous FIRGs are indeed buried quasars. In most cases there is strong evidence that merging activity is intimately linked with starbursts and the galaxy Arp 220 provides a prime example. Indeed, the latest HST data strongly supports the link between powerful AGN activity and mergers. Quasars are shown to be much more disturbed systems than originally expected, indicative of merging and interaction activity. However, the link between quasar luminosity and host galaxy luminosity remains unclear from to the recent HST data but the infrared imaging data fit in comfortably with the idea that the more luminous galaxies host the more luminous AGNs and that below a certain threshold galaxies do not house quasars as we define them.

The observational evidence for supermassive black holes is growing with time. This gives us confidence we are on the right track in that gravity is the ultimate power source for AGNs. The very asymmetric Fe K-α emission line discovered by the ASCA X-ray satellite is very hard to explain by anything other than gravitational broadening. The red wing is too wide for any reasonable level of Compton scattering, pure doppler broadening would produce a stronger blue line and other lines (such as Fe K-β) would lie to the blue side of 6.4 keV (which they do not). So this is very powerful evidence for supermassive black holes. When we add the supporting cases of NGC4258 and even the local galaxies of M31, M32 and M87, the evidence is overwhelming.

9.7 CONCLUDING WORDS

By now the reader should have a good overview of where we are in active galaxy research. Extraordinary progress has been made over the past two decades. Perhaps the major pieces of the puzzle are now already in place, a great number of the details are also there. But just as in a jigsaw puzzle, some of our remaining pieces can still be equally well fitted in a number of places, we have not yet managed to reduce the parameter space enough, but probably most of the edge and the four corner pieces are correct. We are now into the time-consuming phase of filling in the details to see if the picture makes sense. Some of the conclusions we have presented in this book may turn out to be incorrect with more data, others will be reinforced. As we close the loop on the big picture, we will be refining the small but important topics of the detailing of the models; this is the progress of science.

AGN research is probably the most active area of astronomical pursuit today and now that we have reached the end of the story the reader will understand why. I hope I have managed to minimize the impact of the hard work that a full comprehension of AGNs entails, but the same time convey to the reader the sheer excitement of the topic. We have not resorted to a significant amount of mathematical analysis in this text; for the undergraduate who might require this, the books listed in the further reading section will provide the tools to fill the gaps. In closing I sincerely hope that the reader has enjoyed learning about AGNs and is left with a sense of wonder that through the close link of theory and observation, we have made so much progress in understanding active galactic nuclei.

9.8 FURTHER READING

General and reviews

Bicknell,G.V., Dopita,M.A., & Quinn,P.J., Eds., *The first Mount Stromlo Symposium: the physics of active galaxies*, Astronomical Society of the Pacific Conference Series, **54**, 1994.

Filippenko,A.V., Ed., *Relationship between active galaxies and starburst galaxies*, Astronomical Society of the Pacific Conference Series., **31**, 1992. (Up to date and contains some very good reviews.)

Holt,S.S., Ed., *Testing the AGN Paradigm*, College Park, MD, American Institute of Physics Conference Series, 254, 1992. (Very comprehensive as the title suggests.)

Mushotzky,R.F., Done,C., & Pounds,K.A., 'X-ray spectra and time variability of active galactic nuclei' *Ann.Rev.Astron.Astrophys*, **31**, 717, 1993. (The big picture from an X-ray perspective.)

Specialized

Unification, models and buried quasars

Armus,L., *et al.*, 'Near-infrared (Fe II) and Paβ imaging and spectroscopy of ARP 220', *Astrophys.J.*, **440**, 200, 1995.

Antonucci,R., & Barvainis,R., 'Narrow-line radio galaxies as quasars in the sky plane', *Astrophys.J.*, **363**, L17, 1990.

Antonucci,R., Hurt,T., & Kinney,A., 'Evidence for a quasar in the radio galaxy Cygnus A from observation of broad-line emission', *Nature*, **371**, 313, 1994.

Baum,S.A. & Heckman,T., 'Extended optical line emitting gas in powerful radio galaxies: what is the radio emission-line connection?', *Astrophys.J.*, **336**, 702, 1989.

Baum,S.A., Zirbel.L. & O'Dea,C.P., 'Toword understanding the Fanaroff-Riley dichotemy in radio source morphology and power', *Astrophys.J.*, **451**, 88, 1995.

Blanco,G.S., Ward,M.J., & Wright,G.S., 'Broad infrared line emission from the nuclei of Seyfert 2 galaxies', *Mon.Not.R.Astron.Soc.*, **242**, L4, 1990.

Boksenberg.A., *et al.* 'Faint object camera imaging and spectroscopy of NGC4151' *Astrophys.J.*, **440**, 151, 1995

Courvoisier,T.J-L. and Camenzind,M., 'The wind and shock model for quasars: confrontation with observations of 3C273', *Astron.Astrophys.*, **224**, 10, 1989.

DeYoung,D.S., 'On the relation between Fanaroff-Riley types I and II radio galaxies', *Astrophys.J.*, **405**, L13, 1993.

Djorgovski,S. *et al.*, 'Discovery of an infrared nucleus in Cygnus A: An obscured quasar revealed?', *Astrophys.J.*, **372**, L67, 1991.

Eales,S.A., *et al.*, 'An optically luminous radio galaxy at z = 3.22 and the K-z diagram at high redshift', *Astrophys.J.*, **409,** 578, 1993.

Economou,F., *et al.*, 'Broad Hα emission in the high-redshift radio galaxy 3C 22', *Mon.Not.R.Astron.Soc.*, **272**, L5, 1995.

Efstathiou,A. & Rowan-Robinson,M. 'Dusty disks in active galactic nuclei', *Mon.Not. R.Astron.Soc.*, **273**, 649, 1995.

Evans,I.N. *et al.* 'Hubble Space Telescope imaging of the narrow-line region of NGC4151', *Astrophys.J.*, **417**, 82, 1993.

Goodrich,R,W., Veilleux,S., & Hill,G.J., 'Infrared spectroscopy of Seyfert 2 galaxies: a look through the obscuring torus', *Astrophys.J.*, **422**, 521, 1994.

Graham,J.R., *et al.*, 'Infrared observations of the z = 3.8 radio galaxy, 4C 41.17, with the W.M. Keck telescope', *Astrophys.J.*, **420**, L5, 1994.

Graham,J.R., *et al.*, 'The double nucleus of ARP 220 unveiled', *Astrophys.J.*, **354**, L5, 1990.

Graham,J.R., & Liu, M.C., 'High-resolution infrared imaging of FSC 10214 + 4724: Evidence for gravitational lensing', *Astrophys.J.*, **449**, L29, 1995.

Hes,R., Barthel, P.D., & Fosbury,R.A.E., 'Support for a unified model of radio galaxies and quasars from isoptropic [O III] emission', *Nature*, **362**, 326, 1993.

Hines,D.C., & Wills,B., 'The polarized spectrum of the FeII-rich broad absorption line QSO IRAS 07598+6508', *Astrophys.J.*, **448**, L69, 1995.

Jarvis, B.J., *et al.*, 'The nucleus of M87: Starburst or monster ?' *Astron.Astrophys.* **244**, L1, 1991.

Lawrence,A.. 'The relative frequency of broad-lines and narrow-lined active galactic nuclei: implications for unified schemes', *Mon.Not.R.Astron.Soc.*, **252**, 586, 1993

Lonsdale,C.J., *et al.*, 'Compact OH megamase and probable quasar activity in the galaxy Arp 220', *Nature*, **370**, 117, 1994.

Matthews,K., *et al.*, 'Near-infrared imaging of FSC 10214 + 4724 with the W.M. Keck telescope', *Astrophys.J.*, **420**, L13, 1994.

Owen,F.N. & Laing,R.A., 'CCD surface photometry of radio galaxies—I. FR class I and II sources', *Mon.Not.R.astron.Soc.,* **238**, 1989.

Puxley,P.J. *et al.* 'Spectroscopy of luminous IRAS galaxies at 1.25 μm', *Mon.Not.R. Astron.Soc.*, **270**, L7, 1994.

Rees,M.J. 'The radio/optical alignment of high-z radio galaxies: triggering star formation in radio lobes', *Mon.Not.R.Astron.Soc.*, **239**, 1P, 1989.

Rowan-Robinson,M., *et al.*, 'A high-redshift IRAS galaxy with huge luminosity - hidden quasar or protogalaxy ?', *Nature*, **351**, 719, 1991.

Rowan-Robinson, M., 'On the unity of activity in galaxies', *Astrophys.J.*, **213**, 635, 1977.

Serjeant, S. *et al.*, 'Spectroscopic evidence that the extreme properties of IRAS F 10214 + 4724 are due to gravitational lensing', *Mon.Not.R.Astron.Soc.*, in press, 1995.

Shaya,E.J., *et al.*, 'Hubble Space telescope planetary camera observations of ARP 220', *Astronom.J.*, **107**, 1675, 1994.

Ulvestad,J., & Wilson,A., 'Radio structure of Seyfert galaxies VII: extension of a distance limited sample', *Astrophys.J.*, **343**, 659, 1989.

Vestergaard,M. & Barthel,P.D., 'The nucleus of the Cygnus A galaxy', *Astronom.J.*, **105**, 456, 1993.

Young,S. *et al.*, 'Spectropolarimetry of the ultraluminous infrared galaxy IRAS 110548 - 1131', *Mon.Not.R.Astron.Soc.*, **260**, L1, 1993.

The central engine

Done,C., *et al*. 'The complex variable soft X-ray spectrum of NGC5548', *Mon.Not. R.Astron.Soc.*, **275**, 417, 1995.

Fabian,A.C., *et al*., 'Pair induced spectral changes and variability in compact X-ray sources', *Mon.Not.R.Astron.Soc.*, **221**, 931, 1986.

Pier,E.A., *et al*., 'The intrinsic nuclear spectrum of NGC 1068', *Astrophys.J.*, **428**, 124, 1994.

Prestwich,A.H., Joseph,R.D., & Wright,G.S., 'Starburst model of merging galaxies', *Astrophys.J.*, **422**, 73, 1994.

Rawlings,S., & Saunders,R., 'Evidence for a common central-engine mechanism in all extragalactic radio sources', *Nature*, **349**, 138, 1991

Rowan-Robinson,M., 'A new model for the infrared emission of quasars', *Mon.Not.R.Astron.Soc.*, **272**, 737, 1995.

Shlosman,I., Begelman,M.C., & Frank,J., 'The fuelling of active galactic nuclei', *Nature*, **345**, 679, 1990.

Smith,D.A., Done,C., & Pounds,K.A., 'Unified theories of active galactic nuclei: the hard X-ray spectrum of NGC1068', *Mon.Not.R.Astron.Soc.*, **263**, 54, 1993.

Turner,E.L., 'Quasars and Galaxy Formation I, the $z > 4$ objects.' *Astronom.J.*, **101**, 5, 1991.

APPENDIX 1

PHYSICAL CONSTANTS (S I units)

Velocity of light in vacuum	c	=	3.00×10^{8}	$m\ s^{-1}$
Gravitational constant	G	=	6.67×10^{-11}	$N\ m^{2}\ kg^{-2}$
Planck's constant	h	=	6.63×10^{-34}	$J\ Hz^{-1}$
Electron charge	e	=	1.60×10^{-19}	C
Electron rest mass	m_e	=	9.11×10^{-31}	kg
Electron rest mass energy	m_e	=	0.511	MeV
Proton rest mass	m_p	=	1.67×10^{-27}	kg
Boltzmann's constant	k	=	1.38×10^{-23}	$J\ K^{-1}$
Stefan–Boltmann constant	σ	=	5.67×10^{-8}	$W\ m^{-2}\ K^{-4}$
Electron volt	eV	=	1.60×10^{-19}	J
Compton wavelength of the electron	λ_C	=	2.43×10^{-12}	m

ASTRONOMICAL CONSTANTS

Astronomical unit	AU	=	1.496×10^{8}	km
Parsec	pc	=	3.086×10^{13}	km
Parsec	pc	=	3.262	light-year
Light year		=	9.46×10^{12}	km
Solar mass	$M_{\odot}$	=	1.99×10^{30}	kg
Solar radius	$R_{\odot}$	=	6.96×10^{5}	km
Solar luminosity	$L_{\odot}$	=	3.83×10^{26}	W
Earth mass	M_E	=	5.98×10^{24}	kg
Earth equatorial radius	R_E	=	6,378	km
Jansky	Jy	=	1.0×10^{-26}	$W\ m^{-2}\ Hz^{-1}$
V-band absolute magnitude of the Sun	$M_{V_{\odot}}$	=	+4.8	
Year	y	=	3.16×10^{7}	s

Index

WILEY-PRAXIS SERIES IN ASTRONOMY AND ASTROPHYSICS

Forthcoming titles

THE VICTORIAN AMATEUR ASTRONOMER: Independent Astronomical Research in Britain 1820–1920
Allan Chapman, Wadham College, University of Oxford, UK

TOWARDS THE EDGE OF THE UNIVERSE: A Review of Modern Cosmology
Stuart G. Clark, Lecturer in Astronomy, University of Hertfordshire

LARGE-SCALE STRUCTURES IN THE UNIVERSE
Anthony P. Fairall, Professor of Astronomy, University of Cape Town, South Africa

MARS AND THE DEVELOPMENT OF LIFE, Second edition
Anders Hansson, Ph.D.

ASTEROIDS: Their Nature and Utilization, Second edition
Charles T. Kowal, Computer Sciences Corp., Space Telescope Science Institute, Baltimore, Maryland, USA

ASTRONOMICAL OBSERVATIONS OF ANCIENT EAST ASIA
Richard Stephenson, Department of Physics, University of Durham, UK; Zhentao Xu, Purple Mountain Observatory, Academia Sinica, Nanjing, China; Yaotiao Tiang, Department of Astronomy, Nanjing University, China

EXPLORATION OF TERRESTRIAL PLANETS FROM SPACECRAFT, Second edition
Yuri Surkov, Chief of the Planetary Exploration Laboratory, Russian Academy of Sciences, Moscow, Russia